James Alexander McLellan

The teacher's hand-book of algebra

Containing methods, solutions and exercises, illustrating the lates and best

treatment of the elements of algebra. Second Edition

James Alexander McLellan

The Teacher's hand-book of algebra

Containing methods, solutions and exercises, illustrating the lates and best treatment of the elements of algebra. Second Edition

ISBN/EAN: 9783337276393

Printed in Europe, USA, Canada, Australia, Japan

Cover: Foto ©Paul-Georg Meister /pixelio.de

More available books at **www.hansebooks.com**

W. J. Gage's Mathematical Series.

THE TEACHER'S
HAND-BOOK OF ALGEBRA;

CONTAINING

METHODS, SOLUTIONS AND EXERCISES

ILLUSTRATING

THE LATEST AND BEST TREATMENT OF THE ELEMENTS
OF ALGEBRA.

BY

J. A. McLELLAN, M.A., LL.D.,
HIGH SCHOOL INSPECTOR FOR ONTARIO.

"*The object of pure Mathematics, which is another name for Algebra, is the unfolding of the laws of the human intelligence.*"—SYLVESTER.

SECOND EDITION—REVISED AND ENLARGED.

TORONTO:
W. J. GAGE & COMPANY.
1881.

Entered according to Act of Parliament of Canada in the year 1880 by W. J GAGE & COMPANY, in the office of the Minister of Agriculture.

PREFACE.

This book—embodying the substance of Lectures at Teachers' Associations—has been prepared at the almost unanimous request of the teachers of Ontario, who have long felt the need of a work to supplement the elementary text-books in common use. The following are some of its special features:

It gives a large number of solutions in illustration of the best methods of algebraic resolution and reduction, some of which are not found in any text-book.

It gives, classified under proper heads and preceded by type-solutions, a great number of exercises, many of them illustrating methods and principles which are unaccountably ignored in elementary Algebras.

It presents these solutions and exercises in such a way that the student not only sees how Algebraic transformations are effected, but also perceives how to form for himself as many additional examples as he may desire.

It shows the student how simple principles with which he is quite familiar, may be applied to the solution of questions which he has thought beyond their reach.

It gives complete explanations and illustrations of important topics which are strangely omitted or barely touched upon in the ordinary books, such as the Principle of Symmetry, Theory of Divisors, Factoring, Applications of Horner's Division, &c.

A few of the exercises are chiefly supplementary to those proposed in the text-books, but the intelligent student will find that even these examples have not been selected in the usual apparently aimless fashion; he will recognise that they are really expressions of certain laws; they are in fact proposed with a view to lead him to investigate these laws for himself as soon as he has sufficiently advanced in his course. Nos. 8, 9, 10 and 11 afford instances of such exercises.

Others of the questions proposed are preparatory or interpretation exercises. These might well have been omitted, were it not that they are generally omitted from the text-books and too often neglected by teachers. Practice in the interpretation of a new notation and in expression by means of it, should always precede its use as a symbolism itself subject to operations. Nos. 23 to 36 of Ex. iii., and nearly the whole of Ex. xv. may serve for instances.

By far the greater number of the exercises are intended for practice in the methods exhibited in the solved examples.* As many as possible of these have been selected for their intrinsic value. They have been gathered from the works of the great masters of analysis, and the student who proceeds to the higher branches of mathematics will meet again with these examples and exercises, and he will find his progress aided by his familiarity with them, and will not have to interrupt his advanced studies to learn processes properly belonging to elementary Algebra. In making this selection, it has been found that the most widely useful transformations are, at the same time, those that best exhibit the methods of reduction here explained, so that they have thus a double advantage. A great part of the exercises have, of necessity, been prepared specially for this work.

Articles and exercises have been prepared on the theory of substitutions, on Elimination, &c., but it has finally been decided to hold these over for Pt. ii., which will probably appear if the present work be favorably received.

CONTENTS.

CHAPTER I.—SUBSTITUTION, HORNER'S DIVISION &C.

	PAGE.
SECT. 1.—Numerical and Literal Substitution.....................	1
SECT. 2.—Fundamental Formulas and their Applications............	10
SECT. 3.—Horner's Methods of Multiplication and Division, and their Applications	21

CHAPTER II.—PRINCIPLE OF SYMMETRY, &C.

SECT. 1.—The Principle of Symmetry and its Applications	33
SECT. 2.—The Theory of Divisors and its Applications................	39

CHAPTER III.—FACTORING.

SECT. 1.—Direct Application of the Fundamental Formulas	62
SECT. 2.—Extended Application of the Formulas	71
SECT. 3.—Factoring by Parts	79
SECT. 4.—Application of the Theory of Divisors.....................	83
SECT. 5.—Factoring a Polynome by Trial Divisors...................	90

CHAPTER IV.—MEASURES AND MULTIPLES, &C.

SECT. 1.—Division, Measures and Multiples	101
SECT. 2.—Fractions	109
SECT. 3.—Ratios	122
SECT. 4.—Complete Squares, &c...................................	130

CHAPTER V.

SIMPLE EQUATIONS OF ONE UNKNOWN QUANTITY..................... 138
Preliminary Equations Resolution by Factors. Fractional Equations. Application of Ratios. Equations involving Surds, Higher Equations, &c.

CHAPTER VI.

SIMULTANEOUS EQUATIONS... 170
Equations of Two Unknown Quantities. Systems of Equations. Application of Symmetry. Equations of Three Unknowns. Systems of Equations.

CHAPTER VII.

EXAMINATION PAPERS... 207

CHAPTER I.

Section I.—Substitution.

Exercise i.

1. If $a = 1$, $b = 2$, $c = 3$, $d = 4$, $x = 9$, $y = 8$, find the value of the following expressions :—

$1 - \{1 - (1 - \overline{1-x})\}$.

$a - (x-y) - (b-c)(d-a) - (y-b)(x+c)$.

$x - y[y - (y-a)\{d + c(b-c)\}]$.

$(x+d)(y+b+c) + (x-d)(a-b-d) + (y+d)(a-x\quad d)$.

$(d-x)^3 + (c+y)^3$.

$(a-b)(c^3 - b^2x) - (c-d)(b^3 - a^y x) + (d-b-c)(d^3 - c^3)$

$\dfrac{d-a}{d+a} + \dfrac{d+c}{d-c} - 2\dfrac{d+b}{d-b}$.

2. If $a = 8$, $b = -4$, $c = -9$, and $2s = a+b+c$, find the value of the following expressions :—

$s(s-a)(s-b)(s-c)$.

$s^2 + (s-a)^2 + (s-b)^2 + (s-c)^2$.

$s^2 - (s-a)(s-b) - (s-b)(s-c) - (s-c)(s-a)$.

$2(s-a)(s-b)(s-c) + a(s-b)(s-c) + b(s-c)(s-a) + c(s-a)(s-b)$.

3. If $a = 2$, $b = -3$, $c = 1$, $x = 4\frac{1}{2}$, find the value of the following expressions :—

$\dfrac{a^2 - b^2}{a^3 + b^3}$, $\dfrac{a^2 + b^2}{a^3 - b^3}$, $\dfrac{(a-b)^2}{(a+b)^3}$, $\dfrac{(a-b)^3}{(a+b)^2}$,

$\dfrac{a^2 + ab + b^2}{a^2 - ab + b^2}$, $\dfrac{a^2 - b^3}{a^3 - b^2}$, $\dfrac{x}{2}\left\{\dfrac{2x-3}{3} - \dfrac{3x-1}{4}\right\}\dfrac{x-1}{2}$,

$\dfrac{(a+b)\{(a+b)^2 - c^2\}}{4b^2c^2 - (a^2 - b^2 - c^2)^2}$,

$\dfrac{a^2(b-c) + b^2(c-a) + c^2(a-b)}{(a-b)(b-c)(c-a)}$.

SUBSTITUTION.

4. If $a = 6$, $b = 5$, $c = -.4$, $d = -3$, find the value of the following expressions:—

$\sqrt{(b^2 + ac)} + \sqrt{(c^2 - 2ac)}$. $\sqrt{\{b^2 + ac + \sqrt{(c^2 - 2ac)}\}}$.

$\dfrac{a^2 - \sqrt{(b^2 + ac)}}{2a - \sqrt{(b^2 - ac)}}$, $\dfrac{c + \sqrt{(d^2 + c^2)}}{c^3 + 2d(d^2 - c^2)}$.

5. If $x = 3$, $y = 4$, $z = 0$, find the value of :—

$\{3x - \sqrt{(x^2 + y^2)}\}^2 \{2x + \sqrt{(x^2 + y^2 + z)}\}$.

$x^y + y^z + z^x$, $(x - y)^{x-y} + (y - z)^{y-z} + (z - x)^{z-x}$.

$(x^3 - y^3) \div \sqrt[y]{\{3x^3 + 3(3x^2 + 3xy + y^2)y\}}$.

6. Calculate the values of $\dfrac{(x + y + z)^3 - 3(x^3 + y^3 + z^3)}{xyz}$ when

    (a) $x = 1$, $y = 2$, $z = 3$.
    (b) $x = 2$, $y = 3$, $z = 4$.
    (c) $x = 3$, $y = 4$, $z = 5$.
    (d) $x = 10$, $y = 11$, $z = 12$.

7. Given $x = 3$, $y = 4$, $z = -5$, calculate the values of

$(x + y + z)^3 - 3(x + y + z)(xy + yz + zx)$.

$x^2(y + z) + y^2(z + x) + z^2(x + y) + 2xyz$.

$x^2(y - z) + y^2(z - x) + z^2(x - y)$.

$(5x - 4z)^2 + 9(4x - z)^2 - (13x - 5z)^2$.

$(3x + 4y + 5z)^2 + (4x + 3y + 12z)^2 - (5x + 5y + 13z)^2$.

8. If $s = a + b + c$, find the value of

$(2s - a)^2 + (2s - b)^2 - (2s + c)^2$, given

(1) $a = 3$, $b = 4$, $c = 5$,   (2) $a = 21$, $b = 20$, $c = 29$,
(3) $a = 119$, $b = 120$, $c = 169$,   (4) $a = 3$, $b = -4$, $c = 5$,
(5) $a = 5$, $b = 12$, $c = -13$.

9. If $a = 1$, $b = 3$, $c = 5$, $d = 7$, $e = 9$, $f = 11$, prove that

$a + b + c + d + e + f = \left(\dfrac{a + f}{2}\right)^2$.

$\dfrac{1}{ab} + \dfrac{1}{bc} + \dfrac{1}{cd} + \dfrac{1}{de} + \dfrac{1}{ef} = \dfrac{1}{2}\left(\dfrac{1}{a} - \dfrac{1}{f}\right)$.

$\dfrac{1}{abc} + \dfrac{1}{bcd} + \dfrac{1}{cde} + \dfrac{1}{def} = \dfrac{1}{4}\left(\dfrac{1}{ab} - \dfrac{1}{ef}\right)$.

$\dfrac{1}{abcd} + \dfrac{1}{bcde} + \dfrac{1}{cdef} = \dfrac{1}{6}\left(\dfrac{1}{abc} - \dfrac{1}{def}\right)$.

$$a^2+b^2+c^2-ab-bc-ca = b^2+c^2+d^2-bc-cd-db =$$
$$c^2+d^2+e^2-cd-de-ec = d^2+e^2+f^2-de-ef-fd.$$

10. If $a=1$, $b=2$, $c=3$, $d=4$, $e=5$, $f=6$, $g=7$, prove that
$a+b+c = \frac{1}{2}cd$, $a+b+c+d = \frac{1}{2}de$,
$a+b+c+d+e = \frac{1}{2}ef$, $a+b+c+d+e+f = \frac{1}{2}fg$,
$a^2+b^2+c^2 = \dfrac{cd(c+d)}{ab(a+b)}$, $a^2+b^2+c^2+d^2 =$
$\dfrac{de(d+e)}{ab(a+b)}$, $a^2+b^2+c^2+d^2+e^2 = \dfrac{ef(e+f)}{ab(a+b)}$,
$a^2+b^2+c^2+d^2+e^2+f^2 = \dfrac{fg(f+g)}{ab(a+b)}$.
$a^3+b^3+c^3 = (a+b+c)^2$,
$a^3+b^3+c^3+d^3 = (a+b+c+d)^2$,
$a^3+b^3+c^3+d^3+e^3 = (a+b+c+d+e)^2$,
$a^3+b^3+c^3+d^3+e^3+f^3 = (a+b+c+d+e+f)^2$
$a^4+b^4+c^4 = \dfrac{cd(c+d)(c^2d-1)}{bc(b+c)}$,
$a^4+b^4+c^4+d^4 = \dfrac{de(d+e)(cde-1)}{bc(b+c)}$,
$a^4+b^4+c^4+d^4+e^4 = \dfrac{ef(e+f)(cef-1)}{bc(b+c)}$,
$a^4+b^4+c^4+d^4+e^4+f^4 = \dfrac{fg(f+g)(cfg-1)}{bc(b+c)}$,
$c^2+d^2 = e^2$, $c^3+d^3+e^3 = f^3$

11. Assume *any* numerical values for x, y, and z, and calculate the values of the following expressions:—

$(x^5 - 10x^3 + 5x)^2 + (5x^4 - 10x^2 + 1)^2 - (x^2+1)^5$.
$(x+1)^3 - 2(x+5)^3 - (x+9)^3 + 2(x+11)^3 + (x+12)^3 - (x+16)^3$.
$(x^2 - y^2)^2 + (2xy)^2 - (x^2+y^2)^2$
$(x^3 - 3xy^2)^2 + (3x^2y - y^3)^3 - (x^2+y^2)^3$.
$(3x^2+4xy+y^2)^2 + (4x^2+2xy)^2 - (5x^2+4xy+y^2)^2$.
$(x-y)^3 + (y-z)^3 + (z-x)^3 - 3(x-y)(y-z)(z-x)$.

Art. I. If x = any number, as, for example, 3, then x^2 (which = $x.x$) = $3x$, x^3 (which = $x.x^2$) = $3x^2$, x^4 (which = $x.x^3$) = $3x^3$, &c. Or $3 = x$, $3x = x^2$, $3x^2 = x^3$, $3x^3 = x^4$, $3x^4 = x^5$, &c. Hence prob lems like the following may be solved like ordinary arithmetical problems in "Reduction Descending."

EXAMPLES.

1. Find the value of $x^2 - 2x - 9$ when $x = 5$.

$$x^2 - 2x - 9$$
$$5$$
$$\overline{}$$
$$5x$$
$$-2x$$
$$\overline{}$$
$$3x$$
$$5$$
$$\overline{}$$
$$15$$
$$-9$$
$$\overline{}$$
$$6$$

Explanation.
$x^2 = 5x$,
$\therefore x^2 - 2x = 3x = 15$, and
$\therefore x^2 - 2x - 9 = 15 - 9 = 6$.

2. Find the value of $x^4 - x^3 - 4x^2 - 3x - 5$ when $x = 3$.

$$x^4 - x^3 - 4x^2 - 3x - 5$$
$$3$$

p_1 .... $3x^3$
$-x^3$

r_1 .... $2x^3$
3

p_2 .... $6x^2$
$-4x^2$

r_2 .... $2x^2$
3

p_3 .... $6x$
$-3x$

r_3 .... $3x$
3

p_4 9
-5

r_4 .... 4.

$\therefore x^4 - x^3 - 4x^2 - 3x - 5 = 4$
if $x = 3$.

SUBSTITUTION.

Explanation.

$x^4 = 3x^3$,

$\therefore x^4 - x^3 = 2x^3 = 6x^2$,

$\therefore x^4 - x^3 - 4x^2 = 2x^2 = 6x$,

$\therefore x^4 - x^3 - 4x^2 - 3x = 3x = 9$,

$\therefore x^4 - x^3 - 4x^2 - 3x - 5 = 4$.

8. Find the value of $2x^4 + 12x^3 + 6x^2 - 12x + 10$. Using coefficients only, we have

$$2 + 12 + 6 - 12 + 10$$
$$-5$$

p_1 ……… -10
$+12$

r_1 ……… $+\ 2$
$-\ 5$

p_2 ……… -10
$+\ 6$

r_2 ……… $-\ 4$
$-\ 5$

p_3 ……… 20
-12

r_3 ……… 8
-5

p_4 ……… -40
$+10$

r_4 ……… -30

$\therefore$ the quantity $= -30$ if $x = -5$.

Art. II. If the coefficients, and also the values of x are small numbers, much of the above may be done mentally, and the work will then be very compact. Thus, performing mentally the multiplications and additions (or subtractions) of the coefficients, and merely recording the partial reductions r_1, r_2, r_3, and the result r_4, the last example would appear as follows :—

-5) 2 $+12$ $+6$ -12 $+10$
         2
        -4
         8
        -30

Art. III. In the above examples, the coefficients are "brought down" and written below the products p_1, p_2, p_3, p_4, and are added or subtracted, as the case may require, to get the partial reductions r_1, r_2, r_3, and the result r_4. Instead of thus "bringing down" the coefficients, we may "carry up" the products p_1, p_2, p_3, p_4, writing them beneath their corresponding coefficients, and thus get r_1, r_2, r_3, r_4 in a third (horizontal) line. Arranged in this way Ex. 2 will appear

$$3 \,\bigg|\, \begin{array}{ccccc} 1 & -1 & -4 & -8 & -5 \\ & +3 & +6 & +6 & +9 \\ \hline 1 & +2 & +2 & +3; & 4; \end{array}$$

and Ex. 3 will appear

$$-5 \,\bigg|\, \begin{array}{ccccc} 2 & +12 & +6 & -12 & +10 \\ & -10 & -10 & +20 & -40 \\ \hline 2 & +2 & -4 & +8; & -30 \end{array}$$

Comparing these arrangements with those first given (Ex. 2 and 3), it will be seen that they are figure for figure the same, except that the multiplier is not repeated.

Art. IV. When there are several figures in the value of x, they may be arranged in a column, and each figure used separately, as in common multiplication. Where only approximate values are required, "contracted multiplication" may be used.

4. Find the value of $3x^5 - 160x^4 + 344x^3 + 700x^2 - 1910x + 1200$, given $x = 51$.

$$\begin{array}{c|cccccc} & 3 & -160 & +344 & +700 & -1910 & +1200 \\ 1 & & 3 & -7 & -13 & 37 & -23 \\ 50 & & 150 & -350 & -650 & 1850 & -1150 \\ \hline & 3 & -7 & -13 & +37 & -23; & +27 \end{array}$$

∴ result is 27.

SUBSTITUTION. 7

5. Given $x = 1·183$, find the value of $64x^4 - 144x + 45$ correct to three decimal places.

```
     | 64    0        0         -144        +45
   1 |       64       75·712    89·5673     -38·0419
   1.|       6·4      7·5712    8·9567      -3·8042
   8 |       5·12     6·0570    7·1654      -3·0434
   3 |       ·192     ·2271     ·2687       -·1141
     |_____
     | 64,   75·712,  89·5673,  -38·0419,   -·0036.
```
∴ result is $-·004$.

Exercise ii.

Find the value of

1. $x^4 - 11x^3 - 11x^2 - 13x + 11$, for $x = 12$.
2. $x^4 + 50x^3 - 16x^2 - 16x - 61$, for $x = -17$.
3. $2x^4 + 249x^3 - 125x^2 + 100$, for $x = -125$.
4. $2x^3 - 473x^2 - 234x - 711$, for $x = 200$.
5. $x^5 - 3x^2 - 8$, for $x = 4$.
6. $x^6 - 515x^5 - 3127x^4 + 525x^3 - 2090x^2 + 3156x - 15792$, for $x = 521$.
7. $2x^5 + 401x^4 - 199x^3 + 399x^2 - 602x + 211$, for $x = -201$.
8. $1000x^4 - 81x$, for $x = ·1$.
9. $99x^4 + 117x^3 - 257x^2 - 325x - 50$, for $x = 1\frac{3}{4}$.
10. $5x^5 + 497x^4 + 200x^3 + 196x^2 - 218x - 2000$, for $x = -99$.
11. $5x^5 - 620x^4 - 1030x^3 + 1045x^2 - 4120x + 9000$, for $x = 205$.

Calculate, correct to three places of decimals,—

12. $x^3 + 3x^2 - 13x - 38$ for $x = 3·58443$, for $x = -3·77931$, and for $x = -2·80512$.
13. $y^4 - 14y^2 + y + 38$ for $y = 3·13131$, for $y = -1·84813$, and for $y = -3·28319$.

Exercise iii.

What do the following expressions become (1) when $x = a$, (2) when $x = -a$?

1. $x^4 - 4ax^3 + 6a^2x^2 - 4a^3x + a^4$.
2. $\sqrt{(x^2 - ax + a^2)}$.   3. $\sqrt{(x^2 + 2ax + a^2)}$.
4. $(x^2 + ax + a^2)^3 - (x^2 - ax + a^2)^3$.

If $x = y = z = a$, find the value of the following expressions:

5. $(x-y)(y-z)(z-x)$.

6. $(x+y)^2(y+z-a)(x+z-a)$.

7. $x(y+z)(y^2+z^2-x^2)+y(z+x)(z^2+x^2-y^2)+z(x+y)(x^2+y^2-z^2)$.

8. $\dfrac{x}{y+z}+\dfrac{y}{x+z}+\dfrac{z}{x+y}$.

Find the value of

9. $\dfrac{x}{a}+\dfrac{x}{b}$ when $x=\dfrac{abc}{a+b}$.

10. $\dfrac{1}{a(b-x)}+\dfrac{1}{b(c-x)}+\dfrac{1}{a(x-c)}$, when $x=\dfrac{b}{a}(a-b+c)$.

11. $\dfrac{x}{a}+\dfrac{x}{b-a}$, when $x=\dfrac{a^2(b-a)}{b(b+a)}$.

12. $(a+x)(b+x)-a(b+c)+x^2$, when $x=\dfrac{ac}{b}$.

13. $bx+cy+az$, when $x=b+c-a$, $y=c+a-b$, $z=a+b-c$.

14. $\dfrac{a(1+b)+bx}{a(1+b)-bx}-\dfrac{a}{a-2bx}$, when $x=-a$.

15. $\left(\dfrac{x+a}{x+b}\right)^3-\dfrac{x+2a+b}{x-a-2b}$, when $x=\frac{1}{2}(b-a)$.

16. $(p-q)(x+2r)+(r-x)(p+q)$, when $x=\dfrac{r(3p-q)}{2q}$.

17. $a^2(b-c)+b^2(c-a)+c^2(a-b)$, when $a-b=0$.

18. $(a+b+c)(bc+ca+ab)-(a+b)(b+c)(c+a)$, when $a=-b$.

19. $(a+b+c)^3-(a^3+b^3+c^3)$, when $a+b=0$.

20. $(x+y+z)^4-(x+y)^4-(y+z)^4-(z+x)^4+x^4+y^4+z^4$, when $x+y+z=0$.

21. $a^3(c-b^2)+b^3(a-c^2)+c^3(b-a^2)+abc(abc-1)$, when $b-a^2=0$.

22. $a^5\left(\dfrac{a^5+5b^5}{a^5-b^5}\right)^5+b^5\left(\dfrac{5a^5+b^5}{b^5-a^5}\right)^5$, when $a^5+b^5=0$.

23. Express in words the fact that
$$(a-b)^2=a^2-2ab+b^2.$$

24. Express algebraically the fact "that the sum of two quantities multiplied by their difference is equal to the difference of the squares of the numbers."

25. The area of the walls of a room is equal to the height multiplied by twice the sum of the length and breadth: what are the areas of the walls in the following cases:

(1) length l, height h, breadth b.

(2) height x, length b feet more than the height, and breadth b feet less than the height.

26. Express in words the statement that
$$(x+a)(x+b) = x^2 + (a+b)x + ab.$$

27. Express in symbols the statement that "the square of the sum of two quantities exceeds the sum of their squares by twice their product."

28. Express in words the algebraic statement,
$$(x+y)^3 = x^3 + y^3 + 3xy(x+y).$$

29. Express algebraically the fact that "the cube of the difference of two quantities is equal to the difference of the cubes of the quantities diminished by three times the product of the quantities multiplied by their difference"

30. If the sum of the cubes of two quantities be divided by the sum of the quantities, the quotient is equal to the square of their difference increased by their product; express this algebraically.

31. Express in words the following algebraic statement:
$$\frac{x^3-y^3}{x-y} = (x+y)^2 - xy.$$

32. The square on the diagonal of a cube is equal to three times the square on the edge; express this in symbols, using l for length of the edge, and d for length of the diagonal.

33. Express in symbols that "the length of the edge of the greatest cube that can be cut from a sphere is equal to the square root of one-third the square of the diameter."

34. Express in symbols that any "rectangle is half the rectangle contained by the diagonals of the squares upon two adjacent sides." [The square on the diagonal of a square is double the square on a side.]

35. The area of a circle is equal to π multiplied into the square

of the radius; express this in symbols. Also express in symbols the area of the ring between two concentric circles.

86. The volume of a cylinder is equal to product of its height into the area of the base, that of a cone is one-third of this, and that of a sphere is two-thirds of the volume of the circumscribing cylinder; express these facts in symbols, using h for the height of the cylinder, and r for the radius of its base.

Exercise iv.

Perform the additions in the following cases:

1. $(b-a)x+(c-b)y$, and $(a+b)x+(b+c)y$.
2. $ax-by$, $(a-b)x-(a+b)y$, and $(a+b)x-(b-a)y$.
3. $(y-z)a^2+(z-x)ab+(x-y)b^2$, and $(x-y)a^2-(z-y)ab-(z-z)b^2$.
4. $ax+by+cz$, $bx+cy+az$, and $cx+ay+bz$.
5. $(a+b)x^2+(b+c)y^2+(a+c)z^2$, $(b+c)x^2+(a+c)y^2+(a+b)z^2$, $(a+c)x^2+(a+b)y^2+(b+c)z^2$, and $-(a+b+c)(x^2+y^2+z^2)$.
6. $x(a-b)^2+y(b-c)^2+z(c-a)^2$, $y(a-b)^2+z(b-c)^2+x(c-a)^2$, and $z(a-b)^2+x(b-c)^2+y(c-a)^2$.
7. $(a-b)x^2+(b-c)y^2+(c-a)z^2$, $(b-c)x^2+(c-a)y^2+(a-b)z^2$, and $(c-a)x^2+(a-b)y^2+(b-c)z^2$.
8. $(a+b)x+(b+c)y-(c+a)z$, $(b+c)z+(c+a)x-(a+b)y$, and $(a+c)y+(a+b)z-(b+c)x$.
9. $a^2-3ab-1\frac{4}{11}b^2$, $2b^2-\frac{3}{8}b^3+c^2$, $ab-\frac{1}{8}b^2+b^3$, and $2ab-\frac{1}{8}b^3$.
10. ax^n-3bx^n, $-9ax^n+7bx^n$, and $-8bx^n+10ax^n$.
11. What will $(ax-by+cz)+(bx+cy-az)-(cx+ay+bz)$ become when $x-y-z=1$?

SECTION II.—FUNDAMENTAL FORMULAS AND THEIR APPLICATION.

4. By Multiplication we get

$(x+r)(x+s) = x^2+(r+s)x+rs$ ....A.
$(x+r)(x+s)(x+t) = x^3+(r+s+t)x^2+(rs+st+tr)x+rst$ ......B.

From A we immediately get

$(x \pm y)^2 = x^2 \pm 2xy + y^2$ .... [1]

$(x+y+z)^2 = x^2 + 2xy + 2xz + y^2 + 2yz + z^2$ [2]

$(\Sigma a)^2 = \Sigma a^2 + 2\Sigma ab$ .... [3]

$(x+y)(x-y) = x^2 - y^2$ .... [4]

From B we derive

$(x \pm y)^3 = x^3 \pm 3x^2 y + 3xy^2 \pm y^3$ .... [5]

$\quad\quad\quad = x^3 \pm y^3 \pm 3xy(x \pm y)$ .... [6]

$(x+y+z)^3 = x^3 + y^3 + z^3 + 3x^2(y+z) + 3y^2(z+x) + 3z^2(x+y)$
$\quad + 6xyz$ .... [7]

$\quad\quad = x^3 + y^3 + z^3 + 3(x+y)(y+z)(z+x)$ .... [8]

$\quad\quad = x^3 + y^3 + z^3 + 3(x+y+z)(xy+yz+zx) - 3xyz$ ... [9]

$(\Sigma a)^3 = \Sigma a^3 + 3\Sigma a^2 b + 6\Sigma abc$ .... [10]

[The symbol Σ means THE SUM OF ALL SUCH TERMS AS]

FORMULA [1].—EXAMPLES.

1. We have at once $(x+y)^2 + (x-y)^2 = 2(x^2 + y^2)$, and $(x+y)^2 - (x-y)^2 = 4xy$.

2. $(a+b+c+d)^2 + (a-b-c+d)^2$ may be written $\{(a+d)+(b+c)\}^2 + \{(a+d)-(b+c)\}^2$, which (Ex. 1) $= 2\{(a+d)^2 + (b+c)^2\}$; similarly
$(a-b+c-d)^2 + (a+b-c-d)^2 = \{(a-d)-(b-c)\}^2 + \{(a-d)+(b-c)\}^2 = 2\{(a-d)^2 + (b-c)^2\}$;
$\therefore (a+b+c+d)^2 + (a-b-c+d)^2 + (a-b+c-d)^2 + (a+b-c-d)^2 = 2\{(a+d)^2 + (b+c)^2 + (a-d)^2 + (b-c)^2\} =$
(again by Ex. 1) $4(a^2 + b^2 + c^2 + d^2)$.

3. Simplify $(a+b+c)^2 - 2(a+b+c)c + c^2$;

This is the square of a binomial of which the first term is $(a+b+c)$ and the second $-c$; the given quantity $\therefore =$ $\{(a+b+c)-c\}^2 = (a+b)^2$.

4. Simplify $(a+b)^4 - 2(a^2+b^2)(a+b)^2 + 2(a^4+b^4)$.

By Ex. 1. $2(a^4+b^4) = (a^2+b^2)^2 + (a^2-b^2)^2$; $\therefore$ given quantity $= (a+b)^4 - 2(a^2+b^2)(a+b)^2 + (a^2+b^2)^2 + (a^2-b^2)^2 = \{(a+b)^2 - (a^2+b^2)\}^2 + (a^2-b^2)^2 = a^4 + 2a^2b^2 + b^4 = (a^2+b^2)^2$.

Exercise v.

1. $(x+3y^2)^2 + (x-3y^2)^2$, $(\frac{1}{2}a^2+3b^2)^2 - (\frac{1}{2}a^2-3b^2)^2$.
2. Shew that $(mx+ny)^2 + (nx-my)^2 = (m^2+n^2)(x^2+y^2)$.
3. "   " $(mx-ny)^2 - (nx-my)^2 = (m^2-n^2)(x^2-y^2)$.
4. Simplify $(a+3b)^2 + 2(a+3b)(a-b) + (a-b)^2$; $\{a-b\}^2$.
5. " $(x+3)^2 + (x+4)^2 - (x+5)^2$, and $(\frac{1}{2}x^2-2y^2)^2 - (\frac{1}{4}y^2+2x^2)^2$.
6. Simplify $(a+b+c)^2 + (b+c)^2 - 2(b+c)(a+b+c)$
7. Shew that $(ax+by)^2 + (cx+dy)^2 + (ay-bx)^2 + (cy-dx)^2 = (a^2+b^2+c^2+d^2)(x^2+y^2)$.
8. Simplify $(x-3y^2)^2 + (3x^2-y)^2 - 2(3x^2-y)(x-3y^2)$.
9. " $(x^2+xy-y^2)^2 - (x^2-xy-y^2)^2$, and $(1+2x+4x^2)^2 + (1-2x+4x^2)^2$.
10. If $a+b = -\frac{3}{4}c$, shew that $(2a-b)^2 + (2b-c)^2 + (2c-a)^2 + 2(2a-b)(2b-c) + 2(2b-c)(2c-a) + 2(2c-a)(2a-b) = \frac{1}{16}c^2$.
11. Simplify $2(a-b)^2 - (a-2b)^2$; $(a^2+4ab+b^2)^2 - (a^2+b^2)^2$.
12. " $(a+b)^2 - (b+c)^2 + (c+d)^2 - (d+a)^2$.
13. " $(\frac{1}{2}x-y)^2 + (\frac{1}{2}y-z)^2 + (\frac{1}{2}z-x)^2 + 2(\frac{1}{2}x-y)(\frac{1}{2}z-x) + 2(\frac{1}{2}y-z)(\frac{1}{2}z-x) + 2(\frac{1}{2}x-y)(\frac{1}{2}y-z)$.
14. Prove that $(x-y)^2 + (y-z)^2 + (z-x)^2 = 2(x-y)(z-y) + 2(y-x)(z-x) + 2(z-y)(z-x)$.
15. Simplify $(1+x)^4 - 2(1+x^2)(1+x^2) + 2(1+x^4)$.
16. " $(x+y+z)^2 - (x+y-z)^2 - (y+z-x)^2 - (z+x-y)^2$.
17. " $(x-2y+3z)^2 + (3z-2y)^2 + 2(x-2y+3z)(2y-3z)$.
18. " $(a^2+b^2-c^2)^2 + (c^2-b^2)^2 + 2(b^2-c^2)(a^2+b^2-c^2)$.
19. " $(x+y)^4 + (x-y)^4 - 2(x-y)^2(x+y)^2$.

FUNDAMENTAL FORMULAS.

20. " $(5a+3b)^2+16(3a+b)^2-(13a+5b)^2$.

21. Shew that $(3a-b)^2+(3b-c)^2+(3c-a)^2-2(b-3a)(3b-c)$
$+2(3b-c)(3c-a)-2(a-3c)(3a-b)-4(a+b+c)^2=0$.

22. If $z^2 = 2xy$, prove that $(2x^2-y^2)^2+(z^2-2y^2)^2+(x^2-2z^2)^2$
$-2(2x^2-y^2)(z^2-2y^2)+2(x^2-2z^2)(z^2-2y^2)-$
$2(x^2-2z^2)(2x^2-y^2)=(x+y)^4$.

23. Simplify $(1+x+x^2+x^3)^2+(1-x-x^2+x^3)^2+$
$(1-x+x^2-x^3)^2+(1+x-x^2-x^3)^2$.

24. Simplify $(ax+by)^4-2(a^2x^2+b^2y^2)(ax+by)^2+$
$2(a^4x^4+b^4y^4)$.

FORMULAS [2] AND [3].—EXAMPLES.

1. $(1-2x+3x^2)^2 = 1-4x+6x^2$
$\qquad\qquad\qquad +4x^2-12x^3$
$\qquad\qquad\qquad\qquad\quad +9x^4$
$\qquad\qquad = 1-4x+10x^2-12x^3+9x^4$.

2. $(ab+bc+ca)^2 = a^2b^2+2ab^2c+2a^2bc+b^2c^2+2abc^2+c^2a^2 =$
$a^2b^2+b^2c^2+c^2a^2+2abc(a+b+c)$.

3. $\{(x+y)^2+x^2+y^2\}^2 = (x+y)^4+2(x+y)^2(x^2+y^2)+x^4+2x^2$
$y^2+y^4 = (x+y)^4+(x+y)^2\{(x+y)^2+(x-y)^2\}+x^4+2x^2y^2+y^4$
$= 2(x+y)^4+(x^2-y^2)^2+x^4+2x^2y^2+y^4 = 2\{(x+y)^4+x^4+y^4\}$.

4. $(x^2+xy+y^2)^2 = x^4+2x^3y+2x^2y^2+x^2y^2+2xy^3+y^4 =$
$(x+y)^2x^2+x^2y^2+y^2(x+y)^2$.

5. In Ex. 3, substitute $b-c$ for x, $c-a$ for y, and consequently $b-a$ for $x+y$, then since $(b-a)^2=(a-b)^2$, Ex. 3 gives
$\{(a-b)^2+(b-c)^2+(c-a)^2\}^2 = 2\{(a-b)^4+(b-c)^4+(c-a)^4\}$.

6. Making the same substitutions in Ex. 4, we have
$(a^2+b^2+c^2-ab-bc-ca)^2 = (a-b)^2(b-c)^2+(b-c)^2(c-a)^2+$
$(c-a)^2(a-b)^2$, or, multiplying both sides by 4,
$\{(a-b)^2+(b-c)^2+(c-a)^2\}^2 = 4(a-b)^2(b-c)^2+4(b-c)^2 \times$
$(c-a)^2+4(c-a)^2(a-b)^2$, and $\therefore$ from Ex. 5, $(a-b)^4+(b-c)^4+$
$(c-a)^4 = 2(a-b)^2(b-c)^2+2(b-c)^2(c-a)^2+2(c-a)^2(a-b)^2$.

Exercise vi.

1. $(1-2x+3x^2-4x^3)^2$, $(1-x+x^2-x^3)^2$.
2. $(1-2x+2x^2-3x^3-x^4)^2$, $(1+3x+3x^2+x^3)^2$.
3. $(2a-b-c^2-1)^2$, $(1-x+y+z)^2$, $(\frac{1}{2}x-\frac{1}{3}y+6z)^2$.
4. $(x^3-x^2y+xy^2-y^3)^2$, $(ax+bx^2+cx^3+dx^4)^2$.
5. Shew that $(a^2+b^2+c^2)(x^2+y^2+z^2)-(ax+by+cz)^2 = (ay-bx)^2+(cx-az)^2+(bz-cy)^2$.
6. Prove that $(a+b)x+(b+c)y+(c+a)z$ multiplied by $(a-b)x+(b-c)y+(c-a)z$, is equal to the difference of the squares of two trinomials.
7. Shew that $(a-b)(a-c)+(b-c)(b-a)+(c-a)(c-b) - \frac{1}{2}\{(a-b)^2+(b-c)^2+(c-a)^2\} = 0$.
8. Simplify $\{a-(b-c)\}^2+\{b-(c-a)\}^2+\{c-(a-b)\}^2$.
9. Shew that $(a^2+b^2-x^2)^2+(a_1^2+b_1^2-x^2)^2+2(aa_1+bb_1)^2$
$= (a^2+a_1^2-x^2)^2+(b^2+b_1^2-x^2)^2+2(ab+a_1b_1)^2$.
10. Prove that $\{(a-b)(b-c)+(b-c)(c-a)+(c-a)(a-b)\}^2 = (a-b)^2(b-c)^2+(b-c)^2(c-a)^2+(c-a)^2(a-b)^2$.
11. Square $2a-\frac{1}{2}bx-\frac{1}{4}cx+2dx$.
12. If $x+y+z=0$, shew that $x^4+y^4+z^4 = (x^2-y^2)^2+(y^2-z^2)^2+(z^2-x^2)^2$.
13. Prove that $a^2(b+c)^2+b^2(c+a)^2+c^2(a+b)^2+2abc(a+b+c) = 2(ab+bc+ca)^2$.

Art. V. To apply formula [4] to obtain the product of two factors which differ only in the signs of some of their terms:— group together all the terms whose signs are the same in one factor as they are in the other, and then form into a second group all the other terms.

Examples.

1. Multiply $a+b-c+d$ by $a-b-c-d$; here the first group is $a-c$, the second $b+d$; ∴ we have
$\{(a-c)+(b+d)\}\{(a-c)-(b+d)\} = (a-c)^2-(b+d)^2$.

2. $(1 + 3x + 3x^2 + x^3)(1 - 3x + 3x^2 - x^3.) = \{(1+3x^2) + (3x+x^3)\}\{(1+3x^2) - (3x+x^3)\} = (1+3x^2)^2 - (3x+x^3)^2 = 1 - 3x^2 + 3x^4 - x^6$.

3. Find the continued product of $a+b+c$, $b+c-a$, $c+a-b$ and $a+b-c$.

The first pair of factors gives $\{(b+c)+a\}\{(b+c)-a\} = (b+c)^2 - a^2 = b^2 + 2bc + c^2 - a^2$.

The second pair gives $\{a-(b-c)\}\{a+(b-c)\} = a^2 - b^2 + 2bc - c^2$; the only term whose sign is the same in both these results is $2bc$; hence, grouping the other terms, we have
$\{2bc+(b^2+c^2-a^2)\}\{2bc-(b^2+c^2-a^2)\} = (2bc)^2 - (b^2+c^2-a^2)^2 = 2a^2b^2 + 2b^2c^2 + 2c^2a^2 - a^4 - b^4 - c^4$.

4. Prove $(a^2+ab+b^2)^2 - a^2b^2 = (a^2+ab)^2 + (ab+b^2)^2$.

The expression $= (a^2+b^2)(a^2+2ab+b^2) = (a^2+b^2)(a+b)^2 = a^2(a+b)^2 + b^2(a+b)^2 = (a^2+ab)^2 + (ab+b^2)^2$.

Exercise vii.

1. $(a^2+2ab+b^2)(a^2-2ab+b^2)$.
2. $(\frac{1}{2}x^2-xy+y^2)(\frac{1}{2}x^2+y^2+xy)$.
3. $(a^2-ab+2b^2)(a^2+ab+2b^2)$; $(x^4+4xy)(x^4-4xy)$.
4. $\{(x+y)x-y(x-y)\}\{(x-y)x-y(y-x)\}$.
5. Simplify: $(x+3)(x-3)+(x+4)(x-4)-(x+5)(x-5)$
6. " $(1+x)^4+(1-x)^4-2(1-x^2)^2$.
7. $(x^2+y^2)^2 - (2xy)^2 - (x^2-y^2)^2$.
8. $(2a^2-3b^2+4c^2)(2a^2+3b^2-4c^2)$.
9. $(2a+b-3c)(b+3c-2a)$; $(2a-b-3c)(b-3c-2a)$.
10. $(x^4+y^4)(x^2+y^2)(x+y)(x-y)$.
11. $(x^2+xy+y^2)(x^2-xy+y^2)(x^4-x^2y^2+y^4)$.
12. $(a+b-ab-1)(a+b+ab+1)$.
13. Prove $(a^2+b^2+c^2)(b^2+c^2-a^2)(c^2+a^2-b^2)(a^2+b^2-c^2) = 4b^4c^4$ when $a^4=b^4+c^4$.
14. $(x^2+y^2-\frac{5}{4}xy)(x^2+y^2+\frac{5}{4}xy)$.
15. $(x^4-2x^3+3x^2-2x+1)(x^4+2x^3+3x^2+2x+1)$.

16. Multiply $(2x-y)a^2 - (x+y)ax + x^3$ by $(2x-y)a^2 + (x+y)ax - x^3$.

Prove the following:

17. $(a^2+b^2+c^2+ab+bc+ca)^2 - (ab+bc+ca)^2 = (a+b+c)^2 \times (a^2+b^2+c^2)$.

18. $(a^2+b^2+c^2+ab+bc+ca)^2 - (a^2+ab+ca-bc)^2 = \{(a+b)(b+c)\}^2 + \{(b+c)(c+a)\}^2$.

19. $4(ab+cd)^2 - (a^2+b^2-c^2-d^2)^2 = (a+b+c-d)(a+b-c+d)(c+d+a-b)(c+d-a+b)$.

20. Find the product of $x^2+y^2+z^2-2xy+2xz-2yz$ and $x^2+y^2+z^2-2xy-2xz+2yz$.

21. $(x^2+y^2+xy\sqrt{2})(x^2-xy\sqrt{2}+y^2)(x^4-y^4)$.

22. $(1-6a+9a^2)(1+2a+3a^2)$.

23. $\{(m+n)+(p+q)\}(m-q+p-n)$.

24. Obtain the product of $1+x+x^2$, x^2+x-1, x^2-x+1, and $+x-x^2$.

25. $(a-b^2)^2 (a+b^2)^2 (a^2+b^4)^2 (a^4+b^8)^2$.

26. Shew that $(x^2+xy+y^2)^2 (x^2-xy+y^2)^2 - (x^2y^2)^2 = x^4+x^2y^2)^2+(x^2y^2+y^4)^2$.

FORMULA A.—EXAMPLES.

1. Multiply x^2-x+5 by x^2-x-7: here the common term is x^2-x, the other terms $+5$, and -7, hence the product $=(x^2-x)^2 +(-7+5)(x^2-x)+(-7 \times 5)=(x^2-x)^2-2(x^2-x)-35=x^4-2x^3-x^2+2x-35$.

2. $(x-a)(x-3a)(x+4a)(x+6a)$: taking the first and third factors together, and the second and fourth, we have the product $=(x^2+3ax-4a^2)(x^2+3ax-18a^2) = (x^2+3ax)^2 - (4a^2+18a^2) \times (x^2+3ax) - 72a^4 =$ &c.

Exercise viii.

1. $(x^2+2x+3)(x^2+2x-4)$; $(x-y+3z)(x-y+5z)$.
2. $(x+1)(x+5)(x+2)(x+4)$; $(x^3+a-b)(x^3+2b-a)$.
3. $(a^2-3)(a^2-1)(a^2+5)(a^2+7)$; $(x^4+x^2+1)(x^4+x^2-2)$.
4. $\{(x+y)^2-4xy\}\{(x+y)^2+5xy\}$.

5. $(x^n+a+7)(x^n-a-9)$; $\left(\frac{x}{y}+\frac{y}{x}-1\right)\left(\frac{x}{y}+\frac{y}{x}+3\right)$.

6. $(nx+y+3)(nx+y+7)$.

7. $(x+a-y)(x+a+3y)$.

8. $(x^{2n}+x^n-a)(x^{2n}+x^n-b)$.

9. $(\frac{1}{2}x^4-y^2+2)(\frac{1}{2}x^4-y^2-4)$.

10. $\left(\frac{1}{x}+\frac{1}{y}-\frac{1}{2}\right)\left(\frac{1}{x}+\frac{1}{y}+2\frac{1}{2}\right)$.

11. Multiply together $x-2+\sqrt{2}$, $x-2+\sqrt{3}$, $x-2-\sqrt{2}$, and $x-2-\sqrt{3}$.

12. $(x+a+b)(x+b-c)(x-a+b)(x+b+c)$.

13. $(a+b+c)(a+b+d)+(a+c+d)(b+c+d)-(a+b+c+d)^2$.

14. Prove that
$(2a+2b-c)(2b+2c-a)+(2c+2a-b)(2a+2b-c)+(2b+2c-a)(2c+2a-b) = 9(ab+bc+ca)$.

Formulas [5] and [6].—Examples.

1. We get at once
$(x+y)^3+(x-y)^3 = 2x(x^2+3y^2)$.
$(x+y)^3-(x-y)^3 = 2y(3x^2+y^2)$.

2. Simplify $(a+b+c)^3 - 3(a+b+c)^2c + 3(a+b+c)c^2 - c^3$.

This plainly comes under formula [5], the first term being $a+b+c$, the second $-c$; hence the expression is $\{(a+b+c)-c\}^3 = (a+b)^3$.

3. Shew that $(x^2+xy+y^2)^3+(xy-x^2-y^2)^3 - 6xy(x^4+x^2y^2+y^4) = 8x^3y^3$.

This comes under formula [6], the first term being (x^2+xy+y^2), and the second $-(x^2-xy+y^2)$; we have therefore
$\{(x^2+xy+y^2)-(x^2-xy+y^2)\}^3 = (2xy)^3 = 8x^3y^3$.

Exercise ix.

Simplify

1. $(1-x^2)^3+(1+x^2)^3$, $(x^2+xy^3)^3-(x^2-xy^3)^3$.

2. $(a+2b)^3-(a-b)^3$, $(3a-b)^3-(3a-2b)^3$.

FUNDAMENTAL FORMULAS.

3. $(x+y-z)^3 + 3(x+y-z)^2 z + z^3 + 3(x+y-z)z^2$.

4. $(a-b)^3 + (a+b)^3 + 6a(a^2 - b^2)$.

5. $(x-y)^3 + (x+y)^3 + 3(x-y)^2(x+y) - 3(y-x)(x+y)^2$.

6. $(1+x+x^2)^3 - (1-x+x^2)^3 - 6x(1+x^2+x^4)$.

7. $(a-b-c)^3 + (b+c)^3 + 3(b+c)^2(a-b-c) + 3(a-b-c)^2(b+c)$.

8. $(3x-4y+5z)^3 - (5z-4y)^3 + 3(5z-4y)^2(3x-4y+5z) - 3(3x-4y+5z)^2(5z-4y)$.

9. $(1+x+x^2)^3 + 3(1-x^3)(2+x^2) + (1-x)^3$.

10. Shew that $a(a-2b)^3 - b(b-2a)^3 = (a-b)(a+b)^3$.

11. Shew that $a^3(a^3-2b^3)^3 + b^3(2a^3-b^3)^3 = (a^3-b^3)(a^3+b^3)^3$.

12. $(x^2+xy+y^2)^3 + 6(x^2+y^2)(x^4+xy+y^4) + (x^2-xy+y^2)^3$.

13. Shew that $a^3(a^3+2b^3)^3 + b^3(2a^3+b^3)^3 + (3a^2b^2)^3 = (a^6+7a^3b^3+b^6)^2$.

14. Simplify $(ax+by)^3 + a^3y^3 + b^3x^3 - 3abxy(ax+by)$.

✓ 15. What will $a^3+b^3+c^3 - 3abc$ become when $a+b+c=0$? 0

16. Find the value of $x^6 - y^6 + z^6 + 3x^2y^2z^2$ when $x^2 - y^2 + z^2 = 0$.

FORMULAS [7], [8] AND [9].—EXAMPLES.

1. Simplify $(2x-3y)^3 + (4y-5x)^3 + (3x-y)^3 - 3(2x-3y)(4y-5x)(3x-y)$.

By [8] this is seen to be $\{(2x-3y)+(4y-5x)+(3x-y)\}^3 = (0)^3 = 0$.

✓ 2. Prove that $(a-b)^3 + (b-c)^3 + (c-a)^3 = 3(a-b)(b-c)(c-a)$.

In [8] substitute $a-b$ for x, $b-c$ for y, and $c-a$ for z; for these values $x+y+z=0$, and the identity appears at once.

✓ 3. Prove $(a+b+c)^3 - (b+c-a)^3 - (a+c-b)^3 - (a+b-c)^3 = 24abc$.

In [7] let $x=b+c-a$, $y=c+a-b$, $z=a+b-c$, and therefore $x+y=2c$, $y+z=2a$, $z+x=2b$, and this identity at once appears.

FUNDAMENTAL FORMULAS.

Exercise x.

1. Cube the following: $1-x+x^2$, $a-b-c$, $1-2x+3x^2-4x^3$.

2. Simplify $(x^2 + 2x - 1)^3 + (2x-1)(x^2 + 2x - 2) - (x^3 + 3x^2 - 1)^3$.

3. *Prove that $(x+y)(y+z)(z+x) + xyz = (x+y+z)(xy+yz+zx)$.

4. Prove that $(ax - by)^3 + a^3y^3 - b^3x^3 + 3abxy(ax - by) = (a^3 - b^3)(x^3 + y^3)$.

5. Simplify $(x-2y)^3 + (y-2z)^3 + (z-2x)^3 + 3(x-y-2z) \times (y-z-2x)(z-x-2y) + (x+y+z)^3$.

6. Simplify $(2x^2 - 3y^2 + 4z^2)^3 + (2y^2 - 3z^2 + 4x^2)^3 + (2z^2 - 3x^2 + 4y^2)^3$.

7. Simplify $(2ax - by)^3 + (2by - cz)^3 + (2cz - ax)^3 + 3(2ax+by-cz)(2by+cz-ax)(2cz+ax-by)$.

8. Prove $(x^3 + 3x^2y - y^3)^3 + \{3xy(x+y)\}^3 = \{(x-y)^3 + 9x^2y\} \times \{x^2+xy+y^2\}^3$.

9. Prove $9(x^3+y^3+z^3) - (x+y+z)^3 = (4x+y+z)(x-y)^2 + (4y+4z+x)(y-z)^2 + (4z+4x+y)(z-x)^2$.

10. If $x+y+z=0$, shew that $x^3+y^3+z^3 = 3xyz$.

11. If $x=2y+3z$, shew that $x^3 - 8y^3 - 27z^3 - 18xyz = 0$.

12. Shew that $(x^2+xy+y^2)^3 + (x^2-xy+y^2)^3 + 8z^6 - 6z^2(x^4+x^2y^2+y^4) = 0$, if $x^2+y^2+z^2=0$.

13. Prove that $8(a+b+c)^3 - (a+b)^3 - (b+c)^3 - (c+a)^3 = 3(2a+b+c)(a+2b+c)(a+b+2c)$.

Prove the following:

14 $(ax-by)^3 + b^3y^3 = a^3x^3 + 3abxy(by-ax)$.

---

*Note that the right-hand member is formed from the left-hand one by changing *additions* into *multiplications*, and *multiplications* into *additions*; hence in $(x+y+z).(xy+yz+zx)$ the signs $+$ and $.$ may be interchanged throughout without altering the value of the expression.

15. $a^3+b^3+c^3-3abc = \frac{1}{2}\{(a-b)^2+(b-c)^2+(c-a)^2\} \times (a+b+c)$.

16. $(a+b+c)\{(a+b-c)(b+c-a)+(b+c-a)(c+a-b)+(c+a-b)(a+b-c)\} = (a+b-c)(b+c-a)(c+a-b)+8abc$.

17. $a^3+b^3+c^3+24abc = (a+b+c)^3 - 3\{a(b-c)^2+b(c-a)^2+c(a-b)^2\}$.

18. $(a+b+7c)(a-b)^2+(b+c+7a)(b-c)^2+(c+a+7b)(c-a)^2 = 2(a+b+c)^3 - 54abc$.

19. $(a+b+c)\{(2a-b)(2b-c)+(2b-c)(2c-a)+(2c-a)(2a-b)\} = (2a-b)(2b-c)(2c-a)+(2a+b-c)(2b+c-a)(2c+a-b)$.

20. If $x^2(y+z)=a^3$, $y^2(z+x)=b^3$, $z^2(x+y)=c^3$, and $xyz=abc$, shew that $a^3+b^3+c^3+2abc = (x+y)(y+z)(z+x)$

EXPANSION OF BINOMIALS.

We have from formula [5]

$(a+b)^3 = a^3+3a^2b+3ab^2+b^3$; multiplying by $a+b$ we get
$(a+b)^4 = a^4 + 4a^3b + 6a^2b^2+4ab^3+b^4$; multiplying this by $a+b$ we get
$(a+b)^5 = a^5+5a^4b+10a^3b^2+10a^2b^3+5ab^4+b^5$.

From these examples we derive the following law for the formation of the terms in the expansion of $a+b$ to any required power :—

(1). The *index* of a, in the *first* term, is that of the given power, and *decreases* by unity in each succeeding term; the index of b begins with unity in the *second* term and *increases* by unity in each succeeding term.

(2). The coefficient of the first term is unity, and the coefficient of any other term is found by multiplying the coefficient of the immediately preceding term by the index of a in that term, and dividing the product by the number of that preceding term. It will be observed that the coefficients equally distant from the extremes of the expansion, are equal.

Exercise xi.

1. Expand $(x+y)^6$, $(x+y)^7$, $(x+y)^8$, $(x+y)^{12}$.
2. What will be the law of *signs* if $-y$ be written for y in (1)?
3. Expand $(a-b)^5$, $(a-2b)^4$, $(2b-a)^4$.
4. Expand $(1+m)^6$, $(m+1)^5$, $(2m+1)^6$.
5. What is the coefficient of the 4th term in $(a-b)^{10}$?
6. Expand $(x^2-y)^4$, $(a-2b^2)^5$, $(a^3-2b^3)^6$.
7. In the expansion of $(a-b)^{12}$, the third term is $66a^{10}b^2$, find the 5th and 6th terms.
8. Shew that $(x+y)^5 - x^5 - y^5 = 5xy(x+y)(x^2+xy+y^2)$.
9. From (8) shew that $2\{(a-b)^5 + (b-c)^5 + (c-a)^5\} = 5(a-b)(b-c)(c-a)\{(a-b)^2 + (b-c)^2 + (c-a)^2\}$.

SECTION III.—HORNER'S METHODS OF MULTIPLICATION AND DIVISION.

EXAMPLES.

1. Find the product of $kx^3 + lx^2 + mx + n$ and $ax^2 + bx + c$. Write the multiplier in a column to the left of the multiplicand, placing each term in the same horizontal line with the partial product it gives:

	kx^3	$+lx^2$	$+mx$	$+n$	
ax^2	akx^5	$+alx^4$	$+amx^3$	$+anx^2$	......p_1
$+bx$		$+bkx^4$	$+blx^3$	$+bmx^2 + bnx$	......p_2
$+c$			$+ckx^3$	$+clx^2 + cmx + cn$	......p_3

$akx^5 + (al+bk)x^4 + (am+bl+ck)x^3 + (an+bm+cl)x^2 + (bn+cm)x + cn$. .... P.

Art. VI. The above example has been given in full, the powers of x being inserted; in the following example detached coefficients are used. It is evident that if the coefficient of the first term of the multiplier be unity, the coefficients of the multiplicand will be the same as those of the first partial product, and may be used for them, thus saving the repetition of a line.

MULTIPLICATION AND DIVISION.

2. Multiply $3x^4 - 2x^3 - 2x + 3$ by $x^2 + 3x - 2$.

$$
\begin{array}{r|rrrrrr}
1 & 3 & -2 & +0 & -2 & +3 & \\
+3 & & +9 & -6 & +0 & -6 & +9 \\
-2 & & & -6 & +4 & -0 & +4 & -6 \\
\hline
& 3x^6 & +7x^5 & -12x^4 & +2x^3 & -3x^2 & +13x & -6.
\end{array}
$$

3. Find the product of $(x-3)(x+4)(x-2)(x-5)$.

$$
\begin{array}{r|rrrrr}
 & 1 & -3 & & & \\
+4 & & +4 & -12 & & \\
\hline
 & 1 & +1 & -12 & & \\
-2 & & -2 & -2 & +24 & \\
\hline
 & 1 & -1 & -14 & +24 & \\
-5 & & -5 & +5 & +70 & -120 \\
\hline
 & x^4 & -6x^3 & -9x^2 & +94x & -120.
\end{array}
$$

4. Multiply $x^3 - 4x^2 + 2x - 3$ by $2x^3 - 3$.

$$
\begin{array}{r|rrrrrrr}
 & 1 & -4 & +2 & -3 & & & \\
2 & 2 & -8 & +4 & -6 & & & \quad [x^3 \times x^3 = x^6] \\
0 & & 0 & 0 & 0 & 0 & & \\
0 & & & 0 & 0 & 0 & 0 & \\
-3 & & & & -3 & +12 & -6 & +9 \\
\hline
 & 2x^6 & -8x^5 & +4x^4 & -9x^3 & +12x^2 & -6x & +9
\end{array}
$$

In this example the missing terms of the muitiplier are supplied by zeros; but instead of writing the zeros as in the example, we may, as in ordinary arithmetical multiplication, "skip a line" for every missing term.

5. Multiply $x^4 - 2x^2 + 1$ by $x^4 - x^2 + 3$.

$$
\begin{array}{r|rrrrr}
1 & 1 & +0 & -2 & +0 & +1 & & & \\
-1 & & -1 & -0 & +2 & -0 & -1 & \quad [x^4 \times x^4 = x^8] \\
+3 & & & +3 & +0 & -6 & +0 & +3 \\
\hline
 & x^8 & & -3x^6 & & +6x^4 & -7x^2 & +3
\end{array}
$$

MULTIPLICATION AND DIVISION. 23

6. Find the value of $(x+2)(x+3)(x+4)(x+5) - 9(x+2)(x+3)$
$\times (x+4) + 3(x+2)(x+3) + 77(x+2) - 85$.

$$
\begin{array}{r|rrrrr}
 & 1 & +5 & & & \\
 & & -9 & & & \\
\hline
 & 1 & -4 & & & \\
+4 & & +4 & -16 & & \\
 & & & +8 & & \\
\hline
 & 1 & +0 & -13 & & \\
+3 & & +3 & +0 & -39 & \\
 & & & & +77 & \\
\hline
 & 1 & +3 & -13 & +38 & \\
+2 & & +2 & +6 & -26 & +76 \\
 & & & & & -85 \\
\hline
 & x^4 & +5x^3 & -7x^2 & +12x & -9 \\
\end{array}
$$

7. Find the coefficient of x^4 in the product of $x - ax^3 + bx^2 - cx + d$ and $x^2 + px + q$.

$$
\begin{array}{r|rrrrr}
 & 1 & -a & +b & -c & +d \\
+p & & & -ap & & \\
+q & & & +q & & \\
\hline
 & & & +(b-ap+q) & & \\
\end{array}
$$

Exercise xii.

Find the product of

1. $(1+x+x^2+x^3+x^4)(1-x+x^3-x^7+x^9-x^{12}+x^{13})$.
2. $(1+x^5)(1-x^5+x^6)(1+x+x^2+x^3+x^4)$.
3. $(x-5)(x+6)(x-7)(x+8)$; $(2x^5-x^2+1)(x^4-x+2)$.
4. $(x^3+5x^2-16x-1)(x^3-5x^2-16x+1)$.
5. $(6x^6-x^5+2x^4-2x^3+2x^2+19x+6)(3x^2+4x+1)$.

Obtain the coefficients of x^4 and lower powers in

6. $(1+\tfrac{1}{2}x-\tfrac{1}{8}x^2+\tfrac{1}{16}x^3-\tfrac{5}{128}x^4)(1-\tfrac{1}{2}x-\tfrac{1}{8}x^2-\tfrac{1}{16}x^3-\tfrac{5}{128}x^4)$.
7. Multiply $2x^7-x^3+3x-4$ by $3x^5-2x^2-x-1$.

Simplify the following:

8. $(x+1)(x+2)(x+3)+3(x+1)(x+2)-10(x+1)+9$.
9. $x(x+1)(x+2)(x+3)-3x(x+1)(x+2)-2x(x+1)+2x$.
10. $x(x-1)(x-2)(x-3)+3x(x-1)(x-2)-2x(x-1)-2x$.
11. $(x-1)(x+1)(x+3)(x+5)-14(x-1)(x+1)+1$.

12. Given that the sum of the four following factors is -1, find (1) the product of the first pair; (2) the product of the second pair; and (3) the product of the sum of the first pair by the sum of the second pair.

(1) x $+x^4$ $+x^{13}$ $+x^{16}$
(2) x^2 $+x^6$ $+x^9$ $+x^{15}$
(3) x^3 $+x^5$ $+x^{12}$ $+x^{14}$
(4) x^6 $+x^7$ $+x^{10}$ $+x^{11}$.

13. Given that the sum of the three following factors is equal to -1, find their product.

(1) x $+x^5$ $+x^8$ $+x^{12}$
(2) x^2 $+x^3$ $+x^{10}$ $+x^{11}$
(3) x^4 $+x^6$ $+x^7$ $+x^9$.

Art. VII. Were it required to *divide* the product P in the first of the above examples by ax^2+bx+c, it is evident that could we find and *subtract* from P the partial products p_2, p_3, (or what would give the same result, could we *add* them with the sign of each term changed), there would remain the partial product p_1, which, divided by the monomial ax^2, would give the quotient Q. This is what Horner's method does, the change of sign being secured by changing the signs of b and c, which are factors in each term of p_2, p_3, respectively.

MULTIPLICATION AND DIVISION.

$$
\begin{array}{r|l}
1. & akx^5 + (al+bk)x^4 + (am+bl+ck)x^3 + (an+bm+cl)x^2 + (bn+cm)x + cn \ldots\ldots P \\
-bx & -blx^4 -bmx^3 -bmx^2 -bnx \ldots\ldots p_2 \\
-c & -clx^3 -clx^2 -cmx\ -cn \ldots\ldots p_3 \\
\hline
ax^2 & akx^5 +alx^4 +amx^3 \ldots\ldots p_1 \\
ax^2 & kx^3 +lx^2 +mx +n \ldots\ldots Q
\end{array}
$$

The dividend and divisor are arranged as in the example, the sign of every term in the divisor, except the first, being changed in order to turn the subtractions into additions. The first term of the dividend (akx^5) is brought down into the line of p_1: dividing this by ax^2, the first term of the divisor, we get kx^3, *the first term of the quotient*. Multiplying this term, kx^3, by $-bx$ and $-c$, respectively, and writing the products in the proper *columns* and *rows*, makes all ready to give the *second* term of p_1, which is got by simply adding up the second column of the work, giving alx^4. Dividing this second term of p_1 by ax^2, gives lx^2, *the second term of the quotient*. Multiply lx^2 by $-bx$ and $-c$ respectively, and proceed in the same way as was done in getting the second term of the quotient, and the third will be obtained. Repeating the steps, the complete quotient and the remainder will finally be obtained.

Should the coefficient of the first term of the divisor be *unity*, the coefficients of the line Q will be the same as those of p_1, and the line Q need not be written down, since one line does for both.

MULTIPLICATION AND DIVISION.

2. Divide $3x^6 + 7x^5 - 12x^4 + 2x^3 - 3x^2 + 13x - 6$ by $x^2 + 3x - 2$.

$$\begin{array}{r|rrrrrr} & 3 & +7 & -12 & +2 & -3 & +18 & -6 \\ -3 & & -9 & +6 & -0 & +6 & -9 \\ +2 & & & +6 & -4 & +0 & -4 & +6 \\ \hline & 3x^4 & -2x^3 & +0 & -2x & +3 \end{array} \quad \{x^6 \div x^2 = x^4\}$$

Compare this example with the second example of Horner's Multiplication, performing a step in multiplication, then the corresponding step in division; then another step in multiplication and the second (corresponding) step in division, and so on.

3. Divide $x^7 - 3x^6 + 4x^4 + 18x^3 - 7x + 12$ by $x^3 - 3x^2 + 3x - 1$.

$$\begin{array}{r|rrrrrrr} & 1 & -3 & +0 & -4 & +18 & +0 & -7 & +12 \\ +3 & & +3 & +0 & -9 & -36 & -27 \\ -3 & & & -3 & -0 & +9 & +36 & +27 \\ +1 & & & & +1 & +0 & -3 & -12 & -9 \\ \hline & x^4 & +0 & -3x^2 & -12x & -9; & 6x^2 & +8x & +3 \end{array} \quad [x^7 \div x^3 = x^4].$$

The quotient is therefore $x^4 - 3x^2 - 12x - 9$, and the remainder $6x^2 + 8x + 3$.

4. Divide $x^8 - 3x^7 - 5x^5 + 2x^4 + 5x^3 + 4x^2 + 1$ by $x^3 + 2x - 1$. The zero coefficient in the divisor may be inserted, or it may be omitted and allowance made for it in the $2x$-line. See examples 4 and 5 in multiplication.

$$\begin{array}{r|rrrrrrrr} & 1 & -3 & +0 & -5 & +2 & +5 & +4 & +0 & +1 \\ -2 & & & & -2 & +6 & +4 & -4 & -6 & +2 \\ +1 & & & & & 1 & -3 & -2 & +2 & +3 & -1 \\ \hline & 1 & -3 & -2 & +2 & +3 & & 1; & 0 & +5 & +0 \end{array}$$

$[x^8 \div x^3 = x^5]$. The quotient is therefore $x^5 - 3x^4 - 2x^3 + 2x^2 + 3x - 1$, and the remainder $5x$.

5. Divide $10x^6 - 11x^5 - 3x^4 + 20x^3 + 10x^2 + 2$ by $5x^3 - 3x^2 + 2x - 2$.

MULTIPLICTION AND DIVISION.

Arranging as in the ordinary method, we have

$$
\begin{array}{r|rrrr|rrr}
 & 10 & -11 & -3 & +20 & +10 & +0 & +2 \\
+3 & & 6 & -3 & -\ 6 & +12 & & \\
-2 & & & -4 & +\ 2 & +\ 4 & -8 & \\
+2 & & & & +\ 4 & -\ 2 & -4 & +8 \\
\hline
5 & 2 & -1 & -2 & +\ 4 & 24 & -12 & +10
\end{array}
$$

Quotient $= 2x^3 - x^2 - 2x + 4 + \dfrac{24x^2 - 12x + 10}{5x^3 - 3x^2 + 2x - 2}$.

We first draw a vertical line with as many vertical columns to the right as are less by unity than the number of terms in the divisor. This will mark the point at which the remainder begins to be formed. We then divide 5 into 10, and thus obtain the first coefficient of the dividend. We next multiply the remaining terms of the divisor by the 2 thus obtained. Adding the second vertical column and dividing by 5, we obtain -1; we multiply by the -1, add the next column and divide the sum by 5, and so on for the others.

This method is not, however, always convenient. If the first term of the dividend be not divisible by the first term of the divisor, the work would be embarrassed with fractions. We may then proceed as in the following examples:

6. Divide $x^5 - 3x^4 + x^3 + 3x^2 - x + 3$ by $2x^3 + x^2 - 3x + 1$.

Let $2x = y$, or $x = \dfrac{y}{2}$.

Substitute $\dfrac{y}{2}$ for x in the dividend and divisor, and we have

$\dfrac{y^5}{2^5} - \dfrac{3y^4}{2^4} + \dfrac{y^3}{2^3} + \dfrac{3y^2}{2^2} - \dfrac{y}{2} + 3 \div \dfrac{2y^3}{2^3} + \dfrac{y^2}{2^2} - \dfrac{3y}{2} + 1$

$= \dfrac{y^5 - 2 \times 3y^4 + 2^2 y^3 + 2^3 \times 3y^2 - 2^4 y + 2^5 \times 3}{2^5} \div$

$\dfrac{y^3 + y^2 - 2 \times 3y + 2^3}{2^2}$

$= \dfrac{y^5 - 6y^4 + 4y^3 + 24y^2 - 16y + 96}{2^3} \div y^3 + y^2 - 6y + 4 \ldots\ldots A.$

Dividing $y^5-6y^4+4y^3+24y^2-16y+96$ by y^3+y^2-6y+4 by the ordinary method, and the quotient by 2^3 we have

$$\frac{y^2-7y+17}{2^3} - \frac{1}{2^3}\cdot\frac{39y^2-114y-28}{y^3+y^2-6y+4}.$$

Substituting for y its value $2x$, and simplifying we get

$$\frac{x^2}{2} - \frac{7x}{4} + \frac{17}{8} - \frac{1}{8}\cdot\frac{39x^2-57x-7}{2x^3+x^2-3x+1}.\quad\ldots\ldots\ldots\ldots B.$$

By comparing the dividend of A with the original question, we find that we have multiplied the successive coefficients of the dividend by 2^0, 2^1, 2^2, &c., and, omitting the first term, we have multiplied the successive coefficients of the divisor by the same numbers. Dividing then by Horner's division we get the coefficients 1, -7, 17, and for coefficients of remainder, -39, 114, and 28. The first of these divided by 2, 2^2, 2^3 are the coefficients of x^3 &c.; and, -39, &c., are divided by 1, 2, 2^2. Hence the work will stand as follows:—

```
      x⁵ −3x⁴ +x³ + 3x² −   x+  3 ÷ 2x³ +x² −3x+1
       1    2   4    8     16  32      1   2    4
    ┌─ 1. −6  +4 +24   − 16+96      1  −6  +4
 −1 │    −1  +7 −17
 +6 │        +6 −42   +102
  4 │           − 4   + 28 −68
    └──────────────────────────
      1.−7 +17 − 39   +114 +28
```

$$*\text{Quotient} = \frac{x^3}{2} - \frac{7x}{4} + \frac{17}{8} - \frac{1}{8}\cdot\frac{39x^2-\dfrac{114x}{2}-\dfrac{28}{4}}{2x^3+x^2-3x+1}$$

$$= \frac{x^2}{2} - \frac{7x}{4} + \frac{17}{8} - \frac{1}{8}\cdot\frac{39x^2-57x-7}{2x^3+x^2-3x+1}.$$

---

*It will, in general, be as convenient to multiply the dividend by such a number as will make its first term exactly divisible by the first term of the divisor, and afterwards divide the quotient by this multiplier.

7. Divide $5x^5 + 2$ by $3x^2 - 2x + 3$.

$$\begin{array}{cccccc|cc}
5x^5 & 0 & 0 & 0 & 0 & +2 & \div\ 3x^2 & -2x+3 \\
1 & 3 & 9 & 27 & 81 & 243 & 1 & 3
\end{array}$$

$$\begin{array}{c|cccc|cc}
 & 5 & 0 & 0 & 0 & 0 & +486 \\
+2 & & 10 & +20 & -50 & -280 & \\
-9 & & & -45 & -90 & +225 & +1260 \\
\hline
 & 5 & +10 & -25 & -140 & -55 & +1746
\end{array}$$
$\quad\quad\quad\quad\quad\quad\quad\quad\quad\quad\quad\quad -2\ +9$

Coeffs. of Quotient $= \dfrac{5}{3} + \dfrac{10}{3^2} - \dfrac{25}{3^3} - \dfrac{140}{3^4} - \dfrac{1}{3^4} \cdot \dfrac{55 - \frac{1746}{3}}{3 - 2 + 3}$.

Quotient $= \dfrac{5x^3}{3} + \dfrac{10x^2}{9} - \dfrac{25x}{27} - \dfrac{140}{81} - \dfrac{1}{81} \cdot \dfrac{55x - 582}{3x^2 - 2x + 3}$.

Exercise xiii.

1. Divide $6x^5 + 5x^4 - 17x^3 - 6x^2 + 10x - 2$ by $2x^2 + 3x - 1$.
2. $(5x^6 + 6x^5 + 1) \div (x^2 + 2x + 1)$.
3. $(a^6 - 6a + 5) \div (a^2 - 2a + 1)$.
4. $(x^5 - 4x^3y^2 - 8x^2y^3 - 17xy^4 - 12y^5) \div (x^2 - 2xy - 3y^2)$.
5. $(a^6 - 3a^4x^2 + 3a^2x^4 - x^6) \div (a^3 - 3a^2x + 3ax^2 - x^3)$.

Divide

6. $4x^4 + 3x^2 - 3x + 1$ by $x^2 - 2x + 3$.
7. $10x^5 + 5x^4 - 90x^3 - 44x^2 + 10x + 1$ by $x^2 - 9$.
8. $x^5 - x^4y + x^3y^2 - x^2y^3 + xy^4 - y^5$ by $x^3 - y^3$.
9. Multiply $x^4 - 4x^3a + 6x^2a^2 - 4xa^3 + a^4$ by $x^2 + 2xa + a^2$, and divide the product by $x^4 - 2x^3a + 2xa^3 - a^4$.

Divide

10. $x^5 - ax^4 + bx^3 - bx^2 + ax - 1$ by $x - 1$.
11. $6x^5 + 7x^4 + 7x^3 + 6x^2 + 6x + 5$ by $2x^2 + x + 1$.
12. $60(x^4 + y^4) + 91xy(x^2 - y^2)$ by $12x^2 - 13xy + 5y^2$.

13. $6x^6 - 481x^5 + 79x^4 + 81x^3 - 81x^2 + 86x - 481$ by $x - 80$.

14. $6x^6 - x^5 + 2x^4 - 2x^3 + 2x^2 + 19x + 6$ by $3x^2 + 4x + 1$.

15. $a(a+2b)^3 - b(2a+b)^3$ by $(a-b)^3$.

16. $(x+y)^3 + 3(x+y)^2z + 3(x+y)z^2 + z^3$ by $(x+y)^2 + 2(x+y)z + z^2$.

17. $10x^{10} + 10x^6 + 10x^3 - 200$ by $x^7 + x^3 - x + 1$.

18. $bmx^4 + (bn+cm)x^3 + cnx^2 + abx + ac$ by $bx + c$.

19. Multiply $1 + \tfrac{2}{13}x - 18x^3$ by $1 - \tfrac{1}{4}x^2 + \tfrac{3}{2}x^3$ and divide the product by $1 + \tfrac{1}{2}x - 8x^2$.

Find the remainders in the following cases:

20. $(x^3 + 3x^2 + 4x + 5) \div (x - 2)$.

21. $(x^4 - 3x^2 + x - 3) \div (x - 1)$.

22. $(x^4 + 4x^3 + 6x + 8) \div (x + 2)$.

23. $(27x^4 - y^4) \div (3x - 2y)$.

24. $(3x^5 + 5x^4 - 3x^3 + 7x^2 - 5x + 8) \div (x^2 - 2x)$.

25. $(5x^4 + 90x^3 + 80x^2 - 100x + 500) \div (x + 17)$.

Art. viii. The following are examples of an important use of Horner's Division:

1. Arrange $x^3 - 6x^2 + 7x - 5$ in powers of $x - 2$.

```
  | 1      -6      7      -5
2 |         2     -8      -2
  |_____
  | 1      -4     -1;     -7
2 |         2     -4
  |_____
  | 1      -2;    -5
2 |        +2
  |_____
  | 1;      0
```

Hence, $x^3 - 6x^2 + 7x - 5 = (x-2)^3 - 5(x-2) - 7$, or as it is generally expressed, $x^3 - 6x^2 + 7x - 5 = y^3 - 5y - 7$ if $y = x - 2$.

MULTIPLICATION AND DIVISION. 81

2. Express $x^4+12x^3+47x^2+66x+28$ in powers of $x+3$.

	1	12	47	66	28
−3		−3	−27	−60	−18
	1	9	20	6 ;	10
−3		−3	−18	−6	
	1	6	2 ·	0	
−3		−3	−9		
	1	3 ;	−7		
−3		−3			
	1 ;	0.			

Hence $x^4+12x^3+47x^2+66x+28 = y^4-7y^2+10$ if $y=x+3$.

After a few solutions have been written out in full, as in the above examples, the writing may be lessened by omitting the lines opposite the increments (-2 in Ex. 1, and 3 in Ex. 2), the multiplication and addition being performed mentally. The last example written in this way would appear as follows:

	1	12	47	66	28
−3	1	9	20	6	(10)
	1	6	2	(0)	
	1	3	(−7)		
	1	(0)			

Exercise xiv.

1. Express x^3-5x^2+3x-8 in powers of $x-1$.
2.   " x^3+3x^2+6x+9 " $x+1$.
3.   " $x^4-8x^3+24x^2-32x+97$ in powers of $x-2$.
4.   " $x^4+12x^3+5x^2-7$ " $x+2$.
5.   " $3x^5-x^3+4x^2+5x-8$ " $x-2$.
6.   " $x^4-7x^3+11x^2-7x+10$ " $x-1\frac{3}{4}$.
7.   " x^3-2x^2-4x+9 " $x-\frac{1}{2}$.
8.   " $x^3-9x^2y+6xy^2-8y^3$ " $x-2y$.
9.   " $x^5-5x^4y+5xy^4-y^5$ " $x-y$.

10. " $8x^3+12x^2y+10xy^2+8y^3$ " $2x+y$.
11. " $x^3-\frac{3}{2}x^2+\frac{3}{8}x-\frac{5}{72}$ " $\frac{1}{8}x-\frac{1}{18}$.
12. " $x^4+8x^3-15x-10$ " $x+2$.

CHAPTER II

Section I.—The Principle of Symmetry.

Art. ix. An expression is said to be symmetrical with respect to two of its letters when these can be interchanged without altering the expression:

Thus if in $a^3+a^2x+ax^2+x^3$, we write x for a, and a for x, we get $x^3+x^2a+xa^2+a^3$, which is identical with the given expression. So, in $x^2+b^2x+ba+a^2x$, if we interchange a and b, there results $x^2+a^2x+ab+b^2x$ which is identical with the given expression; but it will be seen that the expression is *not* symmetrical with respect to x and b, or x and a.

An expression is symmetrical with respect to *three* of its letters a, b, c, when a can be changed into b, b into c, and c into a, without altering the expression.

Thus $a^3+b^3+c^3-3abc$ remains unaltered by changing a into b, b into c, and c into a, and is therefore symmetrical with respect to these letters. So, $a^2b+b^2a+a^2c+c^2a+b^2c+bc^2$, and $(a-b)^3+(b-c)^3+(c-a)^3$, are each symmetrical with respect to a, b, c.

Again $(x-a)(a-b)^2+(x-b)(b-c)^2+(x-c)(c-a)^2$ is symmetrical with respect to a, b, c, but not with respect to x and any of the other letters.

Generally, an expression is symmetrical with respect to any number of its letters $a, b, c, \ldots h, k$, when a can be changed into b, b into c, c into $d \ldots \ldots h$ into k, and k into a, without altering the expression.

SYMMETRY.

A symmetric function of several letters is frequently represented by writing each *type-term* once, preceded by the letter Σ; thus for $a+b+c+ \ldots +l$ we write Σa, and for $ab+ac+ad+ \ldots +bc+bd+ \ldots$ (*i. e.* the sum of the products of *every* pair of the letters considered) we write Σab.

Exercise xv.

Write the following in full:

1. $\Sigma a^2 b$, $\Sigma(a-b)^2$, $\Sigma a(b-c)$, $\Sigma ab(x-c)$, $\Sigma a^3 b^2 c$, $\Sigma (a+b) \times (c-a)(c-b)$, $\Sigma \{(a+c)^2 - b^2\}$, and $\Sigma a(b+c)^2$, each for a, b, c.

2. Σabc, $\Sigma a^2 b$, $\Sigma a^2 bc$, $\Sigma(a-b)$, and $\Sigma a^2(a-b)$, each with respect to a, b, c, d.

Shew that the following are symmetrical:

3. $(x+a)(a+b)(b+x)+abx$, with respect to a and b.

4. $(a+b)^2+(a-b)^2$ with respect to a and b, and also with respect to a and $-b$.

5. $(ab-xy)^2 - (a+b-x-y)\{ab(x+y)-xy(a+b)\}$ with respect to a and b, and also with respect to x and y.

6. $a^2(b-c)-b^2(a-c)-c^2(b-a)$ with respect to a, b, c.

7. $(ac+bd)^2+(bc-ad)^2$ with respect to a^2 and b^2, and also with respect to c^2 and d^2.

8. $x^6+y^6+3xy(x^2+xy+y^2)$ with respect to x and y.

9. $\{x^3-y^3+3xy(2x+y)\}^3+\{y^3-x^3+3xy(2y+x)\}^3$ with respect to x and y.

10. $a(a+2b)^3+b(b+2a)^3$ with respect to a and b, and also with respect to a and $-b$.

11. $ab[\{(a+c)(b+c)+2c(a+b)\}^2 - (a-c)^2(b-c)^2]$ with respect to a, b, c.

12. $a^2b^2+b^2c^2+c^2a^2+2abc(a+b+c)$ with respect to ab, bc, ca.

With respect to what letters are the following symmetrical?

13. $xyz+5xy+2(x^2+y^2)$.

14. $2(a^2x^2+b^2y^2)-2ab(xy+yz+zx)$.

15. $(f^2-h^2)^2+4g^2(f+h)^2+(2fh-2g^2)^2$,

16. $(x+y)(x-z)(y-z) - xyz$.
17. $a^2b^2 + b^2c^2 + c^2a^2 - 2abc(a+b-c)$.
18. $x^6 - y^6 + z^6 - 3(x^2-y^2)(y^2-z^2)(z^2+x^2)$.
19. $(a+b)^2 + (a+c)^2 + (b-c)^4$.
20. $(a+b)^4 + (a-c)^4 + (b+c)^4 + (a+c)^4$.
21. $(a+b)^4 + (a-c)^4 + (b+c)^4 + (a+c)^4 + (c-b)^4$.

Select the type-terms in:

22. $a^2 + 2ab + b^2 + 2bc + c^2 + 2ca$
23. $a(b^2-c^2) + b(c^2-a^2) + c(a^2-b^2) + (a+b)(b+c)(c+a)$.
24. $a(b+c)^2 + b(c+a)^2 + c(a+b)^2 - 12abc$.

Write down the *type-terms* in:

25. $(x+y)^5, (x-y)^5, (x+y)^5 - x^5 - y^5$.
26. $(x+y)^7 + (x-y)^7, (x+y)^7 - (x-y)^7$.
27. $(x+y+z)^4, (x-y-z)^4$.
28. $(a+b+c+d)^4, (a^2+b^2+c^2+d^2)^2$.
29. $(a+b)^3 + (b+c)^3 + (c+a)^3$.

Art. x. In reducing an algebraic expression from one form to another, advantage may be taken of the principle of symmetry: for, it will be necessary to calculate only the *type-terms*, and the others may be written down from these.

EXAMPLES.

1. Find the expansion of $(a+b+c+d+e+\&c.)^2$

This expression is symmetrical with respect to a, b, c, &c.; hence the expansion also must be symmetrical, and as it is a product of *two* factors, it can contain only the squares a^2, b^2, c^2, &c., and the products in pairs, $ab, ac, ad \ldots bc, bd$, &c.; so that a^2 and ab are *type-terms*.

Now $(a+b)^2 = a^2 + 2ab + b^2$; and the addition of terms involving a, b, c, &c., will not alter the terms $a^2 + 2ab$, but will merely give additional terms of the same type. Hence from symmetry we get

SYMMETRY.

$$(a+b+c+d+e+\&c.)^2 = a^2 + 2ab + 2ac + 2ad + 2ae + \ldots$$
$$+ b^2 + 2bc + 2bd + 2be + \ldots$$
$$+ c^2 + 2cd + 2ce + \ldots$$
$$+ d^2 + 2de + \ldots$$
$$+ e^2 + \ldots$$

This may be compactly written

$$(\Sigma a)^2 = \Sigma a^2 + 2\Sigma ab.$$

2. Expand $(a+b)^3$.

This has been found by actual multiplication—see formula [5] —but we may also proceed as follows:

(1) The expression is of *three* dimensions, and is symmetrical with respect to a and b.

(2) The type-terms are a^3, a^2b.

Hence $(a+b)^3 = a^3 + b^3 + n(a^2b + b^2a)$, where n is *numerical*.

To find the value of n, put $a = b = 1$, and we have $(1+1)^3 = 1+1+n(1+1)$; $\therefore n = 3$.

3. Expand $(x+y+z)^3$.

This is of three dimensions, and is symmetrical with respect to x, y, z. We have

$$(x+y+z)^3 = \{(x+y)+z\}^3 = (x+y)^3 + \&c.$$

$= x^3 + 3x^2y + \&c.$, which are type-terms, the only other possible type-term being xyz.

Now, since the expression contains $3x^2y$, it must also contain $3x^2z$, that is, it must contain $3x^2(y+z)$. Hence

$$(x+y+z)^3 = \quad x^3 + 3x^2(y+z)$$
$$+ y^3 + 3y^2(z+x)$$
$$+ z^3 + 3z^2(y+x)$$
$$+ n(xyz), \text{ where } n \text{ is numerical, and}$$

may be found by putting $x = y = z = 1$ in the last result, giving

$$(1+1+1)^3 = 1+1+1+3(1+1)+3(1+1)+3(1+1)+n;$$
$$\therefore n = 6.$$

SYMMETRY.

4. Similarly we may shew that
$$(a+b+c+d)^3 = a^3+3a^2(b+c+d)+6bcd$$
$$+b^3+3b^2(c+d+a)+6cda$$
$$+c^3+3c^2(d+a+b)+6dab$$
$$+d^3+3d^2(a+b+c)+6abc.$$

5. Expand $(a+b+c+\&c.)^3$.

The type-terms are a^3, a^2b, abc.

Expanding $(a+b+c)^3$, we get $a^3+3a^2b+6abc+\&c.$

Hence by symmetry we have
$$(\Sigma a)^3 = \Sigma a^3 + 3\Sigma a^2 b + 6\Sigma abc.$$

6. Simplify $(a+b-2c)^2+(b+c-2a)^2+(c+a-2b)^2$.

This expression is symmetrical, involving terms of the types a^2 and ab. Now a^2 occurs with 1 as a coefficient in the first square, with 4 as a coefficient in the second square, and with 1 as a coefficient in the third square, and hence $6a^2$ is one type-term of the result: ab occurs with 2 as a coefficient in the first square, with -4 as a coefficient in the second square, and with -4 as a coefficient in the third square, and hence $-6ab$ is the second type-term in the result: hence the total result is $6(a^2+b^2+c^2-ab-bc-ca)$.

7. Simplify $(x+y+z)^3+(x-y-z)^3+(y-z-x)^3+(z-x-y)^3$.

This is symmetrical with respect to x, y, z; and the type-terms are x^3, $3x^2y$, $6xyz$:

(1) x^3 occurs in each of the first two cubes, and $-x^3$ in each of the second two cubes, $\therefore$ there are no terms of the type x^3 in the result.

(2) $3x^2y$ occurs in the *first* and *third* cubes, and $-3x^2y$ in the second and fourth, $\therefore$ there are no terms of this type in the result.

(3) $6xyz$ occurs in each of the four cubes, $\therefore$ $24xyz$ is the total result.

8. Prove $(a^2+b^2+c^2+d^2)(w^2+x^2+y^2+z^2) - (aw+bx+cy+dz)^2 = (ax-bw)^2+(ay-cw)^2+(az-dw)^2+(by-cx)^2+(bz-dx)^2+(cz-dy)^2$.

The left hand member (considered as given) is symmetrical with respect to the pairs of letters, a and w, b and x, c and y, d and z, that is, any two pairs may be interchanged without affecting the expression. As the expression is only of the second degree in these pairs, no term can involve three pairs as factors; hence the type-terms may be obtained by considering all the terms involving a, b, w, x; these are $a^2w^2, a^2x^2, b^2w^2, b^2x^2, -a^2w^2, -b^2x^2, -2abwx$, and are the terms of $(ax-bw)^2$ which is consequently a type-term. From $(ax-bw)^2$ we derive the five other terms of the second member by merely changing the letters.

9. Prove that
$(x^2-yz)^3+(y^2-zx)^3+(z^2-xy)^3-3(x^2-yz)(y^2-zx)(z^2-xy)$ is a complete square.

The expression will remain symmetrical if $(x^2-yz)(y^2-zx)(z^2-xy)$, instead of being multiplied by -3, be subtracted from each of the preceding terms, thus giving

$$(x^2-yz)\{(x^2-yz)^2-(y^2-zx)(z^2-xy)\}$$
$$+(y^2-zx)\{(y^2-zx)^2-(z^2-xy)(x^2-yz)\}$$
$$+(z^2-xy)\{(z^2-xy)^2-(x^2-yz)(y^2-zx)\}$$
$$=(x^2-yz)x(x^3+y^3+z^3-3xyz)$$
$$+\&c.$$
$$+\&c.$$
$$=(x^3+y^3+z^3-3xyz)(x^3+y^3+z^3-3xyz).$$

Exercise xvi.

Simplify the following:

1. $(a+b+c)^2+(a+b-c)^2+(b+c-a)^2+(c+a-b)^2$
2. $(a-b-c)^2+(b-a-c)^2+(c-a-b)^2$.
3. $(a+b+c-d)^2+(b+c+d-a)^2+(c+d+a-b)^2+(d+a+b-c)^2$.
4. $(a+b+c)^2-a(b+c-a)-b(a+c-b)-c(a+b-c)$.
5. $(x+y+z+n)^2+(x-y-z+n)^2+(x-y+z-n)^2+(x+y-z-n)^2$.
6. $(a+b+c)^3+(a+b-c)^3+(b+c-a)^3+(c+a-b)^3$.

38 SYMMETRY.

7. $(x-2y-3z)^2+(y-2z-3x)^2+(z-2x-3y)^2$.

8. $(ma+nb+rc)^3-(ma+nb-rc)^3-(nb+rc-ma)^3-(rc+ma-nb)^3$.

9. $a(b+c)(b^2+c^2-a^2)+b(c+a)(c^2+a^2-b^2)+c(a+b)(a^2+b^2-c^2)$.

10. $(ab+bc+ca)^2-2abc(a+b+c)$.

Prove the following:

11. $(ax+by+cz)^2+(bx+cy+az)^2+(cx+ay+bz)^2+(ax+cy+bz)^2+(cx+by+az)^2+(bx+ay+cz)^2$
$=2(a^2+b^2+c^2)(x^2+y^2+z^2)+4(ab+bc+ca)(xy+yz+zx)$.

12. $(a+b+c)^4+(b+c-a)^4+(c+a-b)^4+(a+b-c)^4$
$=4(a^4+b^4+c^4)+24(a^2b^2+b^2c^2+c^2a^2)$.

13. $(a+b+c)^4 = \Sigma a^4 + 4\Sigma a^3b + 6\Sigma a^2b^2 + 12\Sigma a^2bc$.

14. $(\Sigma a)^4 = \Sigma a^4 + 4\Sigma a^3b + 6\Sigma a^2b^2 + 12\Sigma a^2bc + 24\Sigma abcd$.

15. $(a^2+b^2+c^2)^3+2(ab+bc+ca)^3-3(a^2+b^2+c^2)\times(ab+bc+ca)^2 = (a^3+b^3+c^3-3abc)^2$.

16. $(a-b)^2(b-c)^2+(b-c)^2(c-a)^2+(c-a)^2(a-b)^2 = (a^2+b^2+c^2-ab-ac-bc)^2$.

17. $(2a-b-c)^2(2b-c-a)^2+(2b-c-a)^2(2c-a-b)^2+(2c-a-b)^2(2a-b-c)^2 = 9(a^2+b^2+c^2-ab-bc-ca)^2$

18. $(ar^2+2brs+cs^2)(ax^2+2bxy+cy^2)-\{arx+b(ry+sx)+csy\}^2 = (ac-b^2)(ry-sx)^2$.

19. $(a^2+ab+b^2)(c^2+cd+d^2) = (ac+ad+bd)^2+(ac+ad+bd)(bc-ad)+(bc-ad)^2$.

20. Shew that there are two ways in which the given product in the last example can be expressed in the form p^2+pq+q^2, and two ways in which it can be expressed in the form p^2-pq+q^2.

21. $6(w^2+x^2+y^2+z^2)^2 = (w+x)^4+(w-x)^4+(w+y)^4+(w-y)^4+(w+z)^4+(w-z)^4+(x+y)^4+(x-y)^4+(x+z)^4+(x-z)^4+(y+z)^4+(y-z)^4$.

22. $\frac{1}{8}\{(a+b+c)^5+(a-b-c)^5+(b-c-a)^5+(c-a-b)^5\} =$
$\frac{1}{3}\{(a+b+c)^3+(a-b-c)^3+(b-c-a)^3+(c-a-b)^3\} \times$
$\frac{1}{4}\{(a+b+c)^2+(a-b-c)^2+(b-c-a)^2+(c-a-b)^2\}$.

Section II.—Theory of Divisors.

Any expression which can be reduced to the form $ax^n + bx^{n-1} + cx^{n-2} + \ldots + \ldots + hx + k$, in which n is a positive integer and $a, b, c, \ldots h, k$ are independent of x, is called a Polynome *in x of degree n.*

The expressions $f(x)^n$, $F(x)^n$, $\varphi(x)^m$, are used as general symbols for polynomes; the index n. m. indicates the degree of the polynome.

Theorem I. If the polynome $f(x)^n$ be divided by $x-a$, the remainder will be $f(a)^n$.

Cor. 1. $f(x)^n - f(a)^n$ is always exactly divisible by $x-a$.

(Particular case: $x^n - a^n$ is always exactly divisible by $x-a$).

Cor. 2. If $f(a)^n = 0$, $f(x)^n$ is exactly divisible by $x-a$, i.e., $f(x)^n$ is an algebraic multiple of $x-a$.

Cor. 3. If the polynome $f(x)^n$ on division by the polynome $\varphi(x)^m$ leave a remainder independent of x, such remainder will be the value of $f(x)^n$ when $\varphi(x)^m = 0$.

Examples.—Theorem I.

1. Find the remainder when $x^5 - 7x^4 + 13x^3 - 16x^2 + 9x - 12$ is divided by $x - 5$.

The remainder will be the value of the given polynome when 5 is substituted for x. (See Art. III.).

$$\begin{array}{r|rrrrr} & 1 & -7 & +13 & 16 & +9 & -12 \\ 5 & & 5 & -10 & 15 & -5 & 20 \\ \hline & 1 & -2 & 3 & -1 & 4\,; & 8 \end{array}$$

Hence the remainder is 8.

2. Find the remainder when $(x-a)^3 + (x-b)^3 + (a+b)^3$ is divided by $x+a$.

For x substitute $-a$, then $(-2a)^3 + (-a-b)^3 + (a+b)^3 = -8a^3$.

3. Find the remainder when $x^3 + a^3 + b^3 + (x+a)(x+b)(a+b)$ is divided by $x + a + b$.

For x substitute $-(a+b)$ and we get
$-(a+b)^3+a^3+b^3+ab(a+b) = -2ab(a+b)$. See Formula [6].

4. Find the remainder when $(x^2+2ax-2a^2)^3(x^2-2ax-2a^2)$
$+32(x-a)^4(x+a)^4$ is divided by x^2-2a^2.

x^2-2a^2 may be struck out wherever it appears.

This reduces the dividend to
$(2ax)^3(-2ax)+32(x-a)^4(x+a)^4 = -16a^4x^4+32(x^2-a^2)^4$.

In this substitute $2a^2$ for x^2 and it becomes
$$-64a^8+32a^8 = -32a^8,$$
which is the required remainder.

Exercise xvii.

1. Find the remainder when $3x^4+60x^3+54x^2-60x+58$ is divided by $x+19$.

2. Find the remainder when $px^3-3qx^2+3rx-s$ is divided by $x-a$.

3. What number added to $4x^5+34x^4+58x^3+21x^2-123x-41$ will give a sum exactly divisible by $2x+13$?

4. What number taken from $10x^{10}-20x^8-10x^6-89x^4-8\cdot9x^2+20$ will leave a remainder exactly divisible by $10x^2-11$?

Find the remainders from the following divisions:

5. $(x+1)^5-x^5 \div x+1$, and $(x+a+3)^3-(x+a+1)^3 \div x+2$.

6. $x^n+y^n \div x-y$; $x^{2n}+y^{2n} \div x+y$; $x^{2n+1}+y^{2n+1} \div x+y$.

7. $(x+1)^3+x^3+(x-1)^3 \div x-2$.

8. $(x-a)^3(x+a)^3+(x^2-2b^2)^3 \div x^2+b^2$.

9. $(x^2+ax+a^2)(x^2-ax+a^2)-(x^2-3ax+2a^2)(x^2+3ax+2a^2)$
$\div x^2+2a^2$.

10. $(9a^2+6ab+4b^2)(9a^2-6ab+4b^2)(81a^4-36a^2b^2+16b^4) \div (3a-2b)^2$.

11. $a^2(x-a)^3+b^2(x-b)^3 \div x-a-b$.

12. $(ax+by)^3+a^3y^3+b^3x^3-3abxy(ax+by) \div (a+b)(x+y)$.

13. $x^3+a^3+b^3-3abx \div x-a+b$; also $\div x+a-b$ also $\div x-a-b$.

14. Any polynome divided by $x-1$ gives for remainder the sum of the coefficients of the terms.

Examples.—Cor. 1.

1. $x^5 + y^5$ is exactly divisible by $x+y$.
In "$x^5 - a^5$ is exactly divisible by $x-a$," substitute $-y$ for a.

2. $mx^3 - px^2 + qx + m + p + q$ is exactly divisible by $x+1$.
This may be written
$\{m x^3 - px^2 + qx\} - \{m(-1)^3 - p(-1)^2 + q(-1)\}$ is exactly divisible by $x - (-1)$.

3. $(x^2 + 6xy + 4y^2)^5 + (x^2 + 2xy + 4y^2)^5$ is exactly divisible by $(x+2y)^2$. For $(x^2 + 6xy + 4y^2)^5 - (-x^2 - 2xy - 4y^2)^5$ is exactly divisible by $(x^2 + 6xy + 4y^2) - (-x^2 - 2xy - 4y^2)$, which is $2(x^2 + 4xy + 4y^2) = 2(x+2y)^2$.

Exercise xviii.

Prove that the following are cases of exact division:

1. $x^{2n+1} + y^{2n+1} \div x+y$; $x^{2n} - y^{2n} \div x+y$.
2. $x^{12} + y^{12} \div x^4 + y^4$; $x^{30} + y^{30} \div x^6 + y^6$; also $\div x^{10} + y^{10}$; also $\div x^2 + y^2$.
3. $(ax+by)^5 + (bx+ay)^5 \div (a+b)(x+y)$.
4. $(ax+by+cz)^3 - (bx+cy+az)^3 \div (a-b)x + (b-c)y + (c-a)z$.
5. $(2y-x)^n - (2x-y)^n \div 3(y-x)$.
6. $(2y-x)^{2n+1} + (2x-y)^{2n+1} \div y+x$.
7. $(my-nx)^5 - (mx-ny)^5 \div (m+n)(y-x)$.
8. $(x+y)^6 + (x-y)^6 \div 2(x^2+y^2)$.
9. $(x^2+xy+y^2)^3 + (x^2-xy+y^2)^3 \div 2(x^2+y^2)$.
10. $(a+b)^9 - (a-b)^9 \div 2b(3a^2+b^2)$.
11. $(x^2+5bx+b^2)^7 + (x^2-bx+b^2)^7 \div 2(x+b)^2$.
12. $(a+b)^{4n+2} + (a-b)^{4n+2} \div 2(a^2+b^2)$.
13. $\{x^3 + 3xy(x-y) - y^3\}^3 + \{x^3 - 9xy(x-y) - y^3\}^3 \div 2(x-y)^3$.
14. $3x^3 - 5x^2 + 4x - 2 \div x - 1$.

42   THEORY OF DIVISORS.

✗ 15. Any polynome in x is divisible by $x-1$ when the sum of the coefficients of the terms is zero.

✓ 16. Any polynome in x is divisible by $x+1$, when the sum of the coefficients of the even powers of x is equal to the sum of the coefficients of the odd powers. (The constant term is included among the coefficients of the even powers).

EXAMPLES.—COR. 2.

1. Show that $a(a+2b)^3 - b(2a+b)^3$ is exactly divisible by $a+b$. By Cor. 2, the substitution of $-b$ for a must cause the polynome to vanish.

Substituting; $a(a-2a)^3 + a(2a-a)^3 = -a^4 + a^4 = 0$.

2. Show that $(ab-xy)^2 - (a+b-x-y)\{ab(x+y) - xy(a+b)\}$ is exactly divisible by $(x-a)(y-a)$, also by $(x-b)(y-b)$.

For x substitute a and the expression becomes
$$(ab-ay)^2 - (b-y)\{ab(a+y) - ay(a+b)\} =$$
$$a^2(b-y)^2 - (b-y)\{a^2(b-y)\} = 0.$$

The expression is, therefore, exactly divisible by $x-a$. But it is symmetrical with respect to x and y, hence it is divisible by $y-a$, and as $x-a$ and $y-a$ are independent factors, the expression is exactly divisible by $(x-a)(y-a)$. Again, the given expression is symmetrical with respect to a and b, hence, making the interchange of a and b, the expression is seen to be divisible by $(x-b)(y-b)$.

3. Show that $6(a^5+b^5+c^5) - 5(a^3+b^3+c^3)(a^2+b^2+c^2)$ is exactly divisible by $a+b+c$.

For a substitute $-(b+c)$ and the result which would be the remainder were the division actually performed, must vanish.
$6\{-(b+c)^5 + b^5 + c^5\} - 5\{-(b+c)^3 + b^3 + c^3\}\{(b+c)^2 + b^2 + c^2\}$
$= 6\{-(b+c)^5 + b^5 + c^5\} + 30bc(b+c)(b^2 + bc + c^2)$. See [1] and [6].

The expansion being of the 5th degree, and symmetrical in b and c, it will be sufficient to show that the coefficients of b^5, b^4c, b^3c^2 vanish, the coefficients of b^2c^3, bc^4, c^5 being the coefficients of the former terms in reverse order. Calculating the coefficients of these type-terms we get

$$6\{-5b^4c - 10b^3c^2 - \ldots\} + 30(b^4c + 2b^3c^2 + \ldots),$$

which evidently vanishes. Hence the truth of the proposition.

4. If $a+b+c=0$, $\frac{1}{5}(a^5+b^5+c^5) = \frac{1}{2}(a^2+b^2+c^2)\cdot\frac{1}{3}(a^3+b^3+c^3)$.

In the last example it has been proved that the *difference* of the quantities here declared to be equal, is a multiple of $a+b+c$, *i.e.*, in this case, a multiple of zero. Hence under the given condition they are equal.

Exercise xix.

Prove that the following are cases of exact division:

1. $(ax-by)^3 + (bx-ay)^3 - (a^3+b^3)(x^3-y^3) \div a, b, x, y, a+b, x-y$.
2. $ax^3 - (a^2+b)x^2 + b^2 \div ax-b$. (Substitute ax for b.)
3. $\begin{cases} (ax+by)^2 - (a-b)(x+z)(ax+by) + (a-b)^2xz \div x+y. \\ (ax-by)^2 - (a+b)(x+z)(ax-by) + (a+b)^2xz \div x+y. \end{cases}$
4. $6a^3x^2 - 4ax^3 - 10axy - 3a^2xy + 2x^2y + 5y^2 \div 2ax-y$.
5. $1{\cdot}2a^4x - 16{\cdot}32a^3x^2 + 4{\cdot}8a^2x^3 + {\cdot}9ax^4 - x^5 \div {\cdot}6ax - 2x^2$.
6. $x^8 + x^6y^2 + x^2y^2 + y^3 \div x^6 + y$.
7. $(c-d)a^2 + 6(bc-bd)a + 9(b^2c-b^2d) \div a+3b$.
8. $x(x - \tfrac{1}{12}y)^5 + y(\tfrac{1}{12}x - y)^5 \div x - y$.
9. $a(a+2b)^3 - b(b+2a)^3 \div a-b$, also $\div a+b$.
10. $a^5 - 2a^4b + a^3b^2 + a^2x^3 - 2ax^3 + b^2x^3 \div (a-b)(x+a)$.
11. $a(b-c)^3 + b(c-a)^3 + c(a-b)^3 \div (a-b), (b-c), (c-a)$.
12. $a^3(b-c) + b^3(c-a) + c^3(a-b) \div (a-b), (b-c), (c-a)$.
13. $a^4(b-c) + b^4(c-a) + c^4(a-b) \div (a-b), (b-c), (c-a)$.
14. $(a-b)^2(c-d)^2 + (b-c)^2(d-a)^2 - (d-b)^2(a-c)^2 \div (a-b), (b-c), (c-d), (d-a)$.
15. $\{(a-b)^2 + (b-c)^2 + (c-a)^2\}\{(a-b)^2c^2 + (b-c)^2a^2 + (c-a)^2b^2\} - \{(a-b)^2c + (b-c)^2a + (c-a)^2b\}^2 \div (a-b), (b-c), (c-a)$.
16. $(x+y)(y+z)(z+x) + xyz \div x+y+z$.

17. $ab(a^2-b^2)+bc(b^2-c^2)+ca(c^2-a^2) \div a+b+c.$

18. $(ab-bc-ca)^2 - a^2b^2 - b^2c^2 - c^2a^2 \div a+b-c.$

19. $(a+2b)^3+(2b-3c)^3-(3c-a)^3+a^3+8b^3-27c^3+a+2b-3c.$

20. $a^3b^3+b^3c^3+c^3a^3-3a^2b^2c^2 \div ab+bc+ca.$

EXAMPLES.—CORS. 3 AND 2.

1. Find the value of $4x^5+9x^3-5x^2+23x+6$ when $2x^2 = 3x-4$.

Since $2x^2-3x+4=0$, we have simply to find the remainder on division by $2x^2-3x+4$, and if it is independent of x, it is the value sought, Cor. 3.

	4	0	9	−5	23	6
3		6	9	15	−3	
−4			−8	−12	−20	4
2	2	3	5	− 1;	0	10

Hence the required value is 10.

2. What value of c will make x^3-5x^2+7x-c exactly divisible by $x-2$.

If 2 be substituted for x, the remainder must vanish, Cor. 2.

	1	−5	7	−c
2		2	−6	2
	1	−3	1;	$2-c$

Hence $2-c=0$, or $c=2$.

3. What value of c will make $6x^5-5x^4+cx^3-20x^2+19x-5$ vanish when $2x^2=3x-1$?

By Cor. 3, the remainder must vanish when the given polynome is divide by $2x^2-3x+1$. We may divide at once and find, if possible, a value of c that will make both terms of the remainder vanish, or we may first express cx^3 in lower terms in x, and then divide and find the required value of c from the remainder.

1st. Method, (see page 28),

	6	−10	$4c$	−160	304	−160
3		18	24	$12c+36$	$36c-420$	
−2			−12	−16	$-8c-24$	$-24c+280$
	6	8	$4c+12$	$12c-140$;	$28c-140$	$-24c+120$

THEORY OF DIVISORS.	45

Hence $28c = 140$ and $24c = 120$. Both of these are satisfied by $c = 5$.

2nd Method. $x^3 = \frac{1}{3}x(3x-1) = \frac{3}{2}x^2 - \frac{1}{3}x = \frac{2}{3}(3x-1) - \frac{1}{3}x = 2\frac{1}{3}x - \frac{2}{3} - \frac{1}{3}x = 1\frac{3}{4}x - \frac{2}{3}$; $\therefore cx^3 = 1\frac{3}{4}cx - \frac{2}{3}c$.

Substituting for cx^3 in the given polynome it becomes
$$6x^5 - 5x^4 - 20x^2 + (1\frac{3}{4}c+19)x - \frac{2}{3}c - 5.$$
Divide and apply Cor. 3.

	6	−10	0	−160	$28c+304$	$-24c-160$
3		18	24	36	−420	
−2			−12	−16	−24	280
	6	8	12	−140 ;	$28c-140$	$-24c+120$

We thus obtain the same remainder as by the former method, and consequently the same result. A comparison of the two methods shews that they are but slightly different in form, but the second method shows rather more clearly that c need not be introduced into the dividend at all, but the proper multiples of it found by the preliminary reduction can be added to or taken from the numerical remainder, and the "true remainder" be thus found, and c determined from it.

Exercise xx.

Find the value of
1. $x^4 - 3x^3 + 4x^2 - 3x + 4$, given $x^2 = x - 1$.
2. $x^5 - 2x^4 - 4x^3 + 13x^2 - 11x - 10$, given $(x-1)^2 = 2$.
3. $2x^5 - 7x^4 + 12x^3 - 11x^2 + 2x - 5$ given $(x-1)^2 + 2 = 0$.
4. $3x^6 + 11x^5 + 10x^3 + 7x^2 + 2x + 3$ given $x^3 + 3x^2 - 2x + 5 = 0$.
5. $6x^7 + 9x^6 - 16x^4 - 5x^3 - 12x^2 - 6x + 60$ given $3x^4 + x - 4 = 0$.

What values of c will make the following polynomes vanish under the given conditions.

6. $x^4 + 13x^3 + 26x^2 + 52x + 8c$, given $x + 11 = 0$.
7. $x^4 - 2x^3 - 9x^2 + 2cx - 14$, given $3x + 7 = 0$.
8. $x^4 - 4x^3 - x^2 + 16x + 6c$, given $x^2 = x + 6$.
9. $2x^4 - 10x^2 + 4cx + 6$, given $x^2 + 3 = 3x$.
10. $2x^4 + x^3 - 7cx^2 + 11x + 10$, given $2x = 5$.

11. $4x^4 + cx^2 + 110x - 105$, given $2x^2 - 5x + 15 = 0$.

12. $3x^5 - 16x^4 + cx^3 - 5x^2 - 114x + 200$, given $x^2 = 3x - 4$.

13. What values of p and q will make $x^4 + 2x^3 - 10x^2 - px + q$ vanish, given $x^2 = 3(x-1)$?

14. What values of p and q will make $a^{12} - 5a^{10} + 10a^8 - 15a^6 + 29a^4 - pa^2 + q$ vanish, given $(a^2 - 2)^2 = a^2 - 3$?

Theorem II. If the polynome $f(x)^n$ vanish on substituting for x each of the n (different) values $a_1, a_2, a_3 \ldots a_n$

$$f(x)^n = A(x-a_1)(x-a_2)(x-a_3) \ldots (x-a_n)$$

in which A is independent of x and consequently is the coefficient of x^n in $f(x)^n$.

Cor. If $f(x)^n$ and $\varphi(x)^m$ both vanish for the same m different values of x, $f(x)^n$ is algebraically divisible by $\varphi(x)^m$.

EXAMPLES.

1. $x^3 + ax^2 + bx + c$ will vanish if 2, or 3, or -4 be substituted for x, determine a, b, c.

The coefficient of the highest power of x is 1;

$\therefore x^3 + ax^2 + bx + c = (x-2)(x-3)(x+4) = x^3 - x^2 - 14x + 24$.

$\therefore a = -1; b = -14; c = 24$.

2. $x^3 + bx^2 + cx + d$ will vanish if -3 or 2, or 5 be substituted for x, determine its value if 3 be substituted for x.

The given polynome $= (x+3)(x-2)(x-5)$;

$\therefore$ the required value is $(3+3)(3-2)(3-5) = -12$.

3. $ax^3 + 3bx^2 + 3cx + d$ will vanish if for x be substituted -3, or $\frac{1}{3}$, or $1\frac{1}{2}$, but it becomes 45 if for x there be substituted 3; determine the values of a, b, c, d.

The coefficient of the highest power of x is a;

$\therefore ax^3 + 3bx^2 + 3cx + d = a(x+3)(x-\frac{1}{3})(x-1\frac{1}{2})$

$\therefore a(3+3)(3-\frac{1}{3})(3-1\frac{1}{2}) = 45; \therefore a = 2$.

$\therefore 2x^3 + 3bx^2 + 3cx + d = 2(x+3)(x-\frac{1}{3})(x-1\frac{1}{2})$

$\therefore b = \frac{2}{3}, c = -3\frac{1}{4}, d = 4\frac{1}{2}$.

4. If x^3+px^2+qx+r vanish for $x=a$ or b, or c, determine p, q, and r in terms of a, b, c.

$$x^3+px^2+qx+r = (x-a)(x-b)(x-c)$$
$$= x^3 - (a+b+c)x^2 + (ab+bc+ca)x - abc$$
$$\therefore p = -(a+b+c) \quad \text{or} \quad -\Sigma a.$$
$$q = ab+bc+ca \quad \text{or} \quad \Sigma ab$$
$$r = -abc \quad \text{or} \quad -\Sigma abc.$$

5. If x^3+px^2+qx+r vanish for $x=a$, or b, or c, determine the polynome that will vanish for $x=b+c$, or $c+a$, or $a+b$.

Since x^3+px^2+qx+r vanishes for $x=a$ or b or c,

x^3-px^2+qx-r will vanish for $x=-a$ or $-b$ or $-c$,

and $-p=a+b+c$;

But the required polynome will vanish for

$x=-p-a$, or $-p-b$, or $-p-c$;

that is, for $x+p=-a$, or $-b$, or $-c$.

Hence it is $(x+p)^3 - p(x+p)^2 + q(x+p) - r =$
$$x^3 + 2px^2 + (p^2+q)x + pq - r.$$

The following is the calculation in the last reduction. (See page 31).

$$\begin{array}{c|cccc}
 & 1 & -p & q & -r \\ \hline
p & 1 & 0 & q\,; & pq-r \\
p & 1 & p\,; & p^2+q & \\
p & 1\,; & 2p & & \\
 & 1 & & &
\end{array}$$

* 6. In any triangle, the square of the area expressed in terms of the lengths of the sides, is a polynome of four dimensions; and the area of the triangle, the lengths of whose sides are 3, 4, and 5, respectively, is 6. Find the polynome expressing the square of the area.

Let a, b, and c be the lengths of the sides, and A the required polynome.

1st. The area vanishes if any two of the sides become together equal to the third side, hence if $a+b=c$, $A=0$, and consequently A is divisible by $a+b-c$. Similarly it is divisible by $b+c-a$ and by $c+a-b$.

2nd. The area vanishes if the three sides vanish together, hence if $a+b+c=0$, $A=0$, and consequently A is divisible by $a+b+c$.

We have thus found four linear factors, but A is of only four dimensions.

$$\therefore A = m(a+b+c)(b+c-a)(c+a-b)(a+b-c),$$

in which m is a numerical constant.

But 6^2 or $36 = m(3+4+5)(4+5-3)(5+3-4)(3+4-5)$

$$= 576m; \therefore m = \tfrac{1}{16}.$$

(The above includes all the ways in which the area of a triangle can vanish, for the vanishing of only one side involves the equality of the other two, or if $a=0$, $b=c$, and $\therefore a+b=c$, which is included in 1st.; if two sides vanish simultaneously, the three must vanish).

Examples on the Corollary.

7. Prove that $(x+1)^{12} - x^{12} - 2x - 1$ is divisible by
$$2x^3 + 3x^2 + x.$$

Factoring the latter expression we find it vanishes for $x=0$, or -1 or $-\tfrac{1}{2}$. Substituting these values in the former polynome, it also vanishes. But these are different values of x, hence the truth of the proposition.

8. $(x+y+z)^5 - x^5 - y^5 - z^5$ is divisible by
$$(x+y+z)^3 - x^3 - y^3 - z^3.$$

The latter expression vanishes if $x = -y$, so also does the former. By symmetry they both vanish if $y = -z$ and if $z = -x$. Hence they are both divisible by $(x+y)(y+z)(z+x)$. But this expression is of three dimensions, as also is the latter of the given polynomes, hence it is a divisor of the former.

9. Prove that $\{(a+b)^5 + (c+d)^5\}(a-b)(c-d) +$
$\{(b+c)^5 + (a+d)^5\}(b-c)(a-d) + \{(b+d)^5 + (c+a)^5\}(b-d)(c-a)$
is algebraically divisible by $(a-b)(c-d)(b-c)(a-d)(b-d)(c-a)$
$\times (a+b+c+d)$, and find the quotient.

Let $a = b$ and the former polynome reduces to
$\{(a+c)^5 + (a+d)^5\}(a-c)(a-d) + \{(a+d)^5 + (c+a)^5\}(a-d)(c-a)$

THEORY OF DIVISORS.

which vanishes, the second complex term differing from the first only in the sign of one factor, having $(c-a)$ instead of $(a-c)$.

Hence the former polynome is divisible by $a-b$, and by symmetry it is also divisible by $a-c$, by $a-d$, by $b-c$, by $b-d$, by $c-d$.

Again, $(a+b)^5+(c+d)^5$ is divisible by $(a+b)+(c+d)$; for, on putting $a+b=-(c+d)$, it becomes $\{-(c+d)\}^5+(c+d)^5$ which $=0$.

Similarly the other terms of the former of the given polnomes are each divisible by $a+b+c+d$, and consequently the whole is so divisible.

Now all these factors are different from each other, hence the former of the given polynomes is divisible by the product of these factors, *i.e.*, by the latter of the given polynomes.

Both of these polynomes are of seven dimensions, hence their quotient must be a number, the same for all values of a, b, c, d.

Put $a=2$, $b=1$, $c=0$, $d=-1$, and divide. The quotient will be found to be -5.

$\therefore \{(a+b)^5+(c+d)^5\}(a-b)(c-d)+\{(b+c)^5+(a+d)^5\} \times (b-c)(a-d)+\{(b+d)^5+(c+a)^5\}(b-d)(c-a) = -5(a-b)(c-d)$
$\times (b-c)(a-d)(b-d)(c-a)(a+b+c+d)$.

N.B.—It is not always necessary to find the factors of the divisor, as the following examples show.

10. Prove that x^2+x+1 is a factor of $x^{14}+x^7+1$.

x^2+x+1 will be a factor of $x^{14}+x^7+1$ provided
$x^{14}+x^7+1=0$ if $x^2+x+1=0$.

If $x^2+x+1\ \ =0$
$\therefore x^3+x^2+x\ \ =0$
$\therefore x^3+x^2+x+1=1$
$\therefore x^3\ \ \ \ \ \ \ \ \ \ \ \ \ =1$
$\therefore x^6=1$ and $x^{12}=1$
$\therefore x^7=x$ and $x^{14}=x^2$
$\therefore x^{14}+x^7+1\ \ =x^2+x+1=0$
$\therefore x^2+x+1$ is a factor of $x^{14}+x^7+1$.

THEORY OF DIVISORS.

Art. X!I   Two other methods of proving this proposition are worthy of notice.

1st. x^2+x+1 will be a factor of $x^{14}+x^7+1$ provided it is a factor of $\{(x^{14}+x^7+1) \pm \text{a multiple of } (x^2+x+1)\}$.

$x^{14}+x^7+1$ differs by a multiple of x^2+x+1 from

$x^{14}+x^{11}(x^2+x+1)+x^8(x^2+x+1)+x^7+x^4(x^2+x+1)+$
$x(x^2+x+1)+1$

$=x^{12}(x^2+x+1)+x^9(x^2+x+1)+x^6(x^2+x+1)+x^3(x^2+x+1)+$
(x^2+x+1)

$=(x^{12}+x^9+x^6+x^3+1)(x^2+x+1).$

Hence x^2+x+1 is a factor of $x^{14}+x^7+1$.

2nd. $\dfrac{x^{14}+x^7+1}{x^2+x+1} = \dfrac{x^{21}-1}{x^7-1} \cdot \dfrac{x-1}{x^3-1} =$

$\dfrac{(x^{21}-1)\{(x^{15}-1)-x(x^{14}-1)\}}{(x^7-1)(x^3-1)} =$

$\dfrac{(x^{21}-1)(x^{15}-1)}{(x^7-1)(x^3-1)} - \dfrac{x(x^{21}-1)(x^{14}-1)}{(x^3-1)(x^7-1)}.$

But we see at once that on reduction both of these fractions give an integral quotient, hence $(x^{14}+x^7+1) \div x^2+x+1$ gives an integral quotient.

11. x^2+x+1 is a factor of $(x+1)^7 - x^7 - 1$.

If $x^2+x+1=0$, $(x+1)^7 - x^7 - 1$ will vanish also, for in such case $x+1 = -x^2$.

∴ $(x+1)^7 - x^7 - 1 = (-x^2)^7 - x^7 - 1 = -x^{14} - x^7 - 1$,
which by the last example vanishes if $x^2+x+1=0$;

∴ x^2+x+1 is a factor of $(x+1)^7 - x^7 - 1$.

For x substitute $\dfrac{x}{y}$ and multiply by y^2 and y^7 respectively, and this example becomes

x^2+xy+y^2 is a factor of $(x+y)^7 - x^7 - y^7$.

Exercise xxi.

Determine the values of a, b, c, d, e, in the following cases:—

1. $x^3 + 3bx^2 + 3cx + d$ vanishes for $x = 2$, or 3, or 4.
2. $x^4 + cx^2 + dx + e$ "   " $x = 1\frac{1}{2}$ or -3 or $4\frac{1}{2}$.
3. $x^3 + bx^2 + cx + 24$ "   " $x = 2$ or -3.
4. $ax^3 + bx^2 + cx + 90$.   "   " $x = 3$ or -5 or 2.
5. $ax^4 + cx^2 - 30x + e$.   "   " $x = 1\frac{1}{2}$ or -4, or $2\frac{1}{2}$.
6. $81x^4 + 6cx^2 + 4dx + e$ "   " $x = 1\frac{2}{3}$ or $-3\frac{1}{3}$ or $1\frac{1}{3}$.
7. $ax^4 + bx^3 + cx^2 - 81$ "   " $x = \frac{4}{5}$ or $\frac{3}{4}$ or 3.
8. $ax^4 + cx^2 + dx + e$ "   " $x = 2$ or $1\frac{1}{2}$ or -1 and becomes 14 for $x = 1$.
9. $ax^3 + cx + d$ vanishes for $x = 1\frac{1}{4}$, or $2\frac{3}{4}$, and becomes 49 for $x = 3$, determine its value for $x = -3$.

Given that $x^3 - px^2 + qx - r$ vanishes for $x = a$, or b, or c, determine the polynome that vanishes for

10. $x = a + 1$, or $b + 1$, or $c + 1$.
11. $x = a - 1$, or $b - 1$, or $c - 1$.
12. $x = 1 - \dfrac{1}{a}$, or $1 - \dfrac{1}{b}$, or $1 - \dfrac{1}{c}$.
13. $x = ab$, or bc, or ca.
14. $x = a^2$, or b^2, or c^2.
15. $x = a(b+c)$, or $b(c+a)$, or $c(a+b)$. $\left\{ a(b+c) = q - \dfrac{r}{a} \right\}$.
16. $x = \dfrac{a+b}{c}$ or $\dfrac{b+c}{a}$ or $\dfrac{c+a}{b}$. $\left\{ \dfrac{a+b}{c} = \dfrac{p}{c} - 1. \right\}$

Prove that the following are cases of exact division:

17. $(x-1)^{12} - x^6 + (x^2 - x + 1)^2 \div x^3 - 2x^2 + 2x - 1$.
18. $(x-1)^{16} - x^8 + (x^2 - x + 1)^8 \div x^3 - 2x^2 + 2x - 1$.
19. $(x-2)^{10}(2x-5)^{10} - x^{10} + 2^{10}(x^2 - 4x + 5)^5 \div x^3 - 6x^2 + 13x - 10$.
20. $(x^2 + 4x + 3)^{18} - x^{18} - x^2 - 5x - 3 \div x^3 + 6x^2 + 8x + 3$.
21. $(9x-4)^{21}(x-1)^{21} - x^{21} - (9x^2 - 14x + 4)^{21} \div (x-1) \times (9x-4)(9x^2 - 14x + 4)$.
22. $\{6(x-1)\}^{13} - (2x^2 + 3x - 4)^{13} + (2x^2 - 3x + 2)^{13} \div (2x^2 + 3x - 4)(2x^2 - 3x + 2)(x - 1)$.

THEORY OF DIVISORS.

23. $\{2(x+1)(x-2)\}^{17}+(x^2-3x+3)^{17}-(3x^2-5x-1)^{17} \div$
$(x+1)(x-2)(x^2-3x+3)(3x^2-5x-1)$.

24. $\{6(x-1)\}^{16}-(2x^2+3x-4)^{16}-(2x^2-3x+2)^{16}+$
$2(2x^2+3x-4)^8(2x^2-3x+2)^4 \div (x-1)(2x^2+3x-4)(2x^2-3x+2)$

25. $\{2(x+1)(x-2)\}^{20}-(x^2-3x+3)^{20}-(3x^2-5x-1)^{20}+$
$2(x^2-3x+3)^9(3x^2-5x-1)^{11} \div (x+1)(x-2)(x^2-3x+3) \times$
$(3x^2-5x-1)$.

26. $1+x^4+x^8 \div 1+x+x^2$.

27. $x^{10}+x^5y^5+y^{10} \div x^2+xy+y^2$.

28. $1+x^3+x^6+x^9+x^{12} \div 1+x+x^2+x^3+x^4$.

29. $1+x^4+x^8+x^{12}+x^{16} \div 1+x+x^2+x^3+x^4$.

30. $x^{15}+x^{10}y^5+x^5y^{10}+y^{15} \div x^3+x^2y+xy^2+y^3$.

31. $x^{17}+x^4+x^3+x+1 \div x^4+x^3+x^2+x+1$.

32. $1+x+x^2+x^3+x^5+x^6+x^{53} \div$
$1+x+x^2+x^3+x^4+x^5+x^6$.

Find the quotient of the following divisions in which D denotes the product

$$(b-c)(c-a)(a-b)(a-d)(b-d)(c-d) ;$$

33. $(b^2c^2+a^2d^2)(b-c)(a-d)+(c^2a^2+b^2d^2)(c-a)(b-d)+$
$(a^2b^2+c^2d^2)(a-b)(c-d) \div D$.

34. $(bc+ad)(b^2-c^2)(a^2-d^2)+(ca+bd)(c^2-a^2)(b^2-d^2)+$
$(ab+cd)(a^2-b^2)(c^2-d^2) \div D$.

35. $(b+c)(a+d)(b^2-c^2)(a^2-d^2)+$ the two similar terms $\div D$.

36. $(b^2+c^2)(a^2+d^2)(b-c)(a-d)+$ " " $\div D$.

37. $\{bc(b+c)^2+ad(a+d)^2\}(b-c)(a-d)+$ " $\div D$.

38. $\{bc(b+c)+ad(a+d)\}(b^2-c^2)(a^2-d^2)+$ " $\div D$.

39. $\{bc(b^3+c^3)+ad(a^3+d^3)\}(b-c)(a-d)+$ " $\div D$.

40. $(b+c-a-d)^4(b-c)(a-d)+$ " $\div D$.

41. The sum of the fractions $\frac{1}{1}, \frac{1}{2}, \frac{1}{3}, \ldots \ldots \frac{1}{n}$, increased by the sum of their products two by two, increased by the sum of their products three by three,............. increased by their product is equal to n.

42. In any trapezium the square of the area expressed in terms of the lengths of the parallel sides and the diagonals, is a polynome of four dimensions, determine that polynome.

43. In any quadrilateral inscribed in a circle, the square of the area expressed in terms of the lengths of the sides, is a polynome of four dimensions, find that polynome.

Theorem III. If the polynome $f(x)^n$ vanish for more than n different values of x, it vanishes identically, the coefficient of every term being zero.

Cor. If a rational integral expression of n dimensions be divisible by more than n linear factors, the expression is identically zero.

EXAMPLES.

1. $\dfrac{(x-a)(x-b)}{(c-a)(c-b)} + \dfrac{(x-b)(x-c)}{(a-b)(a-c)} + \dfrac{(x-c)(x-a)}{(b-c)(b-a)} - 1 = 0$, if a, b, and c are unequal; for this is a polynome of *two* dimensions in x, but it vanishes for $x = a$, and, therefore, by symmetry for $x = b$, and for $x = c$, that is, for *three* different values of x, hence it vanishes identically.

2. $\{(a+b)^2+(c+d)^2\}(a-b)(c-d)+\{(c+b)^2+(b+d)^2\}(b-c)(a-d)+\{(c+a)^2+(b+d)^2\}(c-a)(b-d) = 0$.

Substitute b for a and the expression becomes
$\{(b+c)^2+(b+d)^2\}(b-c)(b-d)+\{(c+b)^2+(b+d)^2\}(c-b)(b-d)$
which vanishes, hence the given expression is divisible by $a-b$, and consequently by symmetry it is divisible by $(a-b)$, $(b-c)$, $(c-d)$, $(a-c)$, $(b-d)$, and $(a-d)$, But the given expression is of only *four* dimensions, while it appears to have *six* linear factors, hence it vanishes identically.

Exercise xxii.

Verify the following:

1. $\dfrac{x^2 y^2 z^2}{b^2 c^2} + \dfrac{(x^2-b^2)(y^2-b^2)(z^2-b^2)}{b^2(b^2-c^2)} + \dfrac{(x^2-c^2)(y^2-c^2)(z^2-c^2)}{c^2(c^2-b^2)}$
$= x^2 + y^2 + z^2 - b^2 - c^2$.

THEORY OF DIVISORS.

2. $\dfrac{y^2 z^2}{b^2 c^2} + \dfrac{(y^2-b^2)(z^2-b^2)}{b^2(b^2-c^2)} + \dfrac{(y^2-c^2)(z^2-c^2)}{c^2(c^2-b^2)} = 1.$

3. $\dfrac{x^2 y^2}{a^2 b^2} + \dfrac{(x^2-a^2)(a^2-y^2)z^2}{(z^2-a^2)(b^2-a^2)a^2} + \dfrac{(x^2-b^2)(b^2-y^2)z^2}{(b^2-z^2)(b^2-a^2)b^2} +$
$+ \dfrac{(z^2-x^2)(z^2-y^2)}{(z^2-a^2)(b^2-z^2)} = 0.$

4. $\dfrac{1}{(x+a)(a-b)(a-c)} + \dfrac{1}{(x+b)(b-c)(b-a)} + \dfrac{1}{(x+c)(c-a)(c-b)}$
$= \dfrac{1}{(x+a)(x+b)(x+c)}.$

5. $bc(b^2-c^2) + ca(c^2-a^2) + ab(a^2-b^2) =$
$(a+b+c)\{a^2(b-c)+b^2(c-a)+c^2(a-b)\}.$

6. $\dfrac{a+x}{x(x-y)(x-z)} + \dfrac{a+y}{y(y-x)(y-z)} + \dfrac{a+z}{z(z-x)(z-y)} = \dfrac{a}{xyz}.$

7. $\dfrac{a^4(b^2-c^2)+b^4(c^2-a^2)+c^4(a^2-b^2)}{a^2(b-c)+b^2(c-a)+c^2(a-b)} =$
$\tfrac{1}{3}\{(a+b+c)^3 - a^3 - b^3 - c^3\}.$

8. $(adf+bcf+bed-ace)^2 + (bce+aed+acf-bdf)^2 =$
$(a^2+b^2)(c^2+d^2)(e^2+f^2).$

9. $\dfrac{(a-b)^5+(b-c)^5+(c-a)^5}{(a-b)(b-c)(c-a)} =$
$\tfrac{5}{2}\{(a-b)^2+(b-c)^2+(c-a)^2\}.$

10. $(-x+y+z)(x-y+z)(x+y-z)+x(x-y+z)(x+y-z)+$
$y(x+y-z)(-x+y+z)+z(-x+y+z)(x-y+z) = 4xyz.$

11. $\dfrac{(a^2-b^2)^3+(b^2-c^2)^3+(c^2-a^2)^3}{(a+b)(b+c)(c+a)} =$
$(a-b)^3+(b-c)^3+(c-a)^3.$

12. $x^2(y+z)^2+y^2(z+x)^2+z^2(x+y)^2+2xyz(x+y+z) =$
$2(xy+yz+zx)^2.$

Theorem IV. If the polynomes $f(x)^n$, $\varphi(x)^m$ (n not less than m) are equal for more than n different values of x, they are equal for *all* values, and the coefficients of equal powers of x in each are equal to one another.

(This is called the Principle of Indeterminate Coefficients. The full use of it cannot be exhibited till the student is able to work simultaneous equations.)

EXAMPLES.

1. $\dfrac{a^2}{(a-b)(a-c)(a-d)} + \dfrac{b^2}{(b-a)(b-c)(b-d)} + \dfrac{c^2}{(c-a)(c-b)(c-d)} + \dfrac{d^2}{(d-a)(d-b)(d-c)} = 0.$

Assume $\dfrac{x^2}{(x-a)(x-b)(x-c)(x-d)} =$

$\dfrac{A}{x-a} + \dfrac{B}{x-b} + \dfrac{C}{x-c} + \dfrac{D}{x-d}$ (α)

in which A, B, C, D are independent of x.

Multiply by $(x-a)(x-b)(x-c)(x-d)$.

$\therefore x^2 = (A+B+C+D)x^3 +$ terms in lower powers of x.

Now this equality holds for more than three values of x, holding in fact for *all* finite values of x.

$\therefore A+B+C+D = 0$ (β)

Again multiply both sides of (α) by $x-a$

$\dfrac{x^2}{(x-b)(x-c)(x-d)} = A + \left(\dfrac{B}{x-b} + \dfrac{C}{x-c} + \dfrac{D}{x-d}\right)(x-a).$

Put $x = a$

$\therefore \dfrac{a^2}{(a-b)(a-c)(a-d)} = A.$

By symmetry $\dfrac{b^2}{(b-a)(b-c)(b-d)} = B$, &c.

Adding

$\dfrac{a^2}{(a-b)(a-c)(a-d)} + \dfrac{b^2}{(b-a)(b-c)(b-d)} + \dfrac{c^2}{(c-a)(c-b)(c-d)}$
$+ \dfrac{d^2}{(d-a)(d-b)(d-c)} = A+B+C+D = 0$ by (β).

2. $\dfrac{a^2(a+b)(a+c)}{(a-b)(a-c)} + \dfrac{b^2(b+c)(b+a)}{(b-c)(b-a)} + \dfrac{c^2(c+a)(c+b)}{(c-a)(c-b)}$
$= (a+b+c)^2.$

Assume $x^3 - px^2 + qx - r = (x-a)(x-b)(x-c)$. ($\alpha$).

$\therefore x^3 + px^2 + qx + r = (x+a)(x+b)(x+c)$. ($\beta$).

$$\frac{x^4 + px^3 + qx^2 + rx}{x^3 - px^2 + qx - r} = x + 2p + \frac{A}{x-a} + \frac{B}{x-b} + \frac{C}{x-c} \quad (\gamma).$$

Multiply by $x^3 - px^2 + qx - r$ and equate the coefficients of the terms in x^2. {In multiplying the fractions in the right-hand member of (γ), use the factor side of (α).}

$$q = q - 2p^2 + A + B + C$$
$$\therefore A + B + C = 2p^2.$$

Multiply both members of (γ) by $x - a$

$$\frac{x(x+a)(x+b)(x+c)}{(x-b)(x-c)} = A + \left\{x + 2p + \frac{B}{x-b} + \frac{C}{x-c}\right\}(x-a).$$

Put $x = a$,

$$\frac{2a^2(a+b)(a+c)}{(a-b)(a-c)} = A.$$

By symmetry

$$\frac{2b^2(b+c)(b+a)}{(b-c)(b-a)} = B \text{ and } \frac{2c^2(c+a)(c+b)}{(c-a)(c-b)} = C;$$

$$\therefore \frac{a^2(a+b)(a+c)}{(a-b)(a-c)} + \frac{b^2(b+c)(b+a)}{(b-c)(b-a)} + \frac{c^2(c+a)(c+b)}{(c-a)(c-b)}$$
$$= \tfrac{1}{2}(A + B + C) = p^2.$$
$$= (a+b+c)^2.$$

3. Extract the square root of $1 + x + x^2 + x^3 + x^4 + \&c$.

Assume the square root to be $1 + ax + bx^2 + cx^3 + dx^4 + \&c$.

$\therefore 1 + x + x^2 + x^3 + x^4 + \&c. = (1 + ax + bx^2 + cx^3 + dx^4 + \&c.)^2$
$= 1 + 2ax + (a^2 + 2b)x^2 + 2(ab+c)x^3 + (2d + 2ac + b^2)x^4 + \&c.$

$\therefore 2a = 1 \qquad \therefore a = \tfrac{1}{2}$

$2b + a^2 = 1 \qquad \therefore b = \tfrac{1}{2}(1 - \tfrac{1}{4}) = \tfrac{3}{8}$

$2(c + ab) = 1 \qquad \therefore c = \tfrac{1}{2} - (\tfrac{1}{2} \times \tfrac{3}{8}) = \tfrac{5}{16}$

$2d + 2ac + b^2 = 1 \qquad \therefore d = \tfrac{1}{2}\{1 - \tfrac{5}{16} - \tfrac{9}{64}\} = \tfrac{35}{128}.$

$\therefore \sqrt{(1 + x + x^2 + \&c.)} = 1 + \tfrac{1}{2}x + \tfrac{3}{8}x^2 + \tfrac{5}{16}x^3 + \tfrac{35}{128}x^4 + \&c.$

(NOTE.—As it is frequently necessary to determine the coefficient of a particular power of x, a few preliminary exercises are given on this subject.)

THEORY OF DIVISORS.

Exercise xxiii.

Determine the coefficient of

1. x^4 in $(1+ax)^3+(1+bx)^5+(1-cx)^6$.
2. x^5 in $(1+x+2x^2+3x^3)(1-x+3x^2+x^3-5x^4)$.
3. x^4 in $(1+x+2x^2+3x^3+4x^4+\&c)(1-x+x^2-x^3+x^4-\&c)$.
4. x^2 in $A(x-b)(x-c)(x-d)+B(x-a)(x-c)(x-d)+C(x-a)(x-b)(x-d)+D(x-a)(x-b)(x-c)$.
5. x^4 in $(1-ax)^3(1+ax)^5$.
6. x^4 in $(1+ax)^3(1-bx)^5$.
7. In the product

$$(1+ax+bx^2+cx^3+\&c.)(1-ax+bx^2-cx^3+\&c.)$$

prove that the coefficients of the odd powers of x must be all zeros.

Determine the value of the following expressions:

8. $\dfrac{1}{(a-b)(a-c)(a-d)} + \dfrac{1}{(b-a)(b-c)(b-d)} + \dfrac{1}{(c-a)(c-b)(c-d)} + \dfrac{1}{(d-a)(d-b)(d-c)}$.

9. $\dfrac{a}{(a-b)(a-c)(a-d)} + \dfrac{b}{(b-a)(b-c)(b-d)} + \&c.$

10. $\dfrac{a^2}{(a-b)(a-c)(a-d)} +$ three similar terms.

11. $\dfrac{a^3}{(a-b)(a-c)(a-d)} +$ " "

12. $\dfrac{a^4}{(a-b)(a-c)(a-d)} +$ three similar terms.

13. $\dfrac{bcd}{(a-b)(a-c)(a-d)} +$ " "

14. $\dfrac{a(a+b)(a+c)}{(a-b)(a-c)} +$ two · "

15. $\dfrac{a^3(a+b)(a+c)}{(a-b(a-c)} +$ " "

16. $\dfrac{a^4(a+b)(a+c)}{(a-b)(a-c)} +$ " "

17. $\dfrac{a(a+b)(a+c)(a+d)}{(a-b)(a-c)(a-d)}$ + three similar terms.

18. $\dfrac{a^2(a+b)(a+c)(a+d)}{(a-b)(a-c)(a-d)} +$ " "

19. $\dfrac{a^3(a+b)(a+c)(a+d)}{(a-b)(a-c)(a-d)} +$ " "

20. $\dfrac{bc(b+c)}{(a-b)(a-c)}$ +two similar terms.

[For numerator use $x^3+2px^2+(p^2+q)x+(pq-r)$.]

21. $\dfrac{(2a+b)(2a+c)}{(a-b)(a-c)}$ + two similar terms.

[For numerator use $x^3-2px^2+(p^2+q)x-(pq-r)$.]

22. $\dfrac{a(b+c)}{(a-b)(a-c)}$ + two similar terms.

[For numerator use $x(x+p)$.]

23. $\dfrac{b+c+d}{(a-b)(a-c)(a-d)}$ + three similar terms.

24. $\dfrac{a^3(bc+cd+db)}{(a-b)(a-c)(a-d)} +$ " " "

25. $\dfrac{bc+cd+db}{(a-b)(a-c)(a-d)} +$ " " "

Extract the square-root of (to 4 terms):

26. $1+x$. | 27. $1-x$. | 28. $1+2x+3x^2+4x^3+$ &c.
29. $1-4x+10x^2-20x^3+35x^4-56x^5+84x^6$.
30. Extract the cube-root of $1+x$. (To 4 terms).

Art. XI. 1. Find the condition that $px^2+2qx+r$ and $p'x^2+2q'x+r'$ shall have a common factor.

Multiply the polynomes by p' and p respectively, and take the difference of the products, also by r' and r respectively, and divide the difference of the products by x.

$$\begin{array}{l|l} p'px^2+2p'qx+p'r & pr'x^2+2qr'x+rr' \\ pp'x^2+2pq'x+pr' & p'rx^2+2q'rx+r'r \\ \hline 2(pq'-p'q)x+(pr'-p'r) & (pr'-p'r)x+2(qr'-r'q). \end{array}$$

Multiply the former of these remainders by $(pr'-p'r)$ and the latter by $2(pq'-p'q)$, and the difference of the products is

$$(pr'-p'r)^2-4(pq'-p'q)(qr'-r'q).$$

But if the given polynomes have a linear factor this remainder must vanish, or
$$(pr' - p'r)^2 = 4(pq' - p'q)(qr' - r'q).$$
If the given polynomes have a quadratic factor, the *linear* remainders must vanish identically, or (Th. III.)
$$pq' - p'q = 0, \; pr' - p'r = 0, \text{ and } qr' - r'q = 0,$$
or, $\dfrac{p}{p'} = \dfrac{q}{q'} = \dfrac{r}{r'}$.

2. Find the condition that $px^3 + 3qx^2 + 3rx + s$ shall have a square factor.

Assume the square factor to be $(x-a)^2$. On division, the remainder must be zero for every finite value of x, and consequently (Th. III.) the co-efficient of each term of the remainder must be zero. Divide by $(x-a)^2$, neglecting the first remainder.

$$\begin{array}{c|cccc}
 & p & 3q & 3r & s \\
a & & pa & pa^2 + 3qa & \\ \hline
 & p & pa + 3q & pa^2 + 3qa + 3r \; ; & R \\
a & & pa & 2pa^2 + 3qa & \\ \hline
 & p & 2pa + 3q \; ; & 3(pa^2 + 2qa + r) &
\end{array}$$

$\therefore pa^2 + 2qa + r = 0$;

$\therefore px^2 + 2qx + r$ is divisible by $x-a$ (Th. I. Cor. 2),

or, $px^3 + 3qx^2 + 3rx + s$ and $px^2 + 2qx + r$ have a common divisor. Multiply the latter polynome by x and subtract the product from the former, and the proposition reduces to

If $px^3 + 3qx^2 + 3rx + s$ have a square factor, $px^2 + 2qx + r$ and $qx^2 + 2rx + s$ will have the square-root of that factor for a common divisor.

3. If $px^3 + 3qx^2 + 3rx + s$ vanish for $x = a$, or b, or c, find in terms of x, p, q, r the value of
$$\frac{1}{x-a} + \frac{1}{x-b} + \frac{1}{x-c}.$$

Reduce to a common denominator and add the numerators
$$\frac{3x^2 - 2(a+b+c)x + (ab+bc+ca)}{(x-a)(x-b)(x-c)}$$

Multiply both numerator and denominator by p and reduce by Th. II., and Ex. 4 of Th. II.

$$\therefore \frac{x^{m+1}}{x-a} + \frac{x^{m+1}}{x-b} + \frac{x^{m+1}}{x-c} = \frac{\dfrac{3(px^2+2qx+r)}{px^3+3qx^2+3rx+s}}{1} = \frac{3(px^{m+3}+2qx^{m+2}+rx^{m+1})}{px^3+3qx^2+3rx+s}.$$

4. If $px^3+3qx^2+3rx+s$ vanish for $x=a$, or b, or c, express in terms of p, q, r, s, the following, $a+b+c$, $a^2+b^2+c^2$, $a^3+b^3+c^3$, ... , $a^m+b^m+c^m$.

Divide x^{m+1} by $x-a$.

a	1						
	a	a^2	a^3			a^m	a^{m+1}
	1	a	a^2	a^3	. . . . .	a^m ;	a^{m+1}

Similarly divide x^{m+1} by $x-b$ and also by $x-c$. add together the quotients

$$\frac{x^{m+1}}{x-a} + \frac{x^{m+1}}{x-b} + \frac{x^{m+1}}{x-c} = 3x^m + (a+b+c)x^{m-1} + (a^2+b^2+c^2)x^{m-2} + (a^3+b^3+c^3)x^{m-3} + \&c.$$

Hence, by the last example, the required expressions are the coefficients taken in order, beginning with the second, of the terms in the quotient of $3(px^{m+3}+2qx^{m+2}+rx^{m+1}) \div (px^3+3qx^2+3rx+s)$. These may now be found by Horner's Division.

5. Writing s_1 for $a+b+c$, s_2 for $a^2+b^2+c^2$, &c., express $(a-b)^4+(b-c)^4+(c-a)^4$ in terms of s_1, s_2, s_3, s_4.

By actual expansion

$$(x-a)^4+(x-b)^4+(x-c)^4 =$$
$$3x^4 - 4(a+b+c)x^3 + 6(a^2+b^2+c^2)x^2 - 4(a^3+b^3+c^3)x + a^4+b^4+c^4 = 3x^4 - 4s_1x^3 + 6s_2x^2 - 4s_3x + s_4.$$

Put $x=a$, $=b$, $=c$ in succession.

$(a-b)^4 \qquad +(c-a)^4 = 3a^4 - 4s_1a^3 + 6s_2a^2 - 4s_3a + s_4$
$(a-b)^4+(b-c)^4 \qquad = 3b^4 - 4s_1b^3 + 6s_2b^2 - 4s_3b + s_4$
$\qquad (b-c)^4 +(c-a)^4 = 3c^4 - 4s_1c^3 + 6s_2c^2 - 4s_3c + s_4$
$\therefore 2\{(a-b)^4+(b-c)^4+(c-a)^4\} = 3s_4 - 4s_1s_3 + 6s_2^2 - 4s_3s_1 + 3s_4$
$\therefore \quad (a-b)^4+(b-c)^4+(c-a)^4 = s_0s_4 - 4s_1s_3 + 3s_2^2$

in which s_0 is written for 3 or $1+1+1$, i.e., $a^0+b^0+c^0$.

THEORY OF DIVISORS.

⋋ Exercise xxiii. (a).

1. Determine the condition necessary in order that $x^2 + px + q$ and $x^3 + p'x + q$ may have a common divisor.

2. The expression $x^6 + 3a^2x^5 + 3bx^4 + cx^3 + 3dx^2 + 3e^2x + f^3$ will be a complete cube if
$$f = \frac{e}{a} = \frac{d}{b} = \frac{c-a^6}{6a^2} = b - a^4.$$

3. Prove that $ax^5 + bx + c$ and $a + bx^4 + cx^5$ will have a common quadratic factor if
$$b^2c^2 = (c^2 - a^2 + b^2)(c^2 - a^2 + ab).$$

4. Prove that $ax^5 + bx^2 + c$ and $a + bx^3 + cx^5$ will have a common quadratic factor if
$$a^3b^2 = (a^2 - c^2)(a^2 - c^3 + bc).$$

5. Prove that $ax^4 + bx^3 + cx + d$ and $a + bx + cx^3 + dx^4$ will have a common quadratic factor if
$$(a+d)^2 = (b-c)(bd - ac).$$

6. $x^3 + px^2 + qx + r$ will be divisible by $x^2 + ax + b$ if
$$a^3 - 2pa^2 + (p^2 + q)a + r - pq = 0, \text{ and } b^3 - qb^2 + rpb - r^2 = 0.$$

7. $x^4 + px + q$ will be divisible by $x^2 + ax + b$ if
$$a^6 - 4qa^2 = p^2 \text{ and } (b^2 + q)(b^2 - q)^2 = p^2b^3.$$

8. Determine the condition necessary in order that $x^4 + 4px^3 + 6qx^2 + 4rx + t$ may have a square factor.

If $x^4 + 4px^3 + 6qx^2 + 4rx + t$ vanish for $x = a$, or b, or c, or d, find in terms of x, p, q, r, t, the value of

9. $\dfrac{x^n}{x-a} + \dfrac{x^n}{x-b} + \dfrac{x^n}{x-c} + \dfrac{x^n}{x-d}.$

10. Σa, Σa^2, Σa^3, Σa^4, Σa^5, Σa^6.

11. $\Sigma(a-b)^4$, $\Sigma(a-b)^6$.

12. Determine the values of the expressions in Ex. 9, 10, 11, for the polynome $x^4 - 14x^2 + x - 38$.

CHAPTER III.

SECTION I.—DIRECT APPLICATION OF THE FUNDAMENTAL FORMULAS

FORMULAS [1] AND [2]. $(x \pm y)^2 = x^2 \pm 2xy + y^2$, &c.

Art. XII. From this it appears that a trinomial of which the extremes are squares, is itself a square if four times the product of the extremes is equal to the square of the mean, and that to factor such a trinomial, we have simply to connect the square root of each of the squares by the sign of the other term, and write the result twice as a factor.

EXAMPLES.

1. $4x^4 - 80x^2y^2 + 400y^4 = (2x^2 - 20y^2)(2x^2 - 20y^2)$
2. $1 - 12x^2y^2 + 36x^4y^4 = (1 - 6x^2y^2)(1 - 6x^2y^2)$.
3. $(a-b)^2 + (b-c)^2 + 2(a-b)(b-c)$. This equals $(a-b+b-c) \times (a-b+b-c) = (a-c)(a-c)$.
4. $x^2 + y^2 + z^2 + 2xy - 2xz - 2yz$.

Here the three squares and the three double products suggest that the expression is the square of a linear *trinomial* in x, y, z.

An inspection of the signs of the double products enables us to determine the signs which are to connect x, y, z: we see that

1st. The signs of x and y must be *alike*.
2nd. The signs of x and z must be *different*.
3rd. The signs of y and z must be *different*. Hence we have $x+y-z$, or $-x-y+z = -(x+y-z)$, and the factors are $(x+y-z)(x+y-z)$.

Exercise xxiv.

1. $9m^2 + 12m + 4$; $c^{2m} - 2c^m + 1$.
2. $y^6 - 2y^3z^3 + z^6$; $16x^2y^2 + 16xy^3 + 4y^4$.
3. $9a^2b^2 + 12abc + 4c^2$; $36x^2y^2 - 24xy^3 + 4y^4$

FACTORING.

4. $\frac{1}{4}x^4 + 16y^2z^2 - 4x^2yz$; $\frac{1}{4}a^4 - \frac{1}{4}a^2b^2c^2 + \frac{1}{8}b^4c^4$.

5. $(a+b)^2 + c^2 - 2c(a+b)$; $9x^8 - \frac{3}{2}x^4y^2 + \frac{1}{16}y^4$.

6. $z^2 + (x-y)^2 - 2z(x-y)$; $\left(\dfrac{a}{b}\right)^{2m} + \left(\dfrac{b}{a}\right)^{2m} - 2$.

7. $(x^2-y)^2 + 2(x^2-y)(y-z^2) + (y-z^2)^2$.

8. $(x^2-xy)^2 - 2(x^2-xy)(xy-y^2) + (xy-y^2)^2$.

9. $(a+b+c)^2 - 2c(a+b+c) + c^2$; $\frac{9}{16}p^6 - 2p^3q^2 + \frac{16}{9}q^4$.

10. $(3x-4y)^2 + (2x-3y)^2 - 2(3x-4y)(2x-3y)$.

11. $(x^2 - xy + y^2)^2 + (x^2 + xy + y^2)^2 + 2(x^4 + x^2y^2 + y^4)$.

12. $(5x^2 + 2xy + 7y^2)^2 + (4x^2 + 6y^2)^2 - 2(4x^2 + 6y^2) \times (5x^2 + 2xy + 7y^2)$.

13. $\left(\dfrac{a}{b}\right)^{2m} + \left(\dfrac{b}{a}\right)^{2n} - 2\left(\dfrac{a}{b}\right)^{m-n}$.

14. $a^2 + b^2 + c^2 - 2ab - 2bc + 2ac$.

15. $a^4 + b^4 + c^4 - 2a^2b^2 - 2a^2c^2 + 2b^2c^2$.

16. $(a-b)^2 + (b-c)^2 + (c-a)^2 + 2(a-b)(b-c) - 2(a-b)(c-a) + 2(b-c)(a-c)$.

17. $4a^4 - 12a^2b + 9b^2 + 16a^2c + 16c^2 - 24bc$.

FORMULA [4]. $x^2 - y^2 = (x+y)(x-y)$.

Art. XIII. In this case we have merely to take the square-root of each of the squares, and connect the results with the sign + for one of the factors, and with the sign − for the other.

EXAMPLES.

1. $(a+b)^2 - (c+d)^2$.
   This $= \{(a+b)+(c+d)\}\{(a+b)-(c+d)\}$
   $= (a+b+c+d)(a+b-c-d)$.

2. Factor $(x^2 + 5xy + y^2)^2 - (x^2 - xy + y^2)^2$.
   Here we have
   $\{(x^2+5xy+y^2)+(x^2-xy+y^2)\}\{(x^2+5xy+y^2)-(x^2-xy+y^2)\}$
   $= 2(x^2+2xy+y^2)(6xy) = 12xy(x+y)^2$.

3. $a^2 - b^2 - c^2 + 2bc$.
   This $= a^2 - (b-c)^2 = (a+b-c)(a-b+c)$.

4. Resolve $(a^2+b^2)^2 - (a^2-b^2)^2 - (a^2+b^2-c^2)^2$.

This $= 4a^2b^2 - (a^2+b^2-c^2)^2$

$= (2ab+a^2+b^2-c^2)(2ab-a^2-b^2+c^2)$.

The former of these factors $= (a+b)^2 - c^2 = (a+b+c)(a+b-c)$; and the latter $= c^2 - (a-b)^2) = (c+a-b)(c-a+b)$.

∴ the given expression

$= (a+b+c)(a+b-c)(c+a-b)(c-a+b)$.

Exercise xxv.

1. $49a^2 - 4b^2$.
2. $9a^2 - \tfrac{1}{4}b^2$.
3. $81a^4 - 16b^4$.
4. $100x^2 - 36y^2$.
5. $5a^2b - 20bx^2y^2$.
6. $9x^6 - 16y^4$.
7. $\tfrac{9}{16}c^2 - 1$.
8. $4y^4 - \tfrac{4}{9}x^2z^2$.
9. $81a^4 - 1$.
10. $a^4 - 16b^4$.
11. $a^{16} - b^{16}$.
12. $a^2 - b^2 + 2bc - c^2$.
13. $(a+2b)^2 - (3x-4y)^2$.
14. $(x^2+y^2)^2 - 4x^2y^2$.
15. $(x+y)^2 - 4z^2$.
16. $(3x+5)^2 - (5x+3)^2$.
17. $4x^2y^2 - (x^2+y^2-z^2)^2$.
18. $(x^2+xy-y^2)^2 - (x^2-xy-y^2)^2$.
19. $(x^2-y^2+z^2)^2 - 4x^2z^2$.
20. $(a+b+c+d)^2 - (a-b+c-d)^2$.
21. $(2+3x+4x^2)^2 - (2-3x+4x^2)^2$.
22. $(a^2+b^2+4ab)^2 - (a^2+b^2)^2$.
23. $(a^2-b^2+c^2-d^2)^2 - (2ac-2bd)^2$.
24. $(x^2-y^2-z^2)^2 - 4y^2z^2$.
25. $(a^6-a^3b^3+b^6)^2 - (a^6-5a^3b^3+b^6)^2$.
26. $a^{12} - b^{12} + 6a^9b^3 - 6b^9a^3 + 8b^9a^3 - 8a^9b^3$.
27. $(x^2+y^2+z^2-xy-yz-zx)^2 - (xy+yz+zx)^2$.
28. $(x^2+y^2+z^2-2xy+2xz-2yz) - (y+z)^2$.
29. $2a^2b^2 + 2b^2c^2 + 2c^2a^2 - a^4 - b^4 - c^4$.
30. $x^4 + y^4 + z^4 - 2x^2y^2 - 2y^2z^2 - 2z^2x^2$.

FACTORING.

FORMULA A. $(x+r)(x+s)=x^2+(r+s)x+rs.$

EXAMPLES

1. $x^6-9x^3+20=(x^3-5)(x^3+4).$
2. $(x-y)^2+x-y-110=(x-y+11)(x-y-10).$
3. $(a^2-ab+b^2)^2+6b(a^2-ab+b^2)-4a^2+9b^2 =$
   $\{(a^2-ab+b^2)+(2a+3b)\}\{(a^2-ab+b^2)-(2a-3b)\}.$
4. $(x^2-5x)^2-6(x^2-5x)-40=(x^2-5x+4)(x^2-5x-10).$
5. $(ax+by+c)^2-(m-n)(ax+by+c)-mn$
   $=(ax+by+c-m)(ax+by+c+n).$

Art. XIV. It will be seen that the first (or *common*) term of the required factors, is obtained by extracting the square root of the first term of the given expression, and that the other terms are determined by observing two conditions:

(1) Their product must equal the *third* term of the given expression.

(2) Their sum (*algebraic*) multiplied into the *common* term already found, must equal the middle term of the given expression. Hence, to make a systematic search for integral factors of an expression of the form $x^2 \pm bx \pm c$, we may proceed as follows:

1st. Write down every pair of factors whose product is c.

2nd. If the sign before c is $+$, select the pair of factors whose *sum* is b, and write *both* factors $x+$, if the sign before b is $+$; $x-$, if the sign before b is $-$.

3rd. But if the sign before c is $-$, select the pair of factors whose *difference* is b, and write before the *larger* factor $x+$ or $x-$, and before the other factor $x-$ or $x+$, according as the sign before b is $+$ or $-$.

EXAMPLES.

1. $x^2+9x+20.$ The factors of 20 in pairs are 1 and 20, 2 and 10, 4 and 5. The sign before 20 is $+$, hence select the factors whose *sum* is 9. These are 4 and 5. The sign before 9 is $+$, hence the required factors are $(x+4)(x+5).$

2. $x^2 - 8x + 12$. Pairs of factors of 12 are 1 and 12, 2 and 6, 3 and 4. Sign before 12 is $+$, therefore take pair whose sum is 8. These are 2 and 6. Sign before 8 is $-$, hence the factors are $(x-2)(x-6)$.

3. $x^2 - 21x - 100$. Pairs of factors of 100 are 1 and 100, 2 and 50, 4 and 25, 5 and 20, 10 and 10. Sign before 100 is $-$, therefore take the pair whose difference is 21. These are 4 and 25. The sign before 21 is $-$, therefore $x-$ goes before 25, the larger factor, and the factors are $(x+4)(x-25.)$

4. $x^2 + 12x - 108$. Pairs of factors of 108 are 1 and 108, 2 and 54, 3 and 36, 4 and 27, 6 and 18, 9 and 12. Sign before 108 is $-$, therefore take the pair whose difference is 12. These are 6 and 18. Sign before 12 is $+$, therefore $x+$ goes before 18, the larger factor, and $x-$ before 6, the other factor; hence the factors are $(x-6)(x+18)$.

NOTE.—It will be found convenient to write the factors in two columns, separated by a short space. Taking Ex. 2 above, proceed thus: Since the sign of the third term is $+$, write the sign of the second term (in this case $-$) above both columns.

$$\begin{array}{cc} - & - \\ 1 & 12 \\ (x-2) & (x-6) \end{array}$$

Ex. 3 above. Since the sign of the third term is $-$, write the sign of the 2nd term (in this case $-$) above the column of larger factors, and the other sign of the pair $\pm$, above the other column.

$$\begin{array}{cc} + & - \\ 1 & 100 \\ 2 & 50 \\ (x+4) & (x-25) \end{array}$$

5. $x^2 - 84x + 64$.

Here we have the factors

$$\begin{array}{cc} - & - \\ 1, & 64 \\ x-2, & x-32 \\ 4, & 16 \end{array}$$

and since the last term has the sign $+$, and the middle term has the sign $-$, we write $-$ over both columns,

6. $x^2 + 12x - 64$.

$$\begin{array}{cc} - & + \\ 1, & 64 \\ 2, & 32 \\ x-4, & x+16. \end{array}$$

Here, since the last term has the sign $-$, we write the sign $(+)$ of the middle term, over the column of larger factors, and the sign $-$ over the other column.

7. $x^4 - 10x^2 - 144$.

Here we have the pairs of factors:

$$\begin{array}{cc} + & - \\ 1, & 144 \\ 2, & 72 \\ 4, & 36 \\ x+8, & x-18. \end{array}$$

And since the sign of the third term is $-$, we write the sign of the second term (in this case $-$) above the column of *larger* factors, and the other sign (of the pair $\pm$) above the other column.

Exercise xxvi.

1. $x^2 - 5x - 14$; $x^2 - 9x + 14$; $x^2 + 7x + 12$.
2. $x^2 - 8x + 15$; $x^2 - 19x + 84$; $x^2 - 7x - 60$.
3. $4x^2 - 2x - 20$; $9x^2 - 150x + 600$.
4. $\frac{1}{4}x^2 + 4\frac{1}{2}x - 36$; $25x^2 + 40x + 15$; $9x^6 - 27x^3 + 20$.
5. $\frac{1}{16}x^2 + 1\frac{3}{4}x + 12$; $16x^4 - 4x^2 - 20$.
6. $x^4 - (a^2 + b^2)x^2 + a^2b^2$; $4(x+y)^2 - 4(x+y) - 99$.
7. $(x^2 + y^2)^2 - (a^2 - b^2)(x^2 + y^2) - a^2b^2$.
8. $(a+b)^2 - 2c(a+b) - 8c^2$.
9. $(x+y)^2 + 2(x^2 + y^2)(x+y) + (x^2 - y^2)^2$.
10. $(a+b)^2 - 4ab(a+b) - (a^2 - b^2)^2$.
11. $(x^2 + xy + y^2)^2 + x^3 - y^3 - 5xy - 2y^2 - 2x^2$.
12. $a^2 - 2a(b-c) - 3(b-c)^2$.

FACTORING.

13. $(x^3+y^3)^2+2a^3(x^3+y^3)+a^6-b^6$.
14. $(x^2-10x)^2-4(x^2-10x)-96$.
15. $(x^2-14x+40)^2-25(x^2-14x+40)-150$.
16. $(x^2-xy+y^2)^2+2xy(x^2-xy+y^2)-3x^2y^2$.
17. z^4-8z^2+2; x^4-2x^2-3; $9x^8+9x^4y^2-10y^4$.
18. $c^{2n}+c^m-2$; x^6-x^3-2; $x^{3m}-2x^my^n-8y^{2n}$.
19. $x^{2m}-(a-b)x^my^n-aby^{2n}$.

Art. XV. Trinomials of the form ax^2+bx+c (a not a square) may sometimes be easily factored from the following considerations:—

The product of two binomials consists of
1st. The product of the *first* terms.
2nd.   "   "   *second*   "
3rd. The sum (*algebraic*) of the products of the terms taken diagonally.

These three conditions guide us in the converse process of resolving a trinomial into its binomial factors.

EXAMPLES.

1. Resolve $6x^2-13xy+6y^2$.

Here the factors of the first term are x and $6x$, or $2x$ and $3x$; those of the third term are y and $6y$, or $2y$ and $3y$. These pairs of factors may be arranged

(1)	(2)	(3)	(4)
x	$2x$	y	$2y$
$6x$	$3x$	$6y$	$3y$

Now, we may take (1) with (3) or (4), or (2) with (3) or (4); but none of these combinations will satisfy the third condition. If, however, in (4) we interchange the coefficients 2 and 3, then (2) and (4) give

$2x$ $3y$, and
$3x$ $2y$, where we can combine the "diagonal" products to make 13, and the factors are

FACTORING.

$2x - 3y$, and
$3x - 2y$.

The coefficients of (2), instead of those of (4), might have been interchanged, giving the same result.

2. $6x^2 - 15xy + 6y^2$.

Here, comparing (2) and (3), Ex. 1, we see that their diagonal products may be combined to give 15, and the factors are

$2x - y$, and $3x - 6y$.

3. $6x^2 - 20xy + 6y^2$.

Here, again referring to Ex. 1, we see at once that it is useless to try *both* (2) *and* (4), since the diagonal products cannot be combined in any way to give a higher result than $13xy$. But comparing (1) and (4), we obtain by interchanging the coefficients in (4) $x - 3y$, and
$6x - 2y$, which satisfy the third condition.

Or, we might interchange the coefficients of (3), and take the resulting terms with (2), getting $2x - 6y$, and
$3x - y$.

4. $6x^2 + 35xy - 6y^2$.

Here the large coefficient of the middle term shows at once that we must take (1) and (3) together. Interchanging the coefficients of (1) we have

$6x - y$, and

$x + 6y$. The same result will be obtained by interchanging the coefficients of (3).

Exercise xxvii.

1. $6x^2 - 37xy + 6y^2$.
2. $6x^2 + 9xy - 6y^2$.
3. $56x^2 - 76xy + 20y^2$.
4. $56x^2 - 36xy - 20y^2$.
5. $56x^2 - 1121xy + 20y^2$.
6. $56x^2 - 68xy + 20y^2$.
7. $56x^2 - 558xy - 20y^2$.
8. $56x^2 + 36xy - 20y^2$.
9. $56x^2 - 67xy + 20y^2$.
10. $56x^2 + 3xy - 20y^2$.

11. $6x^2 - 16xy - 6y^2$.
12. $6x^2 + 5xy - 6y^2$.
13. $56x^2 + 562xy + 20y^2$.
14. $56x^2 - 122xy + 20y^2$.
15. $56x^2 - 102xy - 20y^2$.
16. $56x^2 - 229xy + 20y^2$.
17. $56x^2 - 94xy + 20y^2$.
18. $56x^2 - 276xy - 20y^2$.
19. $36x^2 - 33xy - 15y^2$.
20. $72x^2 - 19xy - 40y^2$.

Art. XVI. *More generally*, trinomials of the form ax^2+bx+c (a not a square) may be resolved by Formula A, thus

Multiplying by a we get $a^2x^2+bax+ac$. Writing z for ax this becomes $z^2+bz+ac$. Factor this trinomial, restore the value of z and divide the result by a.

EXAMPLES.

1. $6x^2+5x-4$. Multiplying by 6, we get $(6x)^2+5(6x)-24$ or $z^2+5z-24$. Factoring, we get $(z-3)(z+8)$, hence the required factors are $\frac{1}{6}(6x-3)(6x+8)=(2x-1)(3x+4)$.

2. $6x^2-13xy+6y^2$. Factoring $z^2-13zy+36y^2$ we get $(z-4y)(z-9y)$, hence the required factors are $\frac{1}{6}(6x-4y)(6x-9y)=(3x-2y)(2x-3y)$.

3. $33-14x-40x^2$. Factoring $1320-14z-z^2$ we get $(30-z)(44+z)$, hence the required factors are $\frac{1}{40}(30-40x) \times (44+40x)=(3-4x)(11+10x)$.

NOTE.—The factors may conveniently be arranged in two columns, each with its appropriate sign above it.

Ex. 1, above

	−	+
	1	24
	2	12

$\frac{1}{6}(6x-3)(6x+8)=(2x-1)(3x+4)$.

Ex. 2, above

	−	−
	1	36
	2	18
	3	12

$\frac{1}{6}(6x-4)(6x-9)=(3x-2)(2x-3)$.

[Another method of factoring trinomials of the form ax^2+bx+c is as follows:

Multiply by $4a$, thus obtaining $4a^2x^2+4abx+4ac$. Add b^2-b^2, which will not change the value, $4a^2x^2+4abx+b^2-b^2+4ac$; by [1] this may be written $(2ax+b)^2-(b^2-4ac)$. Factor this by [4] and divide the result by $4a$.

FACTORING. 71

Ex. Factor $56x^2 + 137x - 27885$. Multiply by 4×56 or 2×112, $112^2 x^2 + 2.137.112x - 6246240$. Add $137^2 - 137^2$, then $112^2 x^2 + 2.137.112x + 137^2 - (137^2 + 6246240) = (112x + 137)^2 - 6265009 = \{(112x + 137) + 2503\}\{(112x + 137) - 2503\} = (112x + 2640)(112x - 2366)$.

We multiplied by 4×56, we must, therefore, now divide by that number. Doing so, we obtain as factors $(7x + 165)(8x - 169)$.]

Exercise xxviii.

1. $10x^2 + x - 21$
2. $10x^2 - 29x - 21$.
3. $10x^2 + 29x - 21$.
4. $6x^2 - 37x + 55$.
5. $12a^2 - 5a - 2$.
6. $12x^2 - 37x + 21$.
7. $12x^2 + 37x + 21$.
8. $15a^6 + 13a^3 b^2 - 20b^4$.
9. $12x^2 - x - 1$
10. $9x^2 y^2 - 3xy^4 - 6y^4$.
11. $4x^2 + 8xy + 3y^2$.
12. $6b^2 x^2 - 7bx^3 - 3x^4$.
13. $6x^4 - x^2 y^2 - 35y^4$.
14. $2x^4 + x^2 - 45$.
15. $4x^4 - 37x^2 y^2 + 9y^4$.
16. $4(x+2)^4 - 37x^2(x+2)^2 + 9x^4$.
17. $6(2x+3y)^2 + 5(6x^2 + 5xy - 6y^2) - 6(3x - 2y)^2$.
18. $6(2x+3y)^4 + 5(6x^2 + 5xy - 6y^2)^2 - 6(3x - 2y)^4$.
19. $6(x^2 + xy + y^2)^2 + 13(x^4 + x^2 y^2 + y^4) - 385(x^2 - xy + y^2)^2$.
20. $21(x^2 + 2xy + 2y^2)^2 - 6(x^2 - 2xy + 2y^2)^2 - 5(x^4 + 4y^4)$.

SECTION II.—EXTENDED APPLICATION OF THE FORMULAS.

Art. XVII. The methods of factoring just explained may be applied to find the rational factors, where such exist, of quadratic multinomials.

EXAMPLES.

1. Resolve $12x^2 - xy - 20y^2 + 8x + 41y - 20$.

In the first place we find the factors of the *first three* terms, which are

$4x + 5y$, and
$3x - 4y$.

Now, to find the *remaining terms* of the required factors, we must observe the following conditions:

1st. Their product must $= -20$.

2nd. The sum (*algebraic*) of the products obtained by multiplying them diagonally into the y's, must $=41y$.

3rd. The sum of the products obtained by multiplying them diagonally into the x's, must $=8x$.

We see at once that -4 with the first pair already found, and $+5$ with the second pair, satisfy the required conditions, and $\therefore$ the factors are
$$4x+5y-4, \text{ and}$$
$$3x-4y+5.$$

2. $p^2+2pr-2q^2+7qr-3r^2+pq$.

Here the factors of $p^2+pq-2q^2$, are
$$p+2q, \text{ and}$$
$$p-q.$$

Now find two factors which will give $-3r^2$, and which multiplied diagonally into the p's and q's respectively, will give $2pr$, and $7qr$; these are found to be $-r$ taken with the *first* pair, and $+3r$ taken with the second pair. Hence the required factors are
$$p+2q-r, \text{ and}$$
$$p-q+3r.$$

Art. XVIII. But the following examples illustrate a surer method.

3. $x^2+xy-2y^2+2xz+7yz-3z^2$.

Reject 1st the terms involving z,
2nd.   "   " y,
3rd.   "   " x,

and factor the expression that remains in each case.

1st. $x^2+xy-2y^2 = (x-y)(x+2y)$.
2nd. $x^2+2xz-3z^2 = (x+3z)(x-z)$.
3rd. $-2y^2+7yz-3z^2 = (-y+3z)(2y-z)$.

Arrange these three pair of factors in two sets of three factors each, by so selecting one factor from each pair that two of each set of three may have the same coefficient of x, two may have the

FACTORING.   73

same coefficient of y, and two the same coefficient of z (*coefficient including sign*). In this example there are
$$x-y,\quad x+3z,\quad -y+3z,$$
and $x+2y,\quad x-z,\quad 2y-z.$

From the first set select the common terms (including signs) and form therewith a trinomial, $x-y+3z.$

Repeat with the second set, and we get $x+2y-z.$

$\therefore x^2+xy-2y^2+2xz+7yz-3z^2 = (x-y+3z)(x+2y-z).$

4. $3x^2-8xy-3y^2+30x+27.$

    1st. $3x^2-8xy-3y^2 = (3x+y)(x-3y).$
    2nd. $3x^2+30x+27 = (3x+3)(x+9).$
    3rd. $-3y^2 \quad +27 = (y+3)(-3y+9).$

$\therefore$ the factors are $(3x+y+3)(x-3y+9).$

5. $6a^2-7ab+2ac-20b^2+64bc-48c^2.$

    1st. $6a^2-7ab-20b^2 = (2a-5b)(3a+4b).$
    2nd. $6a^2+2ac-48c^2 = (2a+6c)(3a-8c).$
    3rd. $-20b^2+64bc-48c^2 = (-5b+6c)(4b-8c).$

$\therefore$ the factors are $(2a-5b+6c)(3a+4b-8c).$

Exercise xxix.

1. $7x^2-xy-6y^2-6x-20y-16.$
2. $20x^2-15xy-5y^2-68x-42y-88.$
3. $3x^4+x^2y^2-4y^4+10x^2-17y^2-13.$
4. $20x^2-20y^2+9xy+28x+35y.$
5. $72x^2-8y^2+55xy+12y-169x+20.$
6. $x^2-xy-12y^2-5x-15y.$
7. $8x^2+18xy+9y^2+2xz-z^2.$
8. $6x^2+6y^2-13xy-8z^2-2yz+8xz.$
9. $6x^4-10y^4+11x^2y^2-25z^2+10y^2+25y^2z^2-15x^2+10x^2z^2.$
10. $15x^4-16y^4-22x^2y^2+15z^4+14y^2z^2+50x^2z^2.$
11. $4a^2-15b^2-4ab-21c^2-36bc-8ac.$
12. $a^4+b^4+c^4-2a^2b^2-2b^2c^2-2c^2a^2.$

Art. XIX. Trinomials of the form $ax^4 + bx^2 + c$ can always be broken up into real factors.

If a and c have different signs, the expression may be factored by Art. XVI.

If a and c are of the same sign, three cases have to be considered: i. $b = 2\sqrt{(ac)}$, ii. $b > 2\sqrt{(ac)}$, iii. $b < 2\sqrt{(ac)}$

Case I. $b = 2\sqrt{(ac)}$. This case falls under Art XII., formula [1]. where examples will be found.

Case II. $b > 2\sqrt{(ac)}$. This case falls under Art XVI., where examples will be found. The following additional examples are resolved by the second method of that article.

EXAMPLES.

1. $4x^4 + 5x^2y^2 + y^4$.

Here we see that $(\frac{5}{4}y^2)^2$ will make, with the first two terms, a perfect square, and we therefore add to the given expression $(\frac{5}{4}y^2)^2 - (\frac{5}{4}y^2)^2$. The expression then becomes

$4x^4 + 5x^2y^2 + (\frac{5}{4}y^2)^2 + y^4 - (\frac{5}{4}y^2)^2$

$= (2x^2 + \frac{5}{4}y^2)^2 - \frac{9}{16}y^4$

$= (2x^2 + \frac{5}{4}y^2 + \frac{3}{4}y^2)(2x^2 + \frac{5}{4}y^2 - \frac{3}{4}y^2)$

$= (2x^2 + 2y^2)(2x^2 + \frac{1}{2}y^2) = (x^2 + y^2)(4x^2 + y^2)$.

2. $3x^4 + 6x^2 + 2$.

Here multiplying by 4×3, and completing the square as in Ex. 1, we have

$36x^4 + 72x^2 + 6^2 + 24 - 6^2 = (6x^2 + 6)^2 - 12$

$= (6x^2 + 6 - \sqrt{12})(6x^2 + 6 + \sqrt{12})$, which divided by 4×3 give the required factors.

3. $ax^4 + bx^2 + c$.

Proceeding as in Ex. 2 we have, by multiplying by $4a$,

$ax^4 + bx^2 + c = \{4a^2x^4 + 4abx^2 + b^2 - b^2 + 4ac\} \div 4a$

$= \{2ax^2 + b + \sqrt{(b^2 - 4ac)}\}\{2ax^2 + b - \sqrt{(b^2 - 4ac)}\} \div 4a$.

Exercise. xxx.

1. x^4+7x^2+1 ; $4x^4+14x^2+1$.
2. $x^4+7x^2y^2+y^4$; $3x^4+5x^2y^2+y^4$.
3. $4x^4+10x^2+3$; $3(x+y)^4+5z^2(x+y)^2+z^4$.
4. $x^4+7x^2y^2+3\frac{1}{4}y^4$; $x^4+7x^2y^2+8\frac{1}{4}y^4$.
5. $4x^4+9x^2y^2+\frac{17}{16}y^4$; $4(a+b)^4+10c^2(a+b)^2+3c^4$.
6. $3x^4+8x^2y^2+4\frac{7}{12}y^4$; $36x^4+96x^2+55$.
7. $5x^4+20x^2+2$; $4a^4+12a^2+1$.
8. $4(x+y)^4+12(x+y)^2z^2+z^4$; $5x^4+20x^2y^2+2y^4$.
9. $9x^4+14x^2+4$; $2x^4+12x^2(y+z)^2+15(y+z)^4$.
10. $2x^4+12x^2+15$; $7x^4+40x^2+45$.
11. $8x^4+30x^2y^2+29y^4$; $7x^4+20x^2y^2-20y^4$.
12. $7(a-b)^4+16(a-b)^2c^2+5c^4$; $\frac{3}{2}a^4+3a^2b^2+b^4$.
13. $3x^4+6x^2y^2+2y^4$; $3(a+b)^4+6(a^2-b^2)^2+2(a-b)^4$.
14. $49a^4-84a^2b^2+22b^4$; $25m^4+60m^2n^2+27n^4$.
15. $49(m+n)^4-84(m^2-n^2)^2+22(m-n)^4$.

Case III. $b<2\sqrt{(ac)}$. This case may be brought under Art. XIII. The following examples illustrate the process of reduction and resolution.

Examples.

1. x^4-7x^2+1.

We have to throw this into the form a^2-b^2 :
$x^4-7x^2+1 = (x^2+1)^2-9x^2 = (x^2+1+3x)(x^2+1-3x)$.

2. $9x^4+3x^2y^2+4y^4 = (3x^2+2y^2)^2-9x^2y^2$
$= (3x^2+2y^2-3xy)(3x^2+2y^2+3xy)$.

3. $x^4+y^4 = (x^2+y^2)^2-2x^2y^2$
$= (x^2+y^2+xy\sqrt{2})(x^2+y^2-xy\sqrt{2})$.

4. $x^4-\frac{1}{4}x^2y^2+y^4 = (x^2+y^2)^2-\frac{9}{4}x^2y^2$
$= (x^2+y^2+\frac{3}{2}xy)(x^2+y^2-\frac{3}{2}xy)$.

5. $ax^4+bx^2+c = (\sqrt{a}\cdot x^2+\sqrt{c})^2-\{2\sqrt{(ac)}-b\}x^2$
$= \{\sqrt{a}\cdot x^2+\sqrt{c}-\sqrt{(2\sqrt{ac}-b)}x\} \times$
$\{\sqrt{a}\cdot x^2+\sqrt{c}+\sqrt{(2\sqrt{ac}-b)}x\}$.

Art. XX. It is seen from these examples that we have merely to add to the given expression what will make with the *first* and *last* terms (arranged as in Ex. 5) a perfect square, and to subtract the same quantity. In Ex. 2, *e. g.*, the square root of $9x^4 = 3x^2$, the square root of $4y^4 = 2y^2$, $\therefore 3x^2 + 2y^2$ is the binomial whose square is required; we need $\therefore 12x^2y^2$; but the expression contains $3x^2y^2$; $\therefore$ we have to *add* and *subtract* $12x^2y^2 - 3x^2y^2 = 9x^2y^2$.

Hence we derive a practical rule for factoring such expressions.

(1). Take the square roots of the two extreme terms and connect them by the proper sign; this gives the first two terms of the required factors.

(2) Subtract the middle term of the given expression from twice the product of these two roots, and the square roots of the difference will be the third terms of the required factors.

6. $x^4 + \tfrac{7}{16}x^2y^2 + y^4$. Here $\sqrt{x^4} = x^2$, $\sqrt{y^4} = y^2$, and the first two terms of the required factors are $x^2 + y^2$; twice the product of these is $+2x^2y^2$, from which subtracting the middle term, $\tfrac{7}{16}x^2y^2$, we get $\tfrac{25}{16}x^2y^2$; the square roots of this are $\pm\tfrac{5}{4}xy$. Hence the factors are $x^2 + y^2 \pm \tfrac{5}{4}xy$.

Note that since $\sqrt{y^4} = +y^2$, or $-y^2$, it may sometimes happen that while the former sign will give irrational factors, the latter will give rational factors, and conversely.

7. $x^4 - 11x^2y^2 + y^4$. Here, taking $+y^2$, we have
$x^2 + y^2 + xy\sqrt{13}$, and $x^2 + y^2 - xy\sqrt{13}$.
But taking $-y^2$, we have
$x^2 - y^2 + 3xy$, and $x^2 - y^2 - 3xy$.

Sometimes *both* signs will give rational factors.

8. $16x^4 - 17x^2y^2 + y^4$. Here we have
$(4x^2 + y^2 + 3xy)(4x^2 + y^2 - 3xy)$, and also
$(4x^2 - y^2 + 5xy)(4x^2 - y^2 - 5xy)$.

FACTORING. 77

Exercise xxxi.

1. $x^4+2x^2y^2+9y^4$, $x^4-x^2y^2+y^4$, $x^4+x^2y^2+y^4$.
2. x^4+4y^4, $16x^4+y^4-x^2y^2$, $\frac{1}{4}x^4+y^4$.
3. x^4+1, x^4+9y^4, $1-12y^2+16y^4$.
4. x^4-7x^2+1, x^4+9, $\frac{1}{4}x^4+y^4-3x^2y^2$.
5. $y^4-x^4+11x^2y^2$, x^8+4y^8, x^4+4x^2+16.
6. $4x^4+y^4-8\frac{1}{4}x^2y^2$, $x^4+y^4-\frac{7}{16}x^2y^2$, $4x^4+1$.
7. $x^{4m}+64y^{4m}$, $x^{4m}+4y^{4m}$, $\frac{1}{4}x^4+\frac{9}{16}y^4-5\frac{3}{4}x^2y^2$.
8. $4x^4-8x^2+1$, $7x^2y^2-\frac{1}{4}x^4-36y^4$, $x^4+a^4y^4$.
9. $m^2x^4+n^2y^4-(2mn+p)x^2y^2$, $x^{4m}+2^{4m-2}y^{4m}$.
10. $16x^4-25x^2+9$, $4x^4-16x^2+4$, $13x^2y^2-9x^4-4y^4$.
11. $4x^4-12\frac{16}{25}x^2y^2+9y^4$, x^4+6x^2+25.
12. $a^4+b^4+(a+b)^4$, $1+a^4+(1+a)^4$.
13. $(x+y)^4-7z^2(x+y)^2+z^4$.
14. $(a+b)^4+7c^2(a+b)^2+c^4$.
15. $16a^4+4(b-c)^4-9a^2(b-c)^2$.
16. $4(a+b)^4+9(a-b)^4-21(a^2-b^2)^2$.
17. $(x^2+y^2-xy)^4-7(x^3+y^3)^2+(x+y)^4$.
18. $(a^2+ab+b^2)^4+7(a^3-b^3)^2+(a-b)^4$.
19. $16a^4+4a^2+1$, x^4-41x^2+16.
20. $x^4+81y^8-63x^2y^4$, $1+z^4+25z^8$.
21. $(a^2+1)^4+4(a^2+1)^2a^2+16a^4$, $(x+1)^4+2(x^2-1)^2+9(x-1)^4$.

Art. XXI. We can apply [4], Art. XIII., to factor expressions of the form $ax^4+bx^3+rbx-r^2a$. This may be written
$$a(x^4-r^2)+bx(x^2+r) = \{a(x^2-r)+bx\}(x^2+r).$$

Examples.

1. $6x^4+4x^3+12x-54$. This
$= 6(x^4-9)+4x(x^2+3) = (x^2+3)\{6(x^2-3)+4x\}$
$= (x^2+3)(6x^2+4x-18).$

2. $11x^4+10x^3-40x-176$. This
$=11(x^4-16)+10x(x^2-4)=(x^2-4)\{11(x^2+4)+10x\}$
$=(x^2-4)(11x^2+10x+44)$.

3. $40x^4+30x^3+60x-160$. This
$=10(4x^4-16)+15x(2x^2+4)=(2x^2+4)\{10(2x^2-4)+15x\}$
$=(2x^2+4)(20x^2+15x-40)$.

Note.—To determine r, take the ratio of the coefficient of x^3 to the coefficient of x.

Exercise xxxii.

Resolve into factors

1. x^4+2x^3+6x-9.
2. $2x^4+2x^3+6x-18$.
3. $x^4+3x^3+12x-16$.
4. $3x^4+x^3-4x-48$.
5. $5x^4+4x^3-12x-45$.
6. $10x^4+5x^3+30x-360$.
7. $\frac{1}{4}x^4+20x^3+4x-\frac{1}{100}$.
8. $25x^4-40x^3+8x-1$.
9. $37\frac{1}{2}x^4-30x^3+48x-96$.
10. $63x^4-39x^3+52x-112$.
11. $810x^4+\frac{81}{4}x^3+\frac{9}{5}x-2\frac{1}{4}$.
12. $242x^4-33x^2-3x-2$.
13. $\frac{1}{4}x^4+\frac{1}{16}x^3-\frac{2}{25}x-\frac{4}{25}$.
14. $80x^4-32x^3y+64xy-320y^4$.
15. $24x^4-12x^3y+30xy^3-150y^4$.
16. $2x^4+\frac{1}{2}x^3y-8xy^3-512y^4$.
17. $11x^4+10x^3-12x-15\frac{21}{25}$.
18. $40x^4+30x^3+60x-160$.
19. $13x^4-12x^3y+72xy^3-468y^4$.
20. $3x^4+3x^3y+12xy^3-48y^4$.
21. $5x^4+4x^3y-12xy^3-45y^4$.
22. $4x^4-14x^3y+28xy^3-16y^4$.
23. $x^4+80x^3y+16xy^3-\frac{1}{25}y^4$.
24. $2x^4-x^3y+6xy^3-72y^4$.

Art. XXII. Formulas [1] and [4] may sometimes be applied to factor expressions of the form
$$ax^4+bx^3+cx^2+rbx+r^2a.$$
This may be put under the form
$$a(x^4+r^2)+bx(x^2+r)+cx^2 = a(x^2+r)^2+bx(x^2+r)+(c-2ar)x^2,$$
which can sometimes be factored.

Examples.

1. $x^4+6x^3+27x^2+162x+729$.

We have $x^4+729+6x(x^2+27)+27x^2$.
$=(x^2+27)^2+6x(x^2+27)+9x^2-36x^2$
$=\{x^2+27+3x\}^2-36x^2$, which gives the factors
$x^2-3x+27$, and $x^2+9x+27$.

FACTORING.

2. $x^4+4x^3+4x^2+20x+25.$ This
$= (x^2+5)^2+4x(x^2+5)-6x^2$
$= (x^2+5)^2+4x(x^2+5)+4x^2-10x^2$
$= \{x^2+5+2x-x\sqrt{10}\}\{x^2+5+2x+x\sqrt{10}\}.$

Exercise xxxiii.

Resolve into factors:

1. $x^4-6x^3+27x^2-162x+729.$
2. $x^4+2x^3+3x^2+8x+16.$
3. $x^4+x^3+x^2+x+1.$
4. $x^4-4x^3+x^2-4x+1.$
5. $4x^4-12x^3-6x^2-12x+4.$
6. $x^4+14x^3-25x^2-70x+25.$
7. $16x^4-24x^3-16x^2+12x+4.$
8. $x^4+5x^3-16x^2+20x+16.$
9. $x^4+6x^3-11x^2-12x+4.$
10. $x^4+4x^3y+x^2y^2+12xy^3+9y^4.$
11. $x^4+6x^3-9x^2-6x+1.$
12. $x^4+4x^3y-19x^2y^2+4xy^3+y^4.$
13. $4x^4+4x^3y-65x^2y^2-10xy^3+25y^4.$
14. $x^4+6x^3y-9x^2y^2-6xy^3+y^4.$
15. $x^4+6x^3y+10x^2y^2+12xy^3+4y^4.$
16. $9x^4+18x^3y-52x^2y^2-12xy^3+4y^4.$
17. $11x^4+10x^3y+39\frac{96}{121}x^2y^2+20xy^3+44y^4.$

SECTION III.—FACTORING BY PARTS.

Art. XXIII. To factor an expression which can be reduced to the form $a.F(x)+b.f(x).$

When the expression is thus arranged, any factor common to a and b, or to $F(x)$ and $f(x)$, will be a factor of the whole expression. The method about to be illustrated will be found useful in cases where only *one* power of some letter is found.

EXAMPLES.

1. Factor $acx^2 - abx - bc^2x + b^2c$.

Here we see that only one power of a occurs, and we therefore group together the terms involving this letter, and those not involving it, getting

$$a(cx^2 - bx) - bc^2x + b^2c$$
$$= ax(cx - b) - bc(cx - b) = (ax - bc)(cx - b).$$

2. Factor $m^2x^2 - mna^2x - mnx + n^2a^2$.

Here we observe that a occurs in only one power (a^2). Therefore we have

$$- a^2(mnx - n^2) + m^2x^2 - mnx$$
$$= -na^2(mx - n) + mx(mx - n)$$
$$= (mx - n)(mx - na^2).$$

3. $2x^2 + 4ax + 3bx + 6ab$.

Here we observe that the expression contains only one power of both a and b. We may, therefore, collect the coefficients in either of the following ways:

$$a(4x + 6b) + (2x^2 + 3bx),$$
$$\text{or, } b(3x + 6a) + (2x^2 + 4ax).$$

Now the expressions in the brackets ought to have a common factor, and we see that this is the case. Hence,

$$a(4x + 6b) + (2x^2 + 3bx)$$
$$= 2a(2x + 3b) + x(2x + 3b) = (2x + 3b)(x + 2a).$$

4. $abxy + b^2y^2 + acx - c^2$
$$= a(bxy + cx) + b^2y^2 - c^2$$
$$= ax(by + c) + (by + c)(by - c) = (by + c)(ax + by - c).$$

5. $y^3 - (2a + b)y^2 + (2ab + a^2)y - a^2b$
$$= -b(y^2 - 2ay + a^2) + y^3 - 2ay^2 + a^2y$$
$$= -b(y^2 - 2ay + a^2) + y(y^2 - 2ay + a^2)$$
$$= (y - b)(y - a)^2.$$

6. $2x^3y + 2bx^4 - bx^3y + 4abx^2y - x^2y^2 + 4axy^2 - 2abxy^2 - 2ay^3$.
$$= b(2x^4 - x^3y + 4ax^2y - 2axy^2) + 2x^3y - x^2y^2 + 4axy^2 - 2ay^3$$
$$= bx(2x^3 - x^2y + 4axy - 2ay^2) + y(2x^3 - x^2y + 4axy - 2ay^2)$$
$$= (y + bx)(2x^3 - x^2y + 4axy - 2ay^2).$$

FACTORING.   81

And $2x^3 - x^2y + 4axy - 2ay^2$
$= a(4xy - 2y^2) + 2x^3 - x^2y$
$= 2ay(2x - y) + x^2(2x - y) = (2ay + x^2)(2x - y)$.

7. $\quad x^3 + (2a - b)x^2 - (2ab - a^2)x - a^2b$
$= b(-x^2 - 2ax - a^2) + x^3 + 2ax^2 + a^2x$
$= -b(x+a)^2 + x(x+a)^2 = (x-b)(x+a)^2$.

8. $\quad px^3 - (p-q)x^2 + (p-q)x + q$
$= q(x^2 - x + 1) + px^3 - px^2 + px$
$= q(x^2 - x + 1) + px(x^2 - x + 1) = (px + q)(x^2 - x + 1)$.

Exercise xxxiv.

1. $x^2y - x^2z - y^2 + yz$.
2. $abxy + b^2y^2 + acx - c^2$.
3. $x^2z^2 + ax^2 - a^2z^2 - a^3$.
4. $2x^2 - ax - 4bx + 2ab$.
5. $x^2 + 2bx + 3ax + 6ab$.
6. $x^3 - b^2x^2 - a^2x + a^2b^2$.
7. $x^5 - a^3x^2 - b^2x^3 + a^3b^2$.
8. $8x^2 + 12ax + 10bx + 15ab$.
9. $a^2 + (ac - b^2)x^2 + bcx^3$.
10. $a^2 + (ac - b^2)x^2 - bcx^3$.
11. $abx^3 + (ac - bd)x^2 - (af + cd)x + df$.
12. $px^3 - (p+q)x^2 + (p+q)x - q$.
13. $a^2 + ab + 2ac - 2b^2 + 7bc - 3c^2$.
14. $x^3 + (a+1)x^2 + (a+1)x + a$.
15. $mpx^3 + (mq - np)x^2 - (mr + nq)x + nr$.
16. $x^3 - (a+b+c)x^2 + (ab + bc + ac)x - abc$.
17. $x^3 + (a-b-c)x^2 - (ab - bc + ca)x + abc$.
18. $x^3 + (a+b-c)x^2 - (bc + ca - ab)x - abc$.
19. $a^2x^3 - a^3x^2y - a^2xy + a^3y^2 - ax^2yz + x^3z - xyz + ay^2z$.
20. $a^2bx^2 + ab^2xy + acdxy + bcdy^2 - aefxz - befyz$.
21. $a^2x^3 - a(b-c)x^2 + c(a-b)x + c^2$.
22. $mx^3 - nx^2y + rx^2z - mxy^2 + ny^3 - ry^2z$.
23. $amx^2 + (mby - nay + mcz)x - nby^2 - ncyz$.
24. $(am - bcm)x^2 + (am - bcn)x + an + nax$.
25. $a^2b^2c^2 - b^2c^2xy - a^2c^2yz + c^2xy^2z - a^2b^2zx + b^2x^2yz + a^2z^2xy - x^2y^2z^2$.
26. $x^5 - m^2x^4 - (n - n^2)x^3 + (m^2n - m^2n^2)x^2 - a(x^2 + n^2 - n)$.
27. $1 - (a-1)x - (a-b+1)x^2 + (a+b-c)x^3 - (b+c)x^4 + cx^5$.
28. $a^3x^3 - a^2(b - c + d)x^2y - (abc - abd + acd)xy^2 + bcdy^3$.
29. $m^2npx^3 - (n^2p - m^2n^2 - m^2pq)x^2 - (n^3 + npq - m^2nq)x - n^2q$.

30. $m^2p^2x^5 + m^3p^2x^4 - (p_*^2n^2 - q^2m^2)x^3y^2 - (p^2n^2 - q^2m^2)x^2y^2 - (n^2q^2 + n^3q^2x)y^4$.

Art. XXIV. Sometimes an expression which does not come directly under the preceding form, may be resolved by first finding the factors of its parts.

EXAMPLES.

1. $abx^2 + aby^2 - a^2xy - b^2xy$.

Here, taking ax out of the first and third terms, and by out of the second and fourth terms, we have

$ax(bx - ay) - by(bx - ay)$, and hence

$(ax - by)(bx - ay)$.

2. $x^4 - (a+b)x^3 + (a^2b + ab^2)x - a^2b^2$.

Here, taking the first and last terms together, and the two middle terms together, we have

$(x^2 + ab)(x^2 - ab) - (a+b)x^3 + ab(a+b)x$

$= (x^2 + ab)(x^2 - ab) - (a+b)x\{x^2 - ab\}$

$= (x^2 - ab)\{x^2 + ab - (a+b)x\} = (x^2 - ab)(x - a)(x - b)$.

3. $x^{3m} - 4x^m + 3$. This equals

$x^{3m} - x^m - 3(x^m - 1) = x^m(x^{2m} - 1) - 3(x^m - 1)$

$= x^m(x^m + 1)(x^m - 1) - 3(x^m - 1)$

$= (x^m - 1)\{x^m(x^m + 1) - 3\}$.

Exercise xxxv.

1. $a^2 - ab + ax - bx$.
2. $abx^2 + b^2xy - a^2xy - aby^2$.
3. $x^4 + ax^3 - a^3x - a^4$.
4. $a^3x + 2a^2x^2 + 2ax^3 + x^4$.
5. $acx^2 + (ad - bc)x - bd$.
6. $25x^4 - 5x^3 + x - 1$.
7. $a^2 - b^2 + ax - ac - bx + bc$.
8. $a^3 + (1+a)ab + b^2$.
9. $x^4 + 2xy(x^2 - y^2) - y^4$.
10. $x^3 - y^3 + x^2 + xy + y^2$.
11. $2b + (b^2 - 4)x - 2bx^2$.
12. $x^3 + 3x^2 - 4$.
13. $p^3 - p^2q - 2pq^2 + 2q^3$.
14. $a^3 + a^2 - 2$.
15. $3a^2b^4 - 2ab^2 - 1$.
16. $y^3 - 3y + 2$.
17. $2a^3 - a^2b - ab^2 + 2b^3$.
18. $b^{3m} + b^{2m} - 2$.
19. $y^{3n} - 2y^{2n}z^n - 2y^nz^{2n} + z^{3n}$.
20. $a^3 - 4ab^2 + 3b^3$.
21. $a^{2m} - 3a^mc^n + 2c^{2n}$.
22. $ax^3 - (a^2 + b)x^2 + b^2$.
23. $35x^{2n} - 6a^2x^n - 9a^4$.
24. $a^2b^2 + 2abc^2 - a^2c^2 - b^2c^2$.
25. $am^2 - ab^2 + b^2m - m^3$.
26. $\frac{1}{x} - 6a^2 + 27a^4$.

27. $(x-y)^3 + (1-x+y)(x-y)z - z^2$.
28. $24m^3 - 28m^2n + 6mn^2 - 7n^3$.
29. $x^{m+n} + x^n y^n + x^m y^m + y^{m+n}$.
30. $x^4 + 2x^3y - a^2x^2 + x^2y^2 - 2axy^2 - y^4$.

Section IV.—Application of the Theory of Divisors.

Art. XXV. By Theorem I. we prove that
$x^n - a^n$ is divisible by $x - a$ always
$x^n - a^n$ " " " $x + a$ when n is *even*
$x^n + a^n$ " " " $x + a$ when n is *odd*.

By actual division we find, in the above cases;—

$$\frac{x^n - a^n}{x - a} = x^{n-1} + x^{n-2}a + \ldots xa^{n-2} + a^{n-1}. \ldots\ldots(1).$$

$$\frac{x^n - a^n}{x + a} = x^{n-1} - x^{n-2}a + \ldots + xa^{n-2} - a^{n-1}. \ldots\ldots(2).$$

$$\frac{x^n + a^n}{x + a} = x^{n-1} - x^{n-2}a + \ldots - xa^{n-2} + a^{n-1}. \ldots\ldots(3).$$

Examples.

1. Resolve into factors $x^3 - y^3$; here $x - y$ is one factor and by (1) the other is $x^2 + xy + y^2$.

2. Resolve $a^3 + (b-c)^3$; here $a + (b-c)$ is one factor; and by (3) the other is $a^2 - a(b-c) + (b-c)^2$.

3. Resolve $x^{15} + 1024y^{10}$. This $= (x^3)^5 + \{(2y)^2\}^5$, one factor of which is $x^3 + (2y)^2$, and by (3) the other factor is
$(x^3)^4 - (x^3)^3(4y^2) + (x^3)^2(4y^2)^2 - x^3(4y^2)^3 + (4y^2)^4$.
$= x^{12} - 4x^9y^2 + 16x^6y^4 - 64x^3y^6 + 256y^8$.

4. Resolve $(x-2y)^3 + (2x-y)^3$ into factors.

Here by (3) we have

$$\frac{(x-2y)^3 + (2x-y)^3}{x-2y + 2x-y} = (x-2y)^2 - (x-2y)(2x-y) + (2x-y)^2$$

∴ the factors are
$$3(x-y)(7x^2 - 13xy + 7y^2).$$

5. Resolve $x^5+x^4y+x^3y^2+x^2y^3+xy^4+y^5$:

By (1) we see that this $= \dfrac{x^6-y^6}{x-y} = \dfrac{(x^3+y^3)(x^3-y^3)}{x-y}$

$=(x+y)(x^2-xy+y^2)(x^2+xy+y^2).$

6. Resolve $x^{11}-x^{10}a+x^9a^2-x^8a^3+x^7a^4-x^6a^5+x^5a^6$

$-x^4a^7+x^3a^8-x^2a^9+xa^{10}-a^{11}$. This $= \dfrac{x^{12}-a^{12}}{x+a}$

$= \dfrac{(x^6+a^6)(x^6-a^6)}{x+a} = \dfrac{(x^6+a^6)(x^3-a^3)(x^3+a^3)}{x+a}$

$=(x^2+a^2)(x^4-x^2a^2+a^4)(x-a)(x^2+xa+a^2)(x^2-xa+a^2).$

Exercise xxxvi.

Factor the following :—

1. x^6-y^6, x^3-1, x^3+8, $8a^3-27x^3$, $8+a^3x^3$.
2. x^5-a^{10}, $27a^3-64$, $a^{12}-b^6$, $x^{10}-32y^5$.
3. Find a factor which, multiplied into

$a^4+a^3b+a^2b^2+ab^3+b^4$, will give a^5-b^5.

4. By what factor must $x^3-4x^2y+16xy^2-64y^3$ be multiplied to give x^4-256y^4 ?

5. Factor $x^7+x^6y+x^5y^2+x^4y^3+x^3y^4+x^2y^5+xy^6+y^7$.

Find the factors of the following :

6. $(3y^2-2x^2)^3-(3x^2-2y^2)^3$, a^8-16b^4.
7. $x^3-y^3-x(x^2-y^2)+y(x-y)$.
8. $b(x^3-a^3)+ax(x^2-a^2)+a^3(x-a)$.
9. $b(m^3+a^3)+am(m^2-a^2)+a^3(m+a)$.
10. $x^6-y^6+2xy(x^4+x^2y^2+y^4)$.
11. $(a^2-bc)^3+8b^3c^3$, $x^{4m}-a^{4n}$.
12. $x^3-3ax^2+3a^2x-a^3+b^3$.
13. $x^3+8y^3+4xy(x^2-2xy+4y^2)$.
14. $8x^3-6xy(2x+3y)+27y^3$.
15. $1-2x+4x^2-8x^3$.
16. $a^5+a^4bc+a^3b^2c^2+a^2b^3c^3+ab^4c^4+b^5c^5$.

FACTORING.

Art. XXVI. The principles illustrated in Section II., chap. II., may be applied to factor various algebraic expressions, as in the following cases :

EXAMPLES.

1. Find the factors of
$$(a+b+c)(ab+bc+ca)-(a+b)(b+c)(c+a).$$

   1st. Observe that the expression is *symmetrical* with respect to a, b, c.

   2nd. If there be any *monomial* factor a must be one. Putting $a=0$, the expression vanishes. ∴ a is a factor, and, by symmetry, b and c are also factors. ∴ abc is a factor.

   3rd. There can be no other *literal* factor, because the given expression is of only *three* dimensions, and abc is of three dimensions.

   4th. But there may be a *numerical* factor, m suppose, so that we have
   $$(a+b+c)(ab+bc+ca)-(a+b)(b+c)(c+a)=mabc.$$
   To find m, put $a=b=c=1$ in this equation, and $m=1$.
   ∴ the expression $=abc$.

2. Resolve $a^2(b-c)+b^2(c-a)+c^2(a-b)$.

   1st. For $a=0$ this does not vanish. ∴ a is not a factor, and by symmetry neither is b nor c.

   2nd. Try a *binomial* factor; this will likely be of the form $b-c$; put $b-c=0$, $i.e.$, $b=c$ in the given expression, and there results
   $$a^2(c-c)+c^2(c-a)+c^2(a-c), \text{ which } = 0,$$
   ∴ $b-c$ is a factor, and by symmetry $c-a$ and $a-b$ are factors. Since the given expression is only of *three* dimensions, there can be no other *literal* factor; but there may be a *numerical* factor, m (say), so that
   $$a^2(b-c)+b^2(c-a)+c^2(a-b)=m(a-b)(b-c)(c-a).$$
   To find the value of m, give a, b, c, in this equation, any values which will not reduce either side to zero; let $a=1$, $b=2$, $c=0$,

and we have $2 = m(-2)$, or $m = -1$; so that the given expression $= -(a-b)(b-c)(c-a)$, or $(a-b)(b-c)(a-c)$.

3. Resolve $a^3(b+c^2)+b^3(c+a^2)+c^3(a+b^2)+abc(abc+1)$.

Here we see at once that there is no *monomial* factor: put $b+c^2 = 0$, *i.e.*, $b = -c^2$, and the expression becomes $a^3(-c^2+c^2)-c^6(c+a^2)+c^3(a+c^4)-c^3a(-c^3a+1)$ which $= 0$; $\therefore b+c^2$ is a factor, and by symmetry $c+a^2$ and $a+b^2$ are also factors; and proceeding as in former examples we find $m = 1$; $\therefore$ the expression $= (b+c^2)(c+a^2)(a+b^2)$.

4. Resolve into factors the expression
$$(a-b)^3+(b-c)^3+(c-a)^3.$$
As before, we find that there are no monomial factors.

Let $a-b = 0$, or $a = b$, and substituting b for a the expression becomes zero; hence

        $a-b$ is a factor.

  By symmetry $b-c$ "

    and $c-a$ "

Hence the factors are
$$m(a-b)(b-c)(c-a).$$
To find m let $a = 0$, $b = 1$, $c = 2$, and we have
$$6 = 2m, \text{ or } m = 3.$$
The factors are, therefore,
$$3(a-b)(b-c)(c-a).$$

5. Resolve into factors
$$a^3(b-c)+b^3(c-a)+c^3(a-b).$$
As before, we find that there are no monomial factors.

Let $a-b = 0$, or $a = b$; substituting b for a, the expression becomes zero;

    therefore $a-b$ is a factor.

    By symmetry $b-c$ "

    and $c-a$ "

Now the product of these three factors is of *three* dimensions, while the expression itself is of *four* dimensions. There must, therefore, be another factor of *one* dimension. It cannot be a monomial factor, for the expression has no such factors. It cannot be a binomial factor, such as $a+b$, for then, by symmetry, $b+c$ and $c+a$ would also be factors, which would give an expression of *six* dimensions. It cannot be a trinomial factor, unless a, b, and c are similarly involved. For instance, if $a-b+c$ were a factor, then, by symmetry, $b-c+a$ and $c-a+b$ would also be factors, and the dimensions would be *six* instead of *four*. The other factor must, therefore, be $a+b+c$. Hence,

$$a^3(b-c)+b^3(c-a)+c^3(a-b) = m(a-b)(b-c)(c-a)(a+b+c).$$

To find m, put $a=0$, $b=1$, and $c=2$, and we have

$$-6 = 6m;$$
$$\therefore m = -1.$$

Hence the factors are

$$-(a-b)(b-c)(c-a)(a+b+c),$$
or, $(a-b)(a-c)(b-c)(a+b+c).$

6. Prove that

$$a^3+b^3+c^3+3(a+b)(b+c)(c+a)$$

is exactly divisible by $a+b+c$, and find all the factors.

Let $a+b+c=0$, or $a = -(b+c)$; substituting this value for a, we have

$$-(b+c)^3+b^3+c^3+3bc(b+c), \text{ or}$$
$$-(b+c)^3+(b+c)^3 \text{ which} = 0, \text{ and}$$

therefore $a+b+c$ is a factor.

As before, we find that there are no monomial factors. Since $a+b+c$, the factor already obtained, is of *one* dimension, the other factor must be of *two* dimensions, and cannot, therefore, be a binomial; for if $a+b$ were a factor, by symmetry $b+c$, and $c+a$ must also be factors. The factors in that case would give a quantity of *four* dimensions, while the expression itself is only of three dimensions. Nor can $a^2+b^2+c^2$ be a factor. For if so, the other factor must involve a numerical multiple of the first power of a, and, therefore, on taking the first power of a out of terms involving first and third powers, we should have left some numerical multiple of $a^2+b^2+c^2$, instead of which we get

$a^2+3(b+c)^2$. Nor can $a_{\cdot}^2+(b+c)^2$ be a factor, for symmetry would require two other factors, viz.: $b^2+(c+a)^2$, and $c^2+(a+b)^2$, thus giving a quantity of *seven* dimensions.

The only factor admissible is, therefore, $(a+b+c)^2$. Hence

$$a^3+b^3+c^3+3(a+b)(b+c)(c+a) = m(a+b+c)(a+b+c)^2$$
$$= m(a+b+c)^3.$$

To find m, let $a=1$, $b=0$, and $c=0$, and we have $1=m$. Hence the factors are

$$(a+b+c)(a+b+c)(a+b+c).$$

7. Simplify
$$a(b+c)^2+b(a+c)^2+c(a+b)^2-(a+b)(a-c)(b-c)$$
$$-(a-b)(a-c)(b+c)+(a-b)(b-c)(a+c).$$

Let $a=0$, and the expression becomes $bc^2+cb^2+bc(b-c)-bc(b+c)-bc(b-c)$, which equals zero; therefore a is a factor; by symmetry, b and c are also factors.

The expression is of three dimensions, and abc is of three dimensions, there cannot therefore be any other literal factor.

Hence the expression $= mabc$.

To find m, let $a=b=c=1$, and we have
$$4+4+4 = m;$$
$$m = 12.$$
∴ the expression $=12abc$.

In the preceding examples the factors have been *linear*, but the principle applies equally well to those of higher dimensions. (See Th. ii. Cor.)

8. Examine whether x^n+1 is a factor of $x^{3n}+2x^{2n}+3x^n+2$.

Let $x^n+1=0$, or $x^n=-1$, and substituting, the expression vanishes, therefore, x^n+1 is a factor.

9. Examine whether a^2+b^2 is a factor of
$$2a^4+a^3b+2a^2b^2+ab^3.$$

Let $a^2+b^2=0$, or $a^2=-b^2$, substituting, we have
$$2b^4-ab^3-2b^4+ab^3 \text{ which} = 0, \text{ and}$$
therefore a^2+b^2 is a factor.

FACTORING. 89

10. Prove that a^3+b^3 is a factor of
$$a^5+a^4b+a^3b^2+a^2b^3+ab^4+b^5.$$
Let $a^3+b^3=0$, or $a^3=-b^3$; substituting, we have
$-a^2b^3-ab^4-b^5+a^2b^3+ab^4+b^5$, which $=0$, and therefore a^3+b^3 is a factor.

Exercise xxxvii.

Resolve into factors

1. $(x+y+z)^3-(x^3+y^3+z^3)$. ✓ $3(x+y)(y+z)(z+x)$
2. $bc(b-c)-ca(a-c)-ab(b-a)$. ✓
3. $(a^2-b^2)^3+(b^2-c^2)^3+(c^2-a^2)^3$. ✓ $3(a^2-b^2)(b^2-c^2)(c^2-a^2)$
4. $x(y+z)^2+y(z+x)^2+z(x+y)^2-4xyz$. $(x+y)(y+z)(z+x)$
✓ 5. $(a+b)^3-(b+c)^3+(c-a)^3$. $3(a+b)(-b-c)(c-a)$
✓ 6. $a(b-c)^3+b(c-a)^3+c(a-b)^3$. $(a+b+c)(a-b)(b-c)(a-a)$
✓ 7. $(a+b+c)(ab+bc+ca)-abc$. $(a+b)(b+c)(c+a)$
8. $a^3(c-b^2)+b^3(a-c^2)+c^3(b-a^2)+abc(abc-1)$. $(a^2-b)(b^2-c)(c^2-a)$
9. $a^2(b+c)+b^2(c+a)+c^2(a+b)+2abc$. $(a+b)(b+c)(c+a)$
10. $(a-b)(c-h)(c-k)+(b-c)(a-h)(a-k)+(c-a)(b-h)(b-k)$. $(a-b)(b-c)$
× 11. $x^4y^2+x^2y^4+x^4z^2+x^2z^4+y^4z^2+y^2z^4+2x^2y^2z^2$. $(x^2+y^2)(y^2+z^2)(z^2$
★ 12. $(a-b)^5+(b-c)^5+(c-a)^5$. $5(a-b)(b-c)(c-a)(a^2+b^2+c^2-ab-bc-ca)$
13. $ab(a+b)+bc(b+c)+ca(c+a)+(a^3+b^3+c^3)$. $(a+b+c)(a^2+b^2+c^2)$
14. $a^4(c-b^3)+b^4(a-c^3)+c^4(b-a^3)+abc(a^2b^2c^2-1)$. $(c-a^3)(a-b^3)(b-$
★ 15. $x^4(y^2-z^2)+y^4(z^2-x^2)+z^4(x^2-y^2)$. $-(x^2-y^2)(y^2-z^2)(z^2-x^2)$
★ 16. $x^4+y^4+z^4-2x^2y^2-2y^2z^2-2z^2x^2$.
17. $(b-c)(x-b)(x-c)+(c-a)(x-c)(x-a)+(a-b)(x-a)(x-b)$. $(a-b)(b-c)(c-a)$
★ 18. $(a+b)^3+(b+c)^3+(c+a)^3+3(a+2b+c)(b+2c+a)(c+2a+b)$. $5(a+b+c)^3$
19. Shew that $a^5+a^2b^2-ab^2-b^3$ has a^2-b for a factor.
20. Shew that $(x+y)^7-x^7-y^7=7xy(x+y)(x^2+xy+y^2)^2$.
21. Examine whether x^2-5x+6 is a factor of
$$x^3-9x^2+26x-24.$$

22. Shew that $a-b+c$ is a factor of
$$a^2(b+c)-b^2(c+a)+c^2(a+b)+abc.$$
23. Shew that a^2+3b is a factor of
$$a^4-4a^3b^3+3a^2b^4+3a^2b-12ab^4+9b^5,$$
and find the other factor.

24. Find the factors of $a^4(b-c)+b^4(c-a)+c^4(a-b)$.

SECTION V.—FACTORING A POLYNOME BY TRIAL DIVISORS.

Art. XXVII. To find, if possible, a rational linear factor of the polynome.
$$x^n + bx^{n-1} + cx^{n-2} + \ldots + hx + k.$$

Substitute successively for x every measure (both positive and negative) of the term k, till one is found, say m, that makes the polynome vanish, then $x-m$ will be a factor of the polynome.

EXAMPLES.

1. Factor $x^3+9x^2+16x+4$.

The measures of 4 are ± 1, ± 2 and ± 4. Since every coefficient of the given polynome is positive, the positive measures of 4 need not be tried. Using the others, it will be found that -2 makes the polynome vanish; thus

$$\begin{array}{r|rrrr} & 1 & 9 & 16 & 4 \\ -2 & & -2 & -14 & -4 \\ \hline & 1 & 7 & 2; & 0 \end{array}$$

Hence the factors are $(x+2)(x^2+7x+2)$.

The labour of substitution may often be lessened by arranging the polynome in ascending powers of x, and using $1 \div$ (measure of k) instead of the measures of k. (This is really substituting $1 \div$ measure of k, for $1 \div x$). Should a fraction occur during the course of the work, further trial of that measure of k will be needless.

FACTORING. 91

EXAMPLES.

2. Factor $x^3 - 10x^2 - 63x + 60$.

The measures of 60 are ± 1, ± 2, ± 3, ± 4, ± 5, &c. Neither $+1$ nor -1 will make the polynome vanish. Try 2; thus

$$\begin{array}{r|rrrr} & 60 & -63 & -10 & 1 \\ 1 & & 30 & & \\ \hline 2 & 30 & -16\tfrac{1}{2} & & \end{array}$$

A fraction occurring we need go no further. -2 will also give a fraction, as may easily be seen. Next try 3; thus

$$\begin{array}{r|rrrr} & 60 & -63 & -10 & 1 \\ 1 & & 20 & & \\ \hline 3 & 20 & -14\tfrac{1}{3} & & \end{array}$$

A fraction again occuring, we may stop. -3 will also give a fraction. Next try 4; thus

$$\begin{array}{r|rrrr} & 60 & -63 & -10 & 1 \\ 1 & & 15 & -12 & \\ \hline 4 & 15 & -12 & -5\tfrac{1}{4} & \end{array}$$

Next try -4.

$$\begin{array}{r|rrrr} & 60 & -63 & -10 & 1 \\ -1 & & -15 & & \\ \hline 4 & 15 & -19\tfrac{1}{4} & & \end{array}$$

Next trying 5 we find it fails, then try -5, thus

$$\begin{array}{r|rrrr} & 60 & -63 & -10 & 1 \\ -1 & & -12 & 15 & -1 \\ \hline 5 & 12 & -15 & 1; & 0 \end{array}$$

The remainder vanishes as required; the factors are, therefore, $(x+5)(x^2-15x+12)$.

Art. XXVIII. When k has a large number of factors, the number that need actually be tried can often be considerably lessened by the following means.

Add together all the coefficients of x (including the constant term k); let the sum be called k_1.

From the sum of the coefficients of the even powers of x (including k) take the sum of the coefficients of the odd powers of x; let the remainder be called k_2. (In the coefficients are included the signs of the terms).

1st. If k_1 vanish, $x-1$ will be a factor of the polynome.

2nd. If k_2 vanish, $x+1$ will be a factor of the polynome.

3rd. If both k_1 and k_2 vanish, x^2-1 will be a factor of the polynome.

4th. If neither k_1 nor k_2 vanish, (writing p for "a positive measure of k greater than 1");

(α) We need not try the substitution of p for x unless $p-1$ be a measure of k_1, and $p+1$ a measure of k_2.

(β) Nor need we try the substitution of $-p$ for x unless $p+1$ be a measure of k_1, and $p-1$ a measure of k_2.

(In trying for measures, the signs of k, k_1, and k_2 may be neglected.

EXAMPLES.

1. Find the factors of $x^3 - 10x^2 - 63x + 60$. (See Ex. 2 above).

Here $k = 60$; $k_1 = 1 - 10 - 63 + 60 = -12$,
$k_2 = -1 - 10 + 63 + 60 = 112$.

Tabulating the trial-measures we get

12	1,	2,	3,	4,			
60	2,	3,	4,	5,	6,	10,	12,
112				4,		7,	

12	3,	4,		6,		
60	2,	3,	4,	5,	6,	10,
112	1,	2,		4,		

(It is evident that 12 is the highest measure of 60 we need try in the upper table, for the next measure, 15, would give 14 as a trial-measure of 12, and higher measures of 60 would give higher trial-measures. Similarly, 10 is the highest measure that need be tried in the lower table.)

FACTORING.

In the upper table, 3 is the only measure of 60 that gives a full column; hence of the positive measures of 60 we need try only the substitution of 3 for x.

In the lower table, 2, 3, and 5 give full columns, hence we must try the substitutions -2, -3, -5 for x.

On trying the four substitutions to which we are thus restricted we find -5 is the only one for which the polynome vanishes. (See Ex. 2 above).

2. Find the factors of $x^4 + 12x^3 - 40x^2 + 67x - 120$.

$$k = -120\,;\ k_1 = 1 + 12 - 40 + 67 - 120 = -80\,;$$
$$k_2 = 1 - 12 - 40 - 67 - 120 = -238.$$

80	1,	2,		4,	5,					
120	2,	3,	4,	5,	6,	8,	10,	12,	15,	&c.
238					7,					

80		4,	5,					16,			
120	2,	3,	4,	5,	6,	8,	10,	15,	20,	24,	&c.
238	1,	2,			7,			14,	21,		

The upper table gives us 6 as a trial-measure, and the lower gives us -3 and -15.

Trying these we get

	-120	67	-40	12	1
1		-20			
6	-20	$7\tfrac{5}{6}$			

Hence $x+15$ and x^3-8x^2+5x-8 are the factors. The latter cannot be resolved, for our tables above tell us we need try only $x-6$, $x+3$, and $x+15$. The first two have been found not to be factors, and 15 will not measure 8.

4. Factor $x^4-27x^2+14x+120$.

$$k=120;\ k_1=1-27+14+120=108$$
$$k_2=1-27-14+120=80.$$

108	1,	2,	3,	4,			9			
120	2,	3,	4,	5,	6,	8,	10,	12,	15,	&c.
80			4,	5,					16,	

108	3,	4,		6,			9			
120	2,	3,	4,	5,	6,	8,	10,	12,	15,	&c.
80	1,	2,		4,	5.					

The upper table gives us 3 and 4, the lower table gives us -2, -3, and -5. Using these in order we get

```
      | 120   14   -27    0    1
  1   |        40   18   -3   -1
 ─────┼──────────────────────────
  3   |  40   18   - 3   -1;   0.    Hence x-3 is a factor.
  1   |        10    7    1
 ─────┼──────────────────────────
  4   |  10    7    1;   0           Hence x-4 is a factor.
 -1   |       - 5   -1
 ─────┼──────────────────────────
  2   |   5    1;   0                Hence x+2 is a factor.
                                    and there remains x+5, a factor.
```

Hence the factors are $(x-3)(x-4)(x+2)(x+5)$.

5. Factor $x^4-px^3+(q-1)x^2+px-q$.

$$k=-q;\ k_1=1-p+(q-1)+p-q=0;$$
$$k_2=1+p+(q-1)-p-q=0.$$

Since both k_1 and k_2 vanish, the polynome is divisible by both $x-1$ and $x+1$.

```
       | 1    -p       q-1       p      -q
   1   |       1      -p+1      q-p      q
  ─────┼──────────────────────────────────
       | 1   -p+1      q-p      q;       0
  -1   |      -1       +p      -q
  ─────┼──────────────────────────────────
       | 1    -p        q;       0
```

FACTORING.

Hence the other factor is $x^2 - px + q$.

6. Factor $x^4 + 2ax^3 + (a^2 + a)x^2 + 2a^2x + a^3$.

$$k = a^3 \; ; \; k_1 = 1 + 2a + (a^2 + a) + 2a^2 + a^3 = (a+1)^3 \; ;$$
$$k_2 = 1 - 2a + (a^2 + a) - 2a^2 + a^3 = a^3 - a^2 - a - 1.$$

The positive measures of k are 1, a, a^2, a^3. Of these 1 may be rejected at once, since neither k_1 nor k_2 vanish, and a^2 and a^3 may also be rejected since k_1 or $(a+1)^3$ is not divisible by either $a^2 \pm 1$ or $a^3 \pm 1$. But k_1 is divisible by $a+1$, and k_2 is divisible by $a-1$; thus we need only try the substitution of $-a$ for x. (See 4 β, page 92)

	1	$2a$	$a^2 + a$	$2a^3$	a^3
$-a$		$-a$	$-a^3$	$-a^2$	$-a^3$
	1	a	a	a^2 ;	0
$-a$		$-a$	0	$-a^2$	
	1	0	a ;	0	

Hence the factors are $(x+a)^2(x^2+a)$.

7. Factor $x^3 - (a+c)x^2 + (b+ac)x - bc$.

$$k = -bc \; ;$$
$$k_1 = 1 - (a+c) + (b+ac) - bc = 1 - a + b - c + ac - bc$$
$$k_2 = -1 - (a+c) - (b+ac) - bc = -(1 + a + b + c + ac + bc).$$

The factors of k_1, other than 1, are b and c. k_1 is not divisible by either $b \pm 1$ nor by $c+1$. However, k_1 is divisible by $c-1$, and k_2 is at the same time divisible by $c+1$, ∴ we need only try the substitution of c for x. (See 4 α, page 86).

	1	$-(a+c)$	$(b+ac)$	$-bc$
c		c	$-ac$	bc
	1	$-a$	b ;	

Hence the factors are $(x-c)(x^2 - ax + b)$.

FACTORING.

Exercise xxxviii.

1. $a^3 - 9a^2 + 16a - 4$.
2. $x^3 - 9x^2 + 26x - 24$.
3. $x^3 - 7x^2 + 16x - 12$.
4. $x^3 - 12x + 16$.
5. $x^3 + 8x^2 + 5x + 3$.
6. $x^4 + 4x^3 + 10x^2 + 12x + 9$.
7. $x^3 - 3x + 2$.
8. $x^4 + 2x^2 + 9$.
9. $m^3 - 3m^2n + 4mn^2 - 2n^3$.
10. $x^3 + 2x^2 + 2$.
11. $m^3 - 5m^2n + 8mn^2 - 4n^3$.
12. $b^3 + b^2c + 7bc^2 + 89c^3$.
13. $m^4 - 4mn^3 + 3n^4$.
14. $a^4 - 7a^3b + 28ab^3 - 16b^4$.
15. $x^3 - 11x^2 + 39x - 45$.
16. $x^3 + 5x^2 + 7x + 2$.
17. $a^3 - 3a^2 - 193a + 195$.
18. $p^3 - 3p^2 - 6p - 8$.
19. $a^4 + 3a^3 - 3a^2 - 7a + 6$.
20. $a^{6n} - 6a^{4n} + 11a^{2n} - 6$.
21. $a^4 - 41a^2b^2 + 16b^4$.
22. $a^4 - a^2b^2 - 2ab^3 + 2b^4$.
23. $p^3 - 4p^2 + 6p - 4$.
24. $x^{3n} + 4x^{2n} - 5$.
25. $y^4 - 5y^3 + 8y^2 - 8$.
26. $a^4 - 2a^3 + 3a^2 - 2a + 1$.
27. $a^3 + a^2b + ab^2 - 3b^3$.
28. $2a^{3n} - a^{2n} - a^n + 2$.

29. $x^4 - 18x^3 + 113x^2 - 288x + 252$.
30. $x^4 - 9x^3y + 29x^2y^2 - 39xy^3 + 18y^4$.

Art. XXIX. To find, if possible, a rational linear factor of the polynome

$$ax^n + bx^{n-1} + cx^{n-2} + \ldots\ldots\ldots + hx + k.$$

First Method. Multiply the polynome by a^{n-1}.

$$(ax)^n + b(ax)^{n-1} + ac(ax)^{n-2} + \ldots\ldots\ldots + a^{n-2}h(ax) + a^{n-1}k ;$$

or writing y for ax,

$$y^n + by^{n-1} + acy^{n-2} + \ldots\ldots\ldots + a^{n-2}hy + a^{n-1}k.$$

Factor this polynome by the method of the last article, replace y by ax, and divide the result by a^{n-1}.

Example.

Factor $3x^4 + 5x^3 - 33x^2 + 43x - 20$.

Multiply by 3^3 and express in terms of $3x$.

$$(3x)^4 + 5(3x)^3 - 99(3x)^2 + 387(3x) - 540 ;$$
or, $y^4 + 5y^3 - 99y^2 + 387y - 540$.

FACTORING.

Here $k = -540$; $k_1 = 1+5-99+387-540 = -246$;
$$k_2 = 1-5-99-387-540 = -1030.$$

$$\begin{array}{c|cccccccc} 246 & 1, & 2, & 3, & 6, & 41, & 82, & 123, & 246. \\ 540 & 2, & 3, & 4, & & & & & \\ 1030 & & & 5, & & & & & \end{array}$$

$$\begin{array}{c|cccc} 246 & 3, & 6, & 41, & \&c. \\ 540 & 2, & 5, & & \\ 1030 & 1, & & & \end{array}$$ (Trying by factors of 246 instead of by factors of 540, for convenience).

The only factors of 540 in full columns are 4 in the upper table and 2 in the lower one; hence we need try only the substitutions 4 and -2.

$$\begin{array}{c|ccccc} & -540 & 387 & -99 & 5 & 1 \\ 1 & & -135 & 63 & -9 & -1 \\ \hline 4 & -135 & 63 & -9 & -1; & 0 \end{array}$$

Hence $y-4$ is a factor. The substitution -2 need not now be tried, since we see that 135 is not a multiple of 2. The other factor is therefore $y^3+9y^2-63y+135$.

Replacing y by $3x$ and dividing by 27;

$$\tfrac{1}{27}(3x-4)(27x^3+81x^2-189x+135)$$
$$= (3x-4)(x^3+3x^2-7x+5),$$

which are the factors.

Art. XXX. *Second Method.* Writing m for "a measure of a," and p for a "measure of k, positive or negative;"

For x substitute every value of $p \div m$ till one, say $p' \div m'$ be found which makes the polynome vanish; then $m'x - p'$ will be a factor. Should a fraction be met with in the course of substitution, further trial of that value of $p \div m$ will be useless.

Should k have more factors than a, it will generally be better to arrange the polynome in ascending powers of x and use values of $m \div p$ instead of $p \div m$, making p positive and m positive or negative.

FACTORING.

To reduce the number of trial-measures, calculate k_1 and k_2, as directed on page 92, then 1, 2, 3 hold as on that page, but in 4 read $p-m$ for $p-1$ and $p+m$ for $p+1$.

EXAMPLES.

1. Factor $36x^3 + 171x^2 - 22x + 480$.

$$k = 480, \quad k_1 = 36 + 171 - 22 + 480 = 665$$
$$k_2 = -36 + 171 + 22 + 480 = 637.$$

m may have any of the values ± 1, ± 2, ± 3, ± 4, ± 6, ± 9, ± 12, ± 18, ± 36.

In forming the table write out the measures of k_1; take each measure in succession and add to it each value of m separately, should the sum measure 480, i.e., k, add to it the same value of m, and should the new sum measure 637, i.e., k_2, keep the measure of 480, writing above it the value of m used. Should the sum in either case not be a measure, another value of m must be tried; when all the values of m have been tried, another measure of 665, i.e., k_1 must be tried till all have been tested. (Measures of k_1 or 665 have been used in this instance because they are much fewer than those of 480; measures of k_2 or 637 would have done equally well).

$m=$	+3,	+1,	+3	−2	−3	−9	−3
665	1,	5,	7	5	7	19	19
480	4,	6,	10	3	4	10	16
637	7,	7,	13	1	1	1	13

Hence the only substitutions that need be tried are

$$\frac{3}{4}, \quad \frac{1}{6}, \quad \frac{3}{10}, \quad \frac{-2}{3}, \quad \frac{-3}{4}, \quad \frac{-9}{10}, \quad \frac{-3}{16}, \quad \text{for } \frac{1}{x}.$$

Arrangement in ascending powers of x.

By actual trial, as below, we find $\mp\frac{3}{16}$ is the only one of these giving a zero remainder.

FACTORING.

	480	− 22	171	36
3		360		
4	120	84½		
1		80		
6	80	9⅔		
3		144		
10	48	12·2		
−2		−320	228	−266
3	160	−114	133;	−230
−3		360		
4	120	− 95½		
−9		−432		
10	48	−45·4		
−3		− 90	21	−36
16	30	− 7	12;	0

(The coefficients are written only once, and understood for the other lines of substitution.)

Hence the factors are $3x + 16$ and $12x^2 - 7x + 30$.

The latter factor cannot be resolved, for 16 will not measure 30, and all the other factors left for trial by the tables above, have been tried and have failed.

2. Factor $10x^4 - x^3(15y+4z) - x^2(40y^2 - 6yz) + x(60y^3 + 16y^2z) - 24y^3z$.

Here $m = \pm 1, \pm 2, \pm 5,$ or ± 10. $k = -24y^3z$.

$k_1 = 10 - (15y+4z) - (40y^2 - 6yz) + (60y^3 + 16y^2z) - 24y^3z$
$= 10 - 15y - 40y^2 + 60y^3 - 2z(2 - 3y - 8y^2 + 12y^3)$
$= (5 - 2z)(2 - 3y - 8y^2 + 12y^3)$.

$k_2 = (5 + 2z)(2 + 3y - 8y^2 - 12y^3)$, as may easily be found by making the calculation.

We get at a glance $2z$ a factor of k, $2z - 5$ a factor of k_1, and $2z + 5$ a factor of k_2; hence taking $m = 5$, we are directed to try the substitution $\dfrac{2z}{5}$ for x.

	10	−(15y+4z)	−(40y² − 6yz)	(60y³ + 16y²z)	−24y³
$\dfrac{2z}{5}$		4z	−6yz	−16y²z	24y³
	2	−3y	−8y²	12y³;	0

Hence $5x-2z$ is a factor, the other being
$$2x^3 - 3x^2y - 8xy^2 + 12y^3.$$
The latter factor being homogeneous, the method of this article may be applied to it.

$m = \pm 1$ or ± 2, $k = 12$, $k_1 = 3$, $k_2 = 15$.

	$m=1,$	$2,$	$1,$	-1	
3	1,	1,	3,	3	The other columns
12	2,	3,	4,	2	are not full.
15	3,	5,	5,	1	

Hence the trial-substitutions (arrangement in ascending powers of x) are $\frac{1}{2}, \frac{2}{3}, \frac{1}{4}, \frac{-1}{2}$.

	12	-8	-3	2
1		6	-1	-2
2	6	-1	-2;	0
2		4	2	
3	2	1;	0	

Final factor is $2y + x$.

Hence the factors are $(x-2y)(2x-3y)(x+2y)$, and these, with the factor $5x-2z$ already found, give the complete resolution of the polynome proposed.

(The factor $5x-2z$, might easily have been got by the method of Art. XXIII., page 79, but the present solution shows we are independent of that article. It may also be obtained by rearranging the polynome in terms of y).

Exercise xxxix.
Factor

1. $2x^3 - 20x^2 + 38x - 20$; $x^3 - 7x^2y + 16xy^2 - 12y^3$.
2. $12x^3 + 5x^2y + xy^2 + 3y^3$; $8x^3 - 14x + 6$.
3. $3x^3 - 15ax^2 + a^2x - 5a^3$; $2x^3 + 9x^2y + 7xy^2 - 3y^3$.
4. $2b^4 - 7b^3c - 4b^2c^2 + bc^3 - 4c^4$; $15a^3 + 47a^2b + 13ab^2 - 12b^3$.
5. $4p^4 + 8p^3q + 7p^2q_n^2 + 8pq^3 + 3q^4$.
6. $150x^4 - 725x^3y + 931x^2y^2 + 920xy^3 - 1152y^4$.
7. $36x^4 - 6(9-7y)x^3 - 7(9+14y)x^2y + 3(49-40y)xy^2 + 180y^3$.
8. $10x^4 - x^3(15y+4z) + x^2(40y^2 + 6yz) + x(60y^3 - 16y^2z) - 24y^3z$

CHAPTER IV.

SECTION I.—DIVISION. MEASURES AND MULTIPLES.

Art. XXXI. When one quantity is to be divided by another the quotient can often be readily obtained by resolving the divisor or dividend, or *both*, into factors.

EXAMPLES.

1. Divide $a^2 - 2ab + b^2 - c^2 + 2cd - d^2$ by $a - b + c - d$. Here we see at once that the dividend $= (a-b)^2 - (c-d)^2$, and $\therefore$ quotient $= a - b - (c-d) = a - b - c + d$.

2. Divide the product of $a^2 + ax + x^2$ and $a^3 + x^3$ by $a^4 + a^2x^2 + x^4$. Here $a^3 + x^3 = (a+x)(a^2 - ax + x^2)$, and the divisor $= (a^2 + ax + x^2)(a^2 - ax + x^2)$ $\therefore$ the quotient is $a + x$.

3. Divide $a^3 + a^2b + a^2c - abc - b^2c - bc^2$ by $a^2 - bc$. The dividend is $a(a^2 - bc) + b(a^2 - bc) + c(a^2 - bc)$ $\therefore$ the quotient $= a + b + c$.

4. $(a^3 + b^3 - c^3 + 3abc) \div (a + b - c)$.

Dividend $= a^3 + b^3 + 3ab(a+b) - c^3 - 3ab(a+b) + 3abc = (a+b)^3 - c^3 - 3ab(a+b-c)$ which is exactly divisible by $a+b-c$; quotient $= a^2 + b^2 + c^2 - ab + bc + ca$.

5. Divide $x^5 - x^4y + x^3y^2 - x^2y^3 + xy^4 - y^5$ by $x^3 - y^3$.

The dividend is (Art. XXV.) evidently $(x^6 - y^6) \div (x+y)$, and this divided by $x^3 - y^3 = (x^3 + y^3) \div (x+y) = x^2 - xy + y^2$.

6. Divide $b(x^3 + a^3) + ax(x^2 - a^2) + a^3(x+a)$ by $(a+b)(x+a)$. Striking the factor $x + a$ out of dividend and divisor we have $b(x^2 - ax + a^2) + ax(x-a) + a^3 = b(x^2 - ax + a^2) + a(x^2 - ax + a^2)$ $= (a+b)(x^2 - ax + a^2)$ $\therefore$ quotient $= x^2 - ax + a^2$.

7. Divide $apx^4 + x^3(aq + bp) + x^2(ar + bq + pc) + x(qc + br) + cr$ by $ax^2 + bx + c$.

Factoring the dividend (Art. XXIII.) we have
$$(ax^2+bx+c)(px^2+qx+r).$$
∴ the quotient = the latter factor.

8. Divide $6x^4-13ax^3+13a^2x^2-13a^3x-5a^4$ by $2x^2-3ax-a^2$. This can be done by Art. XVII. The divisor is $2x^2-a^2-3ax$, and we see at once that $3x^2+5a^2$ must be two terms of the quotient.

Multiplying diagonally into the first two terms of the divisor, and adding the products, we get $+7a^2x^2$; but $+13a^2x^2$ is required. ∴ $+6a^2x^2$ is still required, and as this must come from the third term multiplied into $-3ax$, that third term must be $-2ax$; ∴ the quotient is $3x^2+5a^2-2ax$.

NOTE.—By multiplying the terms $-2ax$, $-3ax$, diagonally into the x^2's and a^2's respectively, we get the remaining terms of the dividend; it is, of course, necessary to test whether the division is exact.

9. Divide $2a^4-a^3b-12a^2b^2-5ab^3+4b^4$ by a^2-b^2-2ab.

Here, as before, one factor is a^2-b^2-2ab; ∴ *two terms of* the other factor are $2a^2-4b^2$. Multiplying, as in the last example, we get $-6a^2b^2$; but $-12a^2b^2$ is required. ∴ $-6a^2b^2$ is still needed, and $+3ab$ is the third term of the required quotient, which is therefore $2a^2-4b^2+3ab$.

Prove that

10. $(1+x+x^2+\ldots+x^{n-1})(1-x+x^2-\ldots+x^{n-1})$
$=1+x^2+x^4+\ldots+x^{2n-2}.$

$$\text{Product} = \frac{1-x^n}{1-x} \cdot \frac{1+x^n}{1+x}$$
$$= \frac{1-x^{2n}}{1-x^2} = 1+x^2+x^4+\ldots+x^{2n-2}.$$

11. Divide $(a^2-bc)^3+8b^3c^3$ by a^2+bc.
$= (a^2-bc)^3+(2bc)^3$ by $(a^2-bc)+2bc$
$= (a^2-bc)^2-(a^2-bc)\times 2bc+(2bc)^2$
$= a^4-4a^2bc+7b^2c^2.$

12. Divide $1+2357947691x^9$ by $1-11x+121x^2$

Dividend $= 1+(11x)^9$

$= \{1-(11x)^3+(11x)^6\}\{1+(11x)^3\}$

Divisor $= \{1+(11x)^3\} \div (1+11x)$.

∴ quotient $= \{1-(11x)^3+(11x)^6\}(1+11x)$.

Exercise xl.

Find the quotients in the following cases:

1. $1-x+x^2-x^3 \div 1-x$.
2. $1-2x^4+x^8 \div x^4+2x^2+1$.
3. $x^{16}+a^8x^8+a^{16} \div x^4-a^2x^2+a^4$.
4. $x^4+4x^2y^2-32y^4 \div x-2y$.
5. $1-4x^2+12x^3-9x^4 \div 1+2x-3x^2$.
6. $(a^2-2ax+x^2)(a^3+3a^2x+3ax^2+x^3) \div a^2-x^2$.
7. $x^3-y^3+z^3+3xyz \div x-y+z$.
8. $6a^4-a^3b+2a^2b^2+13ab^3+4b^4 \div 2a^2-3ab+4b^2$.
9. $4x^4-x^2y^2+6xy^3-9y^4 \div 2x^2+3y^2-xy$.
10. $a^4+b^4-c^4-2a^2b^2 \div a^2-b^2-c^2$.
11. $21a^4-16a^3b+16a^2b^2-5ab^3+2b^4 \div 3a^2-ab+b^2$.
12. $2a^3-7a^2-46a-21 \div 2a^2+7a+3$.
13. $\{a^3(b-c)+b^3(c-a)+c^3(a-b)\} \div a+b+c$.
14. $x^3-3ax^2+3a^2x-a^3+b^3 \div x-a+b$.
15. $x^4-y^4+z^4+2x^2z^2-2y^2-1 \div x^2-y^2+z^2-1$.
16. $x^4-(a+c)x^3+(b+ac)x^2-bcx \div x-c$.
17. $x^3+x^2y+xy^2+y^3 \div x+y$.
18. $x^7-x^6y+x^5y^2-x^4y^3+x^3y^4-x^2y^5+xy^6-y^7 \div x^4+y^4$.
19. $a^4+b^4-c^4-2a^2b^2-2c^2-1 \div a^2-b^2-c^2-1$.
20. $a^4-ab^3-ac^3-2a^3b+2b^4+2bc^3+3a^3c-3b^3c-3c^4$
$\div a+3c-2b$.
21. $a^2b-bx^2+a^2x-x^3 \div (x+b)(a-x)$.
22. $a(b-c)^3+b(c-a)^3+c(a-b)^3 \div a^2-ab-ac+bc$.

23. $a^2b^2+2abc^2-a^2c^2-b^2c^2 \div ab+ac-bc$.

24. $x^3+y^3+3xy-1 \div x+y-1$.

25. $x^6-x^3-2 \div x^2-x+1$.

26. $a^4-29a^2-50a-21 \div a^2-5a-7$.

27. $(2x-y)^2a^4-(x+y)^2a^2x^2+2(x+y)ax^4-x^6 \div (2x-y)a^2+(x+y)ax-x^3$.

28. $(x^3-1)a^3-(x^3+x^2-2)a^2+(4x^2+3x+2)a-3(x+1) \div (x-1)a^2-(x-1)a+3$.

Art. XXXII. The HIGHEST COMMON FACTOR of two algebraic quantities may, in general, be readily found by factoring. The H. C. F. is often discovered by taking the sum or difference (or sum *and* difference) of the given expressions, or of some multiples of them.

EXAMPLES.

1. Find the H. C. F. of $(b-c)x^2+(2ab-2ac)x+a^2b-a^2c$, and $(ab-ac+b^2-bc)x+a^2c+ab^2-a^2b-abc$.

Taking out the common factor $b-c$ we get $(b-c)(x^2+2ax+ab)$ and $(b-c)\{(a-b)x-a^2+ab\}$;

∴ $b-c$ is the H. C. F. of the given expressions.

2. Find the H. C. F. of
$$1-x+y+z-xy+yz-zx-xyz, \text{ and}$$
$$1-x-y-z+xy+yz+zx-xyz.$$

Their difference is $2y+2z-2xy-2zx = 2(1-x)(y+z)$.

Their sum is $2-2x+2yz-2xyz = 2(1-x)(1+yz)$.

∴ the H. C. F. is $(1-x)$.

3. Find the H. C. F. of $x^5+3x^4-8x^2-9x-3$, and
$$x^5-2x^4-6x^3+4x^2+13x+6.$$

The annexed method of finding the H. C. F. depends on the principle, that if a quantity measures two other quantities, it will measure any multiple of their sum or difference.

MEASURES AND MULTIPLES.

```
 1 + 3    0 - 8 - 9 - 3   (a)
 1 - 2 -  6 + 4 +13 + 6   (b)
 ─────────────────────
 5 + 6 - 12 - 22 - 9     (c)       [ = (a) - (b) ]

 2 + 6    0 - 16 - 18 - 6            (a) × 2
 1 - 2 -  6 +  4 + 13 + 6            (b)
 ─────────────────────
 3 + 4 -  6 - 12 -  5     (d)

   15 + 18 - 36 - 66 - 27            (c) × 3
   15 + 20 - 30 - 60 - 25            (d) × 5
   ───────────────────
      -  2 -  6 -  6 -  2
         1 +  3 +  3 +  1   (f)
   ───────────────────
   25 + 30 - 60 - 110 - 45           (c) × 5
   27 + 36 - 54 - 108 - 45           (d) × 9
   ───────────────────
      - 2 -  6 -  6 -  2
        1 +  3 +  3 +  1    (g)
```

H. C. F. = $(x+1)^3$.

The coefficients are written in two lines, (a) and (b). They are then subtracted so as to cancel the first terms. (a) is next multiplied by 2, and added to cancel the last terms. If (c) and (d) had been the same their terms would have been the coefficients of the H. C. F. Since they are not, we proceed with them as with (a) and (b) till they become the same. When (a) and (b) do not contain the same number of terms it is more convenient to find only (c), and then use this with the quantity containing the same number of terms. The general rule is to operate on lines containing the same, or nearly the same number of terms.

4. Find the H. C. F. of $3x^3 + 2x^2 - 14x + 8$, and
$$6x^3 - 11x^2 + 13x - 12.$$

```
   3 +  2 - 14 +  8      (a)
   6 - 11 + 13 - 12      (b)
   ─────────────────
   6 +  4 - 28 + 16                  (a) × 2
   ─────────────────                 (b) - (a),
      15 - 41 + 28       (c)
     (5 - 7)(3 - 4)
   H. C. F. = 3x - 4.    (d)
```

If (a) and (b) have a common factor its first term must measure 3 and 6, and its last term must measure 8 and 12. (c) is not therefore, the H. C. F. Resolve (c) into factors. $5x-7$ is not a factor of (a) and (b). If, therefore, (a) and (b) have a common factor it is $3x-4$. On trial $3x-4$ is found to be a factor of (a) and $\therefore$ it is the H. C. F. of (a) and (b).

5. If x^2+px+q, and x^2+rx+s have a common factor, prove that this factor is

$x+\dfrac{q-s}{p-r}$. If $x-a$ be the common factor then the remainders on dividing the given expressions by $x-a$, must be zero, $i.\ e.$,

$a^2+pa+q=0$, and $a^2+ra+s=0$, or

$(p-r)a=s-q,\ \therefore\ a=\dfrac{s-q}{p-r}$, and

$x-a=x-\dfrac{s-q}{p-r}=x+\dfrac{q-s}{p-r}$.

6. What value of a will make $a^2x^2+(a+2)x+1$, and $a^2x^2+a^2-5$, have a common measure.

They cannot have a monomial factor. Neither can they have one of two dimensions unless $(a+2)$ vanishes, $i.e.$, unless $a=-2$, in which case the expressions become $4x^2+1$, and $4x^2-1$, which have no C. F. Hence if the given quantities have a C. F., it must be of the form $x+m$; dividing $a^2x^2+a^2-5$ by $x+m$, we have for remainder,

$a^2m^2+a^2-5=0$, or $m^2=\dfrac{5-a^2}{a^2}$; $\therefore\ m=\dfrac{1}{a}\sqrt{(5-a^2)}$, in which $\sqrt{(5-a^2)}$ must be possible and integral, $\therefore\ a^2=4$, $(a^2=1$ gives values to m which on *trial* fail) and $a=\pm\ 2$, of which the positive value must be taken, and $\therefore\ 2x+1$ is the C. F.

7. If the H. C. F. of a and b be c, the L. C. M. of $(a+b)(a^3-b^3)$, and $(a-b)(a^3+b^3)$ is $\dfrac{a^6-b^6}{c^2}$.

Let $a=mc$, $b=nc$, and $\therefore\ a^3=m^3c^3$, $b^3=n^3c^3$. Thus
$(a\ +b\ )=c\ (m\ +n\ );\ (a\ -b\ )=c\ (m\ -n\ )$, and
$(a^3+b^3)=c^3(m^3+n^3);\ (a^3-b^3)=c^3(m^3-n^3)$.
$\therefore\ (a+b)(a^3-b^3)=c^4(m+n)(m^3-n^3)$, and
$(a-b)(a^3+b^3)=c^4(m-n)(m^3+n^3)$.

MEASURES AND MULTIPLES.   107

The H. C. F. of the last expressions is $c^4(m^2-n^2)$, ∴ the
L. C. M. $= c^4(m^6-n^6) = \dfrac{c^6(m^6-n^6)}{c^2} = \dfrac{a^6-b^6}{c^2}$.

8. If $(x-a)^2$ measures x^3+qx+r, find the relation between q and r.

Let $x+m$ be the other factor, then
$x^3+qx+r = (x-a)^2(x+m) = x^3+(m-2a)x^2+(a^2-2am)x+ma^2$
equating coefficients,

$m-2a=0$, $a^2-2am=q$, $ma^2=r$

∴ $m=2a$, and ∴ $a^2-4a^2=q$, $2a^3=r$, and

$a^2 = -\dfrac{q}{3}$, or $a^6 = -\dfrac{q^3}{27}$; and $a^3 = \dfrac{r}{2}$ or $a^6 = \dfrac{r^2}{4}$

∴ $\dfrac{r^2}{4} = -\dfrac{q^3}{27}$, or $\dfrac{r^2}{4}+\dfrac{q^3}{27} = 0$.

Or thus:—

Dividing x^3+qx+r by $(x-a)^2$ we find the remainder
$(q+3a^2)x+r-2a^3$
and as this will be the same for *all* values of x, we have, by equating coefficients,

$q+3a^2 = 0$,
and $r-2a^3 = 0$,
or $q^3 = -27a^6$
and $r^2 = 4a^6$;

therefore $\dfrac{r^2}{4}+\dfrac{q^3}{27} = 0$, as before.

Exercise xli.

Find the H. C. F. of the following:

1. $2x^4+3x^3+5x^2+9x-3$, $3x^4-2x^3+10x^2-6x+3$.
2. $x^3+(a+1)x^2+(a+1)x+a$, $x^3+(a-1)x^2-(a-1)x+a$.
3. $px^3-(p+q)x^2+(p-q)x+q$, $px^3-(p+q)x^2+(p+q)x+q$.
4. $ax^3-(a-b)x^2-(b-c)x-c$, $2ax^3+(a+2b)x^2+(b+2c)x+c$.
5. $1-3\frac{2}{5}x-3\frac{1}{2}x^2+\frac{1}{2}x^3-x^4$, $1-1\frac{1}{15}x-3x^2+1\frac{1}{15}x^3+x^4$.
6. $ac^{2a}+bc^{2b}+(a+b)c^{a+b}$, $a^c c^a+a^c c^b+c^a b^c+b^c c^b$.

7. $a^2x^3 + a^5 - 2abx^3 + b^2x^3 + a^3b^2 - 2a^4b$, and
$2a^2x^4 - 5a^4x^2 + 3a^6 - 2b^2x^4 + 5a^2b^2x^2 - 3a^4b^2$.

8. $(ax+by)^2 - (a-b)(x+z)(ax+by) + (a-b)^2xz$, and
$(ax-by)^2 - (a+b)(x+z)(ax-by) + (a+b)^2xz$.

9. $a(b^2-c^2) + b(c^2-a^2) + c(a^2-b^2)$ and
$a(b^3-c^3) + b(c^3-a^3) + c(a^3-b^3)$.

10. $a^{3m} + a^{2m} + a^m + 1$, and $a^{3m} - a^{2m} + a^m - 1$.

11. If $x^3 + ax^2 + bx + c$, and $x^2 + a'x + b'$, have a common factor of one dimension in x, it must be one the factors of
$(a-a')x^2 + (b-b')x + c$.

12 Determine the H. C. F. of $(a-b)^5 + (b-c)^5 + (c-a)^5$, and $(a^2-b^2)^5 + (b^2-c^2)^5 + (c^2-a^2)^5$.

13. Find the H. C. F. of
$2(y^3 - 2y^2 - y + 2)x^3 + 3(y^2-1)x^2 - (2y^3 - y^2 - 2y + 1)$, and
$3(y^3 - 4y^2 + 5y - 2)x^2 + 7(y^2 - 2y + 1)x - (3y^3 - 5y^2 + y + 1)$.

14. If $x^2 + px + q$, and $x^2 + mx + n$ have a common linear factor, shew that
$(n-q)^2 + n(m-p)^2 = m(m-p)(n-q)$.

15. Find the L. C. M. of $x^3 - 3x^2 + 3x - 1$, $x^3 - x^2 - x + 1$, $x^4 - 2x^3 + 2x - 1$, and $x^4 - 2x^3 + 2x^2 - 2x + 1$.

16. Find the L. C. M. of
$x^3 + 6x^2 + 11x + 6$, $x^3 + 7x^2 + 14x + 8$.
$x^3 + 8x^2 + 19x + 12$, and $x^3 + 9x^2 + 26x + 24$.

17. Find the value of y which will make
$2(y^2 + y)x^2 + (11y - 2)x + 4$ and
$2(y^3 + y^2)x^3 + (11y^2 - 2y)x^2 + (y^2 + 5y)x + 5y - 1$, have a common measure.

18. The product of the H. C. F. and L. C. M. of two quantities is equal to half the sum of their squares, one of them is $2x^3 - 11x^2 + 17x - 6$; find the other.

19. If $x + a$ and $x - a$ are both measures of $x^3 + px^2 + qx + r$, shew that $pq = r$.

20. If x^3+qx+r and x^3+mx+n have a common measure $(x-a)^2$, show that $q^3n^3=m^3r^3$.

21. If the H. C. F. of x^3+px+q and x^2+mx+n, be $x+a$, their L. C. M. is
$$x^4+(m-a)x^3+px^2+(a^3+mp)x+a(m-a)(a^2+p).$$

22. If x^2+qx+1, and x^3+px^2+qx+1, have a common factor of the form $x+a$, shew that $(p-1)^2-q(p-1)+1=0$.

23. If x^3+px^2+q, and x^2+mx+n, have $x+a$ for their H. C. F., shew that their L. C. M. is
$$x^4+(m-a+p)x^3+p(m-a)x^2+a^2(a-p)x+a^2(a-p)(m-a).$$

24. If x^2+px+1, and x^3+px^2+qx+1, have $x-a$ for a common factor, shew that $a=\dfrac{1}{1-q}$.

25. Find the H. C. F. of $(a^2-b^2)^3+(b^2-c^2)^3+(c^2-a^2)^3$, and $a^5(b-c)+b^5(c-a)+c^5(a-b)$.

26. If α be the H. C. F. of b and c, β the H. C. F. of c and a, γ the H. C. F. of a and b, and δ the H. C. F. of a, b, and c, then the L. C. M. of a, b, c, is $\dfrac{abc\delta}{\alpha\beta\gamma}$.

27. If $x+c$ be the H. C. F. of x^2+ax+b, and $x^2+a'x+b'$, their L. C. M. will be $x^3+(a+a'-c)x^2+(aa'-c^2)x+(a-c)(a'-c)c$.

28. Shew that the L. C. M. of the quantities in Ex. 2 (solved above) will be a complete square if $x=y^2+z^2-y^2z^2$.

29. Find the H. C. F. of $x^8+2x^6+3x^4-2x^2+1$, and
$$6x^9+x^7+17x^5-7x^3-2.$$

SECTION II.—FRACTIONS.

Art. XXXIII. When required to reduce a fraction to its lowest terms, we can often apply some of the preceding methods of factoring to discover the H. C. F. of the numerator and denominator.

FRACTIONS.

EXAMPLES.

1. $\dfrac{ac+by+ay+bc}{af+2bx+2ax+bf} = \dfrac{c(a+b)+y(a+b)}{f(a+b)+2x(a+b)} = \dfrac{c+y}{f+2x}$.

2. $\dfrac{a^4-ba^3-a^2b^2+ab^3}{a^5-ba^4-ab^4+b^5} = \dfrac{a\{a^3+b^3-ab(a+b)\}}{a(a^4-b^4)-b(a^4-b^4)}$

$= \dfrac{a(a+b)(a-b)^2}{(a-b)(a^4-b^4)} = \dfrac{a}{a^2+b^2}$.

3. $\dfrac{x^5+x^4y+x^3y^2+x^2y^3+xy^4+y^5}{x^5-x^4y+x^3y^2-x^2y^3+xy^4-y^5}$.

Here the numerator is evidently $(x^6-y^6) \div (x-y)$, and the denominator is $\dfrac{x^6-y^6}{x+y}$. The result is $\therefore \dfrac{x+y}{x-y}$.

4. $\dfrac{(x+y)^5-x^5-y^5}{(x+y)^4+x^4+y^4} = \dfrac{5x^4y+10x^3y^2+10x^2y^3+5xy^4}{(x+y)^4-x^2y^2+(x^2+y^2)^2-x^2y^2}$

$= \dfrac{5xy\{x^3+y^3+2xy(x+y)\}}{(x^2+y^2+xy)\{(x+y)^2+xy+x^2+y^2-xy\}}$

$= \dfrac{5xy(x+y)(x^2+xy+y^2)}{2(x^2+xy+y^2)^2} = \dfrac{5xy(x+y)}{2(x^2+xy+y^2)}$

5. $\dfrac{x^2-12x+35}{x^3-10x^2+31x-30}$.

Here we see at once that the numerator $= (x-5)(x-7)$; and it is plain that $x-7$ is not a factor of the denominator; we $\therefore$ try $x-5$ (Horner's division), and find the quotient to be x^2-5x+6.

$\therefore$ the result $= \dfrac{x-7}{x^2-5x+6}$.

6. $\dfrac{x^4+2x^2+9}{x^4-4x^3+8x-21}$.

The factors of the numerator are at once seen to be x^2+2x+3, and x^2-2x+3, of which the latter is one factor of the denominator, the other being (Horner's division) x^2-2x-7 : $\therefore$ the result is $\dfrac{x^2+2x+3}{x^2-2x-7}$.

Exercise xlii

Reduce the following to their lowest terms:

1. $\dfrac{x^2-7x+6}{x^3-2x^2-8x-96}$, $\quad \dfrac{3xy^2-13xy+14x}{7y^3-17y^2+6y}$.

2. $\dfrac{x^4+a^2x^2+a^4}{x^4+ax^3-a^3x-a^4}$, $\quad \dfrac{x^2+x-12}{x^3-5x^2+7x-3}$.

3. $\dfrac{x^3-3x+2}{x^3+4x^2-5}$, $\quad \dfrac{x^4+2x^2+9}{x^4-4x^3+4x^2-9}$.

4. $\dfrac{2+bx}{2b+(b^2-4)x-2bx^2}$, $\quad \dfrac{x^3+2x^2+2x}{x^5+4x}$.

5. $\dfrac{5a^5+10a^4x+5a^3x^2}{a^3x+2a^2x^2+2ax^3+x^4}$, $\quad \dfrac{20x^4+x^2-1}{25x^4+5x^3-x-1}$.

6. $\dfrac{x^7-x^6y+x^5y^2-x^4y^3+x^3y^4-x^2y^5+xy^6-y^7}{x^7+x^6y+x^5y^2+x^4y^3+x^3y^4+x^2y^5+xy^6+y^7}$.

7. $\dfrac{3a^2x^4-2ax^2-1}{4a^3x^6-2a^2x^4-3ax^2+1}$, $\quad \dfrac{x^2+\left(\dfrac{a}{b}+\dfrac{b}{a}\right)xy+y^2}{x^2+\left(\dfrac{a}{b}-\dfrac{b}{a}\right)xy-y^2}$.

8. $\dfrac{a^2(b-c)+b^2(c-a)+c^2(a-b)}{abc(a-b)(b-c)(c-a)}$.

9. $\dfrac{(a+b+c)^2}{a^3(b-c)+b^3(c-a)+c^3(a-b)}$.

10. From Ex. 4 (solved above) show that
$$\dfrac{(a-b)^4+(b-c)^4+(c-a)^4}{(a-b)^5+(b-c)^5+(c-a)^5} = \dfrac{(a-b)^2+(b-c)^2+(c-a)^2}{5(a-b)(b-c)(c-a)}.$$

11. $\dfrac{(x+y)^5-x^5-y^5}{(x+y)^7-x^7-y^7}$.

12. Shew that
$$\dfrac{(a-b)^7+(b-c)^7+(c-a)^7}{(a-b)^5+(b-c)^5+(c-a)^5} = \dfrac{7}{10}\{(a-b)^2+(b-c)^2+(c-a)^2\}.$$

Art. XXXIV. In reducing complex fractions it is often convenient to multiply both terms of the complex fraction by the L. C. M. of all the *denominators* involved.

EXAMPLES.

1. Simplify $\dfrac{\frac{1}{2}(x+1\frac{1}{3})-\frac{2}{3}(1-\frac{1}{4}x)}{1\frac{3}{4}-\frac{1}{3}(x+4\frac{1}{4})}$.

Here the L. C. M. of all the denominators involved is 12;
∴ multiplying both terms of the complex fraction by 12, and removing brackets, we have

$$\frac{6x+8-8+6x}{21-4x-17} = \frac{12x}{4-4x} = \frac{3x}{1-x}.$$

2. $\dfrac{a-\dfrac{a-b}{1+ab}}{1+\dfrac{a(a-b)}{1+ab}}$. Here multiplying both terms by $1+ab$, we get

$$\frac{a(1+ab)-a+b}{1+ab+a(a-b)} = \frac{b(a^2+1)}{a^2+1} = b.$$

3. $\dfrac{1}{x-1+\dfrac{1}{1+\dfrac{c}{4-x}}}$. Here multiplying both terms of the fraction which follows $x-1$ by $4-x$, the given fraction becomes at once $\dfrac{1}{x-1+\dfrac{4-x}{4}}$, and now multiplying both terms by 4, we have $\dfrac{4}{4x-4+4-x} = \dfrac{4}{3x}$.

It may be observed that when the fraction is reduced to the form $\dfrac{a}{b} \div \dfrac{c}{d}$, we may strike out any factor common to the two *denominators*, and also any factor common to the two *numerators*; it is sometimes more convenient to do this than to multiply directly by the L. C. M. of all the denominators.

FRACTIONS. 113

4. Simplify $\left(\dfrac{a+b}{a-b} + \dfrac{a-b}{a+b}\right) \div \left(\dfrac{a^2+b^2}{a^2-b^2} - \dfrac{a^2-b^2}{a^2+b^2}\right)$.

Here the numerator of the first fraction is $(a+b)^2+(a-b)^2$ and the denominator is a^2-b^2; the numerator of second fraction is $(a^2+b^2)^2-(a^2-b^2)^2$, and the denominator is a^4-b^4; the former denominator cancels this to a^2+b^2, which, of course, becomes a multiplier of the first numerator:

$\therefore$ we have $\dfrac{(a^2+b^2)\{(a+b)^2+(a-b)^2\}}{(a^2+b^2)^2-(a^2-b^2)^2} = \dfrac{(a^2+b^2)^2}{2a^2b^2}$.

Occasionally, we at once discover a common complex factor, strike this out, and simplify the result.

5. $\dfrac{\dfrac{1}{a}+\dfrac{1}{b}+\dfrac{1}{c}}{\dfrac{1}{a^2}+\dfrac{1}{b^2}-\dfrac{1}{c^2}+\dfrac{2}{ab}}$: here the den. $= \left(\dfrac{1}{a}+\dfrac{1}{b}\right)^2 - \dfrac{1}{c^2}$

$= \left(\dfrac{1}{a}+\dfrac{1}{b}+\dfrac{1}{c}\right)\left(\dfrac{1}{a}+\dfrac{1}{b}-\dfrac{1}{c}\right)$, and cancelling the common factor we have

$\dfrac{1}{\dfrac{1}{a}+\dfrac{1}{b}-\dfrac{1}{c}}$, and multiplying by abc, this $= \dfrac{abc}{bc+ca-ab}$.

Exercise xliii.

Simplify the following:

1. $\dfrac{1-\frac{1}{2}\{1-\frac{1}{3}(1-x)\}}{1-\frac{1}{3}\{1-\frac{1}{2}(1-x)\}}$, $\dfrac{\dfrac{a+b}{a-b}+\dfrac{a-b}{a+b}}{\dfrac{a+b}{a-b}-\dfrac{a-b}{a+b}}$.

2. $\dfrac{\dfrac{x}{x+y}+\dfrac{x}{x-y}}{\dfrac{2x}{x^2-y^2}}$, $\dfrac{\dfrac{1}{1-a}-\dfrac{1}{1+a}}{\dfrac{a}{1-a}+\dfrac{1}{1+a}}$.

3. $1+\dfrac{a}{1+a+\dfrac{2a^3}{1+a}}$, $x - \dfrac{x-y}{\dfrac{(x-a)(x-z)}{x+y}}$

4. $\dfrac{\dfrac{a^2+b^2}{2a^2} - \dfrac{y^2}{a^2+b^2}}{\dfrac{a^2+b^2}{2b^2} - \dfrac{2a^2}{a^2+b^2}}$; $\quad \dfrac{\dfrac{1}{a} + \dfrac{1}{ab^3}}{b-1+\dfrac{1}{b}}$.

5. $\dfrac{\dfrac{a+b}{c+d} + \dfrac{a-b}{c-d}}{\dfrac{a+b}{c-d} + \dfrac{a-b}{c+d}}$; $\quad \dfrac{a+b+\dfrac{b^2}{a}}{a+b+\dfrac{a^2}{b}}$.

6. $\dfrac{3xyz}{yz - zx - xy} - \dfrac{\dfrac{x-1}{x} + \dfrac{y-1}{y} + \dfrac{z-1}{z}}{\dfrac{1}{x} + \dfrac{1}{y} + \dfrac{1}{z}}$.

7. $\dfrac{\dfrac{2}{a^2} + \dfrac{2}{b^2} + \dfrac{2}{c^2} + \dfrac{a^4+b^4+c^4}{a^2b^2c^2}}{\dfrac{a}{bc} + \dfrac{b}{ac} + \dfrac{c}{ab}}$.

8. $\dfrac{a^3+a^2b+ab^2+b^3}{a^3-a^2b+ab^2-b^3} \div \dfrac{a^2+2ab+b^2}{a^2-b^2}$.

9. $\left(\dfrac{a+b}{a-b} + \dfrac{a^2+b^2}{a^2-b^2}\right) \div \left(\dfrac{a-b}{a+b} - \dfrac{a^3-b^3}{a^3+b^3}\right)$.

10. $\dfrac{\dfrac{1}{a} + \dfrac{1}{b+c}}{\dfrac{1}{a} - \dfrac{1}{b+c}} \left\{ 1 + \dfrac{b^2+c^2-a^2}{2bc} \right\}$.

11. $\dfrac{\dfrac{2(1-x)}{1+x} + \dfrac{(1-x)^2}{(1+x)^2} + 1}{\dfrac{2(1+x)}{1-x} + \left(\dfrac{1+x}{1-x}\right)^2 + 1}$; $\quad \dfrac{\left(\dfrac{x-a}{x+a}\right)^2 + \left(\dfrac{x+a}{x-a}\right)^2 - 2}{\left(\dfrac{x-a}{x+a}\right)^2 + 2 + \left(\dfrac{x+a}{x-a}\right)^2}$.

12. $\dfrac{\dfrac{x}{y} + 1 + \dfrac{y}{x}}{\dfrac{x}{y} - 1 + \dfrac{y}{x}} \div \dfrac{1 - \dfrac{y^3}{x^3}}{1 + \dfrac{y^3}{x^3}}$.

FRACTIONS.

13. $\dfrac{3\left(\dfrac{a-b}{a+b}\right)^2 - \left(\dfrac{a-b}{a+b}\right)^3 - 3\left(\dfrac{a-b}{a+b}\right)+1}{3\left(\dfrac{a+b}{a-b}\right) - 3\left(\dfrac{a+b}{a-b}\right)^2 + \left(\dfrac{a+b}{a-b}\right)^3 - 1}.$

14. $\dfrac{x^5 - x^4y + x^3y^2 - x^2y^3 + xy^4 - y^5}{x^5 + x^4y + x^3y^2 + x^2y^3 + xy^4 + y^5} \div \left(\dfrac{x-y}{x+y}\right)^2.$

15. $\left(\dfrac{1-x^2}{1-x^3} + \dfrac{1-x}{1-x+x^2}\right) \div \left(\dfrac{1+x}{1+x+x^2} - \dfrac{1-x^2}{1+x^3}\right).$

16. Find the value of
$$\dfrac{a}{2na - 2nx} + \dfrac{b}{2nb - 2nx} \text{ when } x = \tfrac{1}{2}(a+b).$$

17. Find the value of $\sqrt{\{1 - \sqrt{(1-x)}\}}$
when $x = 2\left(\dfrac{1-b}{1+b}\right)^2 - \left(\dfrac{1-b}{1+b}\right)^4.$

18. Find value of
$$\dfrac{\sqrt{(a+bx)} + \sqrt{(a-bx)}}{\sqrt{(a+bx)} - \sqrt{(a-bx)}} \text{ when } x = \dfrac{2ac}{b(1+c^2)}.$$

Art. XXXV. When the sum of several fractions is to be found, it is generally best, instead of reducing at once all the fractions to a common denominator, to take two (or more) of them together, and combine the results.

EXAMPLES.

1. Find the sum of
$$\dfrac{x+y}{2x-2y} - \dfrac{y-x}{2x+2y} - \dfrac{x^2-y^2}{x^2+y^2}.$$
Here taking the first two together we have
$$\dfrac{(x+y)^2 + (x-y)^2}{2(x^2-y^2)} = \dfrac{x^2+y^2}{x^2-y^2}; \text{ now add this to } -\dfrac{x^2-y^2}{x^2+y^2}$$
and we get $\dfrac{(x^2+y^2)^2 - (x^2-y^2)^2}{x^4-y^4} = \dfrac{4x^2y^2}{x^4-y^4}.$

2. Find the sum of
$$\frac{1+x}{1-x} + \frac{4x}{1+x^2} + \frac{8x}{1+x^4} - \frac{1-x}{1+x}.$$
Here, taking the first and the last together, we have
$$\frac{(1+x)^2 - (1-x)^2}{1-x^2} = \frac{4x}{1-x^2};$$
taking this result with the second fraction, we have
$$4x\left(\frac{1}{1+x^2} + \frac{1}{1-x^2}\right) = \frac{8x}{1-x^4};$$
now take this result with the remaining fraction and we get
$$8x\left(\frac{1}{1-x^4} + \frac{1}{1+x^4}\right) = \frac{16x}{1-x^8}.$$

3. $\dfrac{x^{3n}}{x^n - 1} - \dfrac{x^{2n}}{x^n + 1} - \dfrac{1}{x^n - 1} + \dfrac{1}{x^n + 1}.$ Taking in pairs those whose denominators are alike, we have
$$\frac{x^{3n}-1}{x^n-1} - \frac{x^{2n}-1}{x^n+1} = x^{2n}+x^n+1 - (x^n-1) = x^{3n}+2.$$

The work is often made easier by *completing the divisions* represented by the fractions.

4. Find the sum of $1 + \dfrac{2x+1}{2(x-1)} - \dfrac{4x+5}{2x+2}.$ By dividing numerators into denominators, this
$$= 1 + 1 + \frac{3}{2x-2} - 2 - \frac{1}{2x+2} = \frac{3}{2x-2} - \frac{1}{2x+2}$$
$$= \frac{3x+3-x+1}{2x^2-2} = \frac{x+2}{x^2-1}.$$

5. $\dfrac{x}{x-2} + \dfrac{x-9}{x-7} - \dfrac{x+1}{x-1} - \dfrac{x-8}{x-6}$: we have, by division
$$1 + \frac{2}{x-2} + 1 - \frac{2}{x-7} - 1 - \frac{2}{x-1} - 1 + \frac{2}{x-6}, \text{ or}$$
$$\frac{2}{x-2} + \frac{2}{x-6} - \frac{2}{x-7} - \frac{2}{x-1} = \frac{2(2x-8)}{(x-2)(x-6)} - \frac{2(2x-8)}{(x-1)(x-7)}.$$
$$= (4x-16)\left\{\frac{1}{x^2-8x+12} - \frac{1}{x^2-8x+7}\right\}$$
$$= (80-20x) \div (x^4 - 16x^3 + 83x^2 - 152x + 84).$$
[denominator $= (x^2-8x)^2 + 19(x^2-8x) + 84$].

6. Find the value of

$$\frac{x+2a}{x-2a} + \frac{x+2b}{x-2b} \text{ when } x = \frac{4ab}{a+b}$$

By division, $1 + \frac{4a}{x-2a} + 1 + \frac{4b}{x-2b}$

$= 2 + 4\left(\frac{a}{x-2a} + \frac{b}{x-2b}\right)$; but the quantity in the brackets

$= \frac{(a+b)x - 4ab}{(x-2a)(x-2b)} = 0$ since $(a+b)x = 4ab$

$\therefore$ the value of the given expression is 2.

Exercise xliv.

Simplify the following:

1. $\dfrac{x-a}{5} + \dfrac{x^2+ax+a^2}{x+a} - \dfrac{x^3-a^3}{x^2-a^2}$.

2. $\dfrac{a^3+b^3}{a^2-ab+b^2} + \dfrac{a^3-3a^2b+3ab^2-b^3}{a^3-b^3} - \dfrac{a(a-b)-b(a-b)}{a^2+ab+b^2}$.

3. $\left(\dfrac{1}{a+x} + \dfrac{1}{a-x} + \dfrac{2a}{a^2+x^2}\right) \times$

$\left(\dfrac{1}{a+x} - \dfrac{1}{a-x} - \dfrac{2x}{a^2+x^2}\right)$.

4. $\dfrac{a}{a+b} + \dfrac{b}{a-b} - \dfrac{ab}{ab-b^2} + \dfrac{ab}{a^2+ab}$.

5. $\dfrac{3+2x}{2-x} - \dfrac{2-3x}{2+x} + \dfrac{16x-x^2}{x^2-4}$.

6. $\dfrac{1}{4a^3(a+x)} + \dfrac{1}{4a^3(a-x)} + \dfrac{1}{2a^2(a^2+x^2)}$.

7. $\dfrac{1}{2}\left(\dfrac{3x+2y}{3x-2y}\right) - \dfrac{1}{2}\left(\dfrac{3x-2y}{3x+2y}\right)$.

8. $\dfrac{x+1}{2x-1} - \dfrac{x-1}{2x+1} - \dfrac{1-3x}{x(1-2x)} + \dfrac{x}{x(4x^2-1)} + \dfrac{1}{x(16x^4-1)}$.

9. $\dfrac{1}{2x+2} - \dfrac{4}{x+2} + \dfrac{9}{2(x+3)} - \dfrac{x-1}{(x+2)(x+3)}$.

10. $\dfrac{2(x+y)}{x-y} - \dfrac{2(y-x)}{x+y} - \dfrac{4(x^2-y^2)}{x^2+y^2} + \dfrac{4(x^4+y^4)}{x^4-y^4}$.

11. $(a-b)\left\{\dfrac{1}{(x+a)^2} + \dfrac{1}{(x+b)^2}\right\} +$

$2\left\{\dfrac{1}{x+a} - \dfrac{1}{x+b}\right\}$

12. $\left\{\dfrac{a+x}{a-x} + \dfrac{4ax}{a^2+x^2} + \dfrac{8a^3x}{a^4+x^4} - \dfrac{a-x}{a+x}\right\} +$

$\left\{\dfrac{a^2+x^2}{a^2-x^2} + \dfrac{4a^2x^2}{a^4+x^4} - \dfrac{a^2-x^2}{a^2+x^2}\right\}$.

13. $\dfrac{5x-4}{9} + \dfrac{12x+2}{11x-8} - \dfrac{10x+17}{18}$.

14. $\dfrac{a}{a^2+b^2} + \dfrac{a}{a^2-b^2} + \dfrac{a^2}{(a-b)(a^2+b^2)} - \dfrac{2a^3-b^3-ab^2}{a^4-b^4}$

15. $\dfrac{12x+10a}{3x+a} + \dfrac{117a+28x}{9a+2x} - 18$.

16. $\dfrac{4x-17}{x-4} - \dfrac{8x-30}{2x-7} + \dfrac{10x-3}{2x-5} - \dfrac{5x-4}{x-1}$.

17. Find the value of $\dfrac{a+b+2c}{a+b-2c} + \dfrac{a+b+2d}{a+b-2d}$

when $a+b = \dfrac{4cd}{c+d}$.

18. $\dfrac{x^{3n}}{x^n-y^n} - \dfrac{y^n x^{2n}}{x^n+y^n} - \dfrac{x^{3n}}{x^n-y^n} + \dfrac{y^{3n}}{x^n+y^n}$.

19. $\dfrac{(a-b)^{3n}}{(a-b)^n-1} - \dfrac{(a-b)^{2n}}{(a-b)^n+1} - \dfrac{1}{(a-b)^n-1} + \dfrac{1}{(a-b)^n+1}$.

20. $\dfrac{1}{(a^2-b^2)(x^2+b^2)} + \dfrac{1}{(b^2-a^2)(x^2+a^2)} - \dfrac{1}{(x^2+a^2)(x^2+b^2)}$

21. $\dfrac{1+x}{1-x^3} + \dfrac{1-x}{1+x^3} - \dfrac{2}{1-x^2} - \dfrac{2x^2}{x^6+1}$.

22. $\dfrac{a^3+a^2b+ab^2+b^3}{a^3-a^2b-ab^2+b^3} \times \dfrac{(a+b)^2-3ab}{(a-b)^2+3ab} \times \dfrac{(a-b)^3-a^3+b^3}{(a+b)^3-a^3-b^3}$.

FRACTIONS.

Art. XXXVI. The following are additional examples in which a knowledge of factoring and of the principle of symmetry is of advantage.

EXAMPLES.

1. $\dfrac{x^2-(y-z)^2}{(x+z)^2-y^2} + \dfrac{y^2-(z-x)^2}{(y+x)^2-z^2} + \dfrac{z^2-(x-y)^2}{(z+y)^2-x^2}.$

Cancelling the common factor $x-y+z$ in the two terms of the first fraction, there results $\dfrac{x+y-z}{x+y+z}$, hence by symmetry, the denominators of the other two fractions will be $x+y+z$, and the numerators will be $y+z-x$, $z+x-y$; $\therefore$ sum of the three numerators $= x+y+z$, and the result $= 1$.

2. Simplify $\dfrac{ab}{(c-a)(c-b)} + \dfrac{bc}{(a-b)(a-c)} + \dfrac{ca}{(b-c)(b-a)}.$

The L. C. M. of denominators is evidently $(a-b)(b-c)(c-a)$. This gives for numerator of first fraction $-ab(a-b)$, and by symmetry the other numerators are $-bc(b-c)$, $-ca(c-a)$.

$\therefore$ we have $-\dfrac{ab(a-b)+bc(b-c)+ca(c-a)}{(a-b)(b-c)(c-a)}.$

$= -\dfrac{(a-b)(b-c)(a-c)}{(a-b)(b-c)(c-a)} = 1.$

2. Reduce the following to a single fraction:

$\dfrac{a}{(a-b)(a-c)(x-a)} + \dfrac{b}{(b-a)(b-c)(x-b)} + \dfrac{c}{(c-a)(c-b)(x-c)}.$

Here the L. C. M. is $(a-b)(b-c)(c-a)(x-a)(x-b)(x-c)$; the numerator of the first fraction is

$-a(b-c)(x-b)(x-c)$, and $\therefore$ by symmetry that of second is $-b(c-a)(x-c)(x-a)$, and that of third is $-c(a-b)(x-a)(x-b)$; and their sum is

$-\{a(b-c)(x-b)(x-c)+b(c-a)(x-c)(x-a)+c(a-b)(x-a)(x-b)\}.$

This vanishes if $a=b$, hence $a-b$ is a factor, and $\therefore$ by symmetry $b-c$ and $c-a$ are also factors.' Now the product of these is of the third degree, while the whole expression rises only to the fourth, *hence x^2 cannot be involved.* The other factor must therefore be of the form $mx+n$, *in which m is a number.*

To determine n put $x=0$, and the expression becomes $abc\{a-b+b-c+c-a\}=0$; ∴ $n=0$, or the other factor is mx.

To determine m put $a=0$, $b=1$, $c=-1$, and m will be found to be 1. The numerator is ∴ $x(a-b)(b-c)(c-a)$, and the result is

$$\frac{x}{(x-a)(x-b)(x-c)}.$$

3. Simplify $\dfrac{a+b}{(b-c)(c-a)} + \dfrac{b+c}{(c-a)(a-b)} + \dfrac{c+a}{(a-b)(b-c)}$.

L. C. M. of denominators is $(a-b)(b-c)(c-a)$;
∴ first numerator is a^2-b^2, and by symmetry
    second   " b^2-c^2, and
    third   " c^2-a^2;
the sum of these $=0$, which is the required result.

4. Reduce

$$\frac{2}{x-y} + \frac{2}{y-z} + \frac{2}{z-x} + \frac{(x-y)^2+(y-z)^2+(z-x)^2}{(x-y)(y-z)(z-x)}.$$

Here the numerator becomes

$2(y-z)(z-x)+2(x-y)(z-x)+2(x-y)(y-z)+$
$(x-y)^2+(y-z)^2+(z-x)^2$, which is evidently
$\{(x-y)+(y-z)+(z-x)\}^2=0$.

5. $a^3\left\{\dfrac{a^3+2b^3}{a^3-b^3}\right\}^3 + b^3\left\{\dfrac{2a^3+b^3}{b^3-a^3}\right\}^3$.

Observe that the denominators become the same by changing the sign between the fractions, and that the expression is symmetrical with respect to a and b. The numerator of the first fraction is $a^{12}+6a^9b^3+12a^6b^6+8a^3b^9$, and by symmetry that of the other is $-b^{12}-6b^9a^3-12b^6a^6-8b^3a^9$. Their sum is ∴
$a^{12}-b^{12}+6a^3b^3(a^6-b^6)-8a^3b^3(a^6-b^6)$
$=(a^6-b^6)\{a^6+b^6+6a^3b^3-8a^3b^3\}=(a^6-b^6)(a^3-b^3)^2$
$=(a^3+b^3)(a^3-b^3)^3$, and since the *denominator* of the given expression is $(a^3-b^3)^3$ ∴ the result is a^3+b^3.

FRACTIONS.   121

Exercise xlv.

Simplify the following:

1. $x\left(\dfrac{x-2y}{x+y}\right)^3 + y\left(\dfrac{2x-y}{x+y}\right)^3.$

2. $a\left(\dfrac{a+2b}{a-b}\right)^3 + b\left(\dfrac{2a+b}{b-a}\right)^3.$

3. $\dfrac{a+b}{(b-c)(c-a)} + \dfrac{b+c}{(c-a)(a-b)} + \dfrac{c+a}{(a-b)(b-c)}.$

4. $\dfrac{1}{(a-b)(a-c)} + \dfrac{1}{(b-a)(b-c)} + \dfrac{1}{(c-a)(c-b)}.$

5. $\dfrac{a-b}{a+b} + \dfrac{b-c}{b+c} + \dfrac{c-a}{c+a} + \dfrac{(a-b)(b-c)(c-a)}{(a+b)(b+c)(c+a)}.$

6. $\dfrac{a^2}{(a+b)(a+c)(x+a)} + \dfrac{b^2}{(a+b)(b-c)(x+b)} - \dfrac{c^2}{(a+c)(b-c)(x+c)}.$

7. $\dfrac{x^2}{(x-y)(x-z)} + \dfrac{y^2}{(y-x)(y-z)} + \dfrac{z^2}{(z-x)(z-y)}.$

8. $\dfrac{a^3}{(a-b)(a-c)} + \dfrac{b^3}{(b-a)(b-c)} + \dfrac{c^3}{(c-a)(c-b)}.$

9. $\dfrac{1}{\left(\dfrac{a}{b}-1\right)\left(\dfrac{a}{c}-1\right)} + \dfrac{1}{\left(\dfrac{b}{a}-1\right)\left(\dfrac{b}{c}-1\right)} + \dfrac{1}{\left(\dfrac{c}{a}-1\right)\left(\dfrac{c}{b}-1\right)}.$

10. $x^3\left(\dfrac{x^3-2y^3}{x^3+y^3}\right)^3 + y^3\left(\dfrac{2x^3-y^3}{x^3+y^3}\right)^3.$

11. $\dfrac{1}{(b+c-2a)(c+a-2b)} + \dfrac{1}{(c+a-2b)(a+b-2c)} + \dfrac{1}{(a+b-2c)(b+c-2a)}.$

12. $\dfrac{b^2-c^2}{(b+c)^2} + \dfrac{c^2-a^2}{(c+a)^2} + \dfrac{a^2-b^2}{(a+b)^2}.$

13. $\dfrac{a^2}{(a-b)(a-c)(x-a)} + \dfrac{b^2}{(b-a)(b-c)(x-b)} + \dfrac{c^2}{(c-a)(c-b)(x-c)}.$

14. $\dfrac{x(y+z)}{(x-y)(z-x)} + \dfrac{y(z+x)}{(y-z)(x-y)} + \dfrac{z(x+y)}{(z-x)(y-z)}.$

15. $\dfrac{(a+b)^2+(b-c)^2+(a+c)^2}{(a+b)(b-c)(a+c)} - \dfrac{2}{a+c} - \dfrac{2}{b-c} + \dfrac{2}{a+b}.$

16. $\dfrac{1}{x(x-a)(x-b)} + \dfrac{1}{a(b-a)(x-a)} + \dfrac{1}{b(b-a)(x-b)}$

SECTION III.—RATIOS.

Art. XXXVII. If $\dfrac{a}{b} = \dfrac{c}{d}$ ∴ $ad = bc$. Now,

dividing $ad = bc$ by ca we have $\dfrac{b}{a} = \dfrac{d}{c}$ . . . . (1).

" $ad = bc$ by cd " $\dfrac{a}{c} = \dfrac{b}{d}$ . . . . (2).

" $ad = bc$ by ab " $\dfrac{d}{b} = \dfrac{c}{a}$ . . . . (3).

Also $\dfrac{ma+nc}{mb+nd}$ = each of the given fractions . . . (4).

For $\dfrac{ma+nc}{mb+nd} = \dfrac{mb\left(\dfrac{a}{b}\right)+nd\left(\dfrac{c}{d}\right)}{mb+nd} = \dfrac{(mb+nd)\dfrac{a}{b}}{mb+nd} = \dfrac{a}{b}$ or $\dfrac{c}{d}.$

A very important case of this is $m = 1$, $n = \pm 1$, hence

$\dfrac{a}{b} = \dfrac{c}{d} = \dfrac{a+c}{b+d} = \dfrac{a-c}{b-d}$ . . . . (5).

Also $\dfrac{a-b}{a+b} = \dfrac{c-d}{c+d}$ . . . . (6).

For by (2) and (5)

$\dfrac{a}{c} = \dfrac{b}{d} = \dfrac{a-b}{c-d} = \dfrac{a+b}{c+d}$ ∴ $\dfrac{a-b}{a+b} = \dfrac{c-d}{c+d}.$

Or thus: $\dfrac{a-b}{a+b} = \dfrac{\dfrac{a}{b}-1}{\dfrac{a}{b}+1} = \dfrac{\dfrac{c}{d}-1}{\dfrac{c}{d}+1} = \dfrac{c-d}{c+d}$

Generally, to prove that if $\dfrac{a}{b} = \dfrac{c}{d}$, any fraction whose numerator and denominator are homogeneous functions of a and b, and are of the same degree, will be equal to a similar fraction formed with c instead of a and d instead of b :—Express the first fraction in terms of $\dfrac{a}{b}$, and for $\dfrac{a}{b}$ substitute its equivalent $\dfrac{c}{d}$, and reduce the result.

By (2) the fractions may be formed of a and c, and b and d.

If $\dfrac{a}{b} = \dfrac{c}{d} = \dfrac{e}{f}$, $\dfrac{ma+nc+pe}{mb+nd+pf} = \dfrac{a}{b}$ or $\dfrac{c}{d}$ or $\dfrac{e}{f}$ ......(7)

$$\dfrac{ma+nc+pe}{mb+nd+pf} = \dfrac{mb\left(\dfrac{a}{b}\right) + nd\left(\dfrac{c}{d}\right) + pf\left(\dfrac{e}{f}\right)}{mb+nd+pf}$$

$$= \dfrac{(mb+nd+pf)\dfrac{a}{b}}{mb+nd+pf} = \dfrac{a}{b}.$$

If $\dfrac{a}{b} = \dfrac{c}{d}$ and $\dfrac{m}{n} = \dfrac{p}{q}$

$\dfrac{ma \pm pc}{nb \pm qd} = \dfrac{pa \pm mc}{qb \pm nd} = \dfrac{ma}{nb}$ or $\dfrac{pa}{qb}$, or &c. . . . . (8)

For $\dfrac{ma}{nb} = \dfrac{pc}{qd} = \dfrac{ma \pm pc}{nb \pm qd}$ by (5)

$\dfrac{pa}{qb} = \dfrac{mc}{nd} = \dfrac{pa \pm mc}{qb \pm nd}$.

But $\dfrac{ma}{nb} = \dfrac{pa}{qb}$, hence the equality stated in (8).

If $\dfrac{a}{b} = \dfrac{c}{d} = \dfrac{e}{f}$ and $\dfrac{m}{n} = \dfrac{p}{q} = \dfrac{r}{s}$,

$\dfrac{ma \pm pc \pm re}{nb \pm qd \pm sf} = \dfrac{pa \pm rc \pm me}{qb \pm sd \pm nf} = $ &c., $= \dfrac{ma}{nb} = $ &c. . (9).

If an upper sign be taken in a numerator, the corresponding upper sign must be taken in the denominator; if a lower sign, the corresponding lower sign, otherwise all the signs are independent of each other.

EXAMPLES.

1. If $\dfrac{a}{b} = \dfrac{c}{d}$, show that $\dfrac{5a - 4b}{7a + 5b} = \dfrac{5c - 4d}{7c + 5d}$.

The given fraction $= \dfrac{5\dfrac{a}{b} - 4}{7\dfrac{a}{b} + 5} = \dfrac{5\dfrac{c}{d} - 4}{7\dfrac{c}{d} + 5} = \dfrac{5c - 4d}{7c + 5d}$.

2. If $\dfrac{a}{b} = \dfrac{c}{d}$ shew that $\dfrac{2a^3 + 3a^2 b}{3a^2 b - 4b^3} = \dfrac{2c^3 + 3c^2 d}{3c^2 d - 4d^3}$.

Dividing the given fraction by b^3 we have

$\dfrac{2\dfrac{a^3}{b^3} + 3\dfrac{a^2}{b^2}}{3\dfrac{a^2}{b^2} - 4}$, and this becomes, on substituting for $\dfrac{a}{b}$ its equal $\dfrac{c}{d}$,

$\dfrac{2\dfrac{c^3}{d^3} + 3\dfrac{c^2}{d^2}}{3\dfrac{c^2}{d^2} - 4} = \dfrac{2c^3 + 3c^2 d}{3c^2 d - 4d^3}$.

3. If $3a = 2b$, find the value of $\dfrac{a^3 + b^3}{a^2 b - ab^2}$. This $= \left(\dfrac{a^3}{b^3} + 1\right) \div \left(\dfrac{a^2}{b^2} - \dfrac{a}{b}\right)$ [by dividing both numerator and denominator by b^3]. But from the given relation $\dfrac{a}{b} = \dfrac{2}{3}$ we have, by substituting for $\dfrac{a}{b}$,

$(\tfrac{8}{27} + 1) \div (\tfrac{4}{9} - \tfrac{2}{3}) = 35 \div (-6) = -5\tfrac{5}{6}$.

4. If $\dfrac{a}{b} = \dfrac{c}{d}$. Prove that $\dfrac{a^3 + b^3}{c^3 + d^3} \times \dfrac{b}{d} = \left(\dfrac{a+b}{c+d}\right)^4$.

We have $\dfrac{a}{c} = \dfrac{b}{d} = \dfrac{a+b}{c+d}$. Also

$\dfrac{a^3 + b^3}{c^3 + d^3} = \dfrac{b^3}{d^3} \left(\dfrac{a^3}{b^3} + 1\right) \div \left(\dfrac{c^3}{d^3} + 1\right) = \dfrac{b^3}{d^3}$, and this multiplied

by $\dfrac{b}{d}$ gives $\dfrac{b^4}{d^4} = \left(\dfrac{a+b}{c+d}\right)^4$.

5. If $\dfrac{x^3+ax^2-bx+c}{x^3-ax^2+bx+c} = \dfrac{x^2+ax-b}{x^2-ax+b}$, shew that $x = \dfrac{b}{a}$.

Multiplying both terms of second fraction by x, it becomes
$\dfrac{x^3+ax^2-bx}{x^3-ax^2+bx}$; now each of the given fractions =
$\dfrac{\text{difference of numerators}}{\text{difference of denominators}}$;

$= \dfrac{c}{c} = 1$ ∴ $x^3+ax-b = x^2-ax+b$

or $2ax = 2b$ ∴ $x = \dfrac{b}{a}$.

6. If $\dfrac{a}{b} = \dfrac{c}{d} = \dfrac{e}{f}$, shew that $\dfrac{ac+ce+ea}{bd+df+fb} = \dfrac{a^2+c^2+e^2}{b^2+d^2+f^2}$.

For $\dfrac{ac}{bd} = \dfrac{ce}{df} = \dfrac{ea}{fb} = \dfrac{ac+ce+ea}{bd+df+fb}$. By (7) making $m=n=p=1$.

Also $\dfrac{a^2}{b^2} = \dfrac{c^2}{d^2} = \dfrac{e^2}{f^2} = \dfrac{a^2+c^2+e^2}{b^2+d^2+f^2}$. By (7).

But $\dfrac{ac}{bd} = \dfrac{a^2}{b^2}$ hence the required equality.

The problem is a particular case of (9), with all the signs $+$ and a for m, b for n, c for p, &c.

(If the fractions *given* equal to one another have not monomial terms, instead of seeking to express the proposed quantity in terms of one fraction and then substituting an equivalent fraction, it is often better to assume a single letter to represent the common value of the fractions given equal, and to work in terms of this assumed letter.)

7. If $\dfrac{a+b}{3(a-b)} = \dfrac{b+c}{4(b-c)} = \dfrac{c+a}{5(c-a)}$,

prove that $32a+35b+27c = 0$.

Assume each of the given fractions $= x$, so that $a+b = 3(a-b)x$, $b+c = 4(b-c)x$, $c+a = 5(c-a)x$, or $\dfrac{a+b}{3} + \dfrac{b+c}{4} + \dfrac{c+a}{5} = x(a-b+b-c+c-a) = 0.$

∴ adding these fractions we have $32a+35b+27c=0.$

This example might also be worked as a particular case of (7), thus

$$\dfrac{a+b}{3(a-b)} = \dfrac{b+c}{4(b-c)} = \dfrac{c+a}{5(c-a)}$$

$$= \dfrac{20(a+b)+15(b+c)+12(c+a)}{60(a-b)+60(b-c)+60(c-a)} = \dfrac{32a+35b+27c}{0},$$

∴ $32a+35b+27c = 0 \times \dfrac{a+b}{3(a-b)} = 0.$

8. If $\dfrac{a^2}{b^2} + \dfrac{c^2}{f^2} = \dfrac{2c}{d}\left\{\dfrac{a}{b} - \dfrac{c}{d} + \dfrac{e}{f}\right\}$, prove that

$$\left(\dfrac{a+c+e}{b+d+f}\right)^2 = \dfrac{a^2+c^2+e^2}{b^2+d^2+f^2}.$$

Transposing terms, &c., we have

$$\dfrac{a^2}{b^2} - \dfrac{2ac}{bd} + \dfrac{c^2}{d^2} + \dfrac{e^2}{f^2} - \dfrac{2ce}{df} + \dfrac{c^2}{d^2} = 0,$$

or $\left(\dfrac{a}{b} - \dfrac{c}{d}\right)^2 + \left(\dfrac{e}{f} - \dfrac{c}{d}\right)^2 = 0;$

that is, the sum of two essentially positive quantities $= 0$;

∴ each of them must $= 0$; hence we have

$$\dfrac{a}{b} - \dfrac{c}{d} = 0, \text{ and } \dfrac{e}{f} - \dfrac{c}{d} = 0;$$

∴ $\dfrac{a}{b} = \dfrac{c}{d} = \dfrac{e}{f}$; ∴ $\dfrac{a^2}{b^2} = \dfrac{a^2+c^2+e^2}{b^2+d^2+f^2}.$

Also $\dfrac{a}{b} = \dfrac{a+c+e}{b+d+f}$; ∴ $\dfrac{a^2}{b^2} = \left(\dfrac{a+c+e}{b+d+f}\right)^2$;

∴ $\left(\dfrac{a+c+e}{b+d+f}\right)^2 = \dfrac{a^2+c^2+e^2}{b^2+d^2+f^2}.$

Exercise xlvi.

1. If $\dfrac{a}{b} = \dfrac{c}{d}$, prove $\dfrac{a^2-ab+b^2}{ab-4b^2} = \dfrac{c^2-cd+d^2}{cd-4d^2}$.

2. If $\dfrac{a}{b} = \dfrac{c}{d}$, prove $\dfrac{a^3-c^3}{b^3-d^3} = \left(\dfrac{a-c}{b-d}\right)^3 = \left(\dfrac{a+c}{b+d}\right)^3$.

3. Given the same, shew that each of these fractions
$$= \sqrt{\left(\dfrac{a^2+c^2}{b^2+d^2}\right)}.$$

4. If $2x = 3y$, write down the value of
$$\dfrac{2x^3 - x^2y + y^3}{x^2y + xy^2 + 2y^3}, \text{ and of } \dfrac{x^4 - 3x^3y + 2y^4}{(x^2-y^2)^2}.$$

5. If $\dfrac{a}{b} = \dfrac{c}{d} = \dfrac{e}{f}$, shew that $\dfrac{a}{b} = \dfrac{ma-nc-pe}{mb-nd-pf}$.

6. From the same relations prove that $\dfrac{a^3}{b^3} = \left(\dfrac{a-mc-ne}{b-md-nf}\right)^3$.

7. If $\dfrac{1+x}{1-x} = \dfrac{b}{a}\left(\dfrac{1+x+x^2}{1-x+x^2}\right)$, then $x^3 = (b-a) \div (b+a)$.

8. If $\dfrac{\sqrt{(a+x)}+\sqrt{(a-x)}}{\sqrt{(a+x)}-\sqrt{(a-x)}} = a$, prove that $x = \dfrac{2a^2}{1+a^2}$.

9. If $\dfrac{mx+a+b}{nx+a+c} = \dfrac{mx-c-d}{nx-b-d}$, prove $x = \dfrac{b-c}{n-m}$.

10. If $\dfrac{a-b}{ay+bx} = \dfrac{b-c}{bz+cx} = \dfrac{c-a}{cy+az} = \dfrac{a+b+c}{ax+by+cz}$,

then each of these fractions $= \dfrac{1}{x+y+z}$, $a+b+c$ not being zero.

11. If $\dfrac{a+b}{a-b} = \dfrac{b+c}{2(b-c)} = \dfrac{c+a}{3(c-a)}$, then $8a + 9b + 5c = 0$.

12. If $\dfrac{\sqrt{a}+\sqrt{(a-x)}}{\sqrt{a}-\sqrt{(a-x)}} = \dfrac{1}{a}$, shew that $\dfrac{a-x}{a} = \left(\dfrac{1-a}{1+a}\right)^2$.

13. If $\dfrac{x^2-yz}{x(1-yz)} = \dfrac{y^2-zx}{y(1-zx)}$, and x, y, z be unequal, shew that each of these fractions is equal to $x+y+z$.

14. If $\dfrac{x^2+2x+1}{x^2-2x+3} = \dfrac{y^2+2y+1}{y^2-2y+3}$, shew that each of these fractions $= (xy-1) \div (xy-3)$.

15. If $\dfrac{25x^2-16}{10x+8} = \dfrac{3(x^2-4)}{2x-4}$, shew that $\dfrac{x-4}{x+5} = \dfrac{3}{5}$.

16. If $y = \dfrac{4bc}{b+c}$ shew that $\dfrac{y+2b}{y-2b} + \dfrac{y+2c}{y-2c} = 2$.

17. If $\dfrac{1}{4}\left(\dfrac{a^2+b^2}{a^2-b^2}\right) = \dfrac{1}{5}\left(\dfrac{b^2+c^2}{b^2-c^2}\right) = \dfrac{1}{6}\left(\dfrac{c^2+a^2}{c^2-a^2}\right)$, prove that $25a^2 + 27b^2 + 22c^2 = 0$.

18. If $\dfrac{a^2}{x^2-yz} = \dfrac{b^2}{y^2-zx} = \dfrac{c^2}{z^2-xy}$, shew that $a^2x + b^2y + c^2z = (a^2+b^2+c^2)(x+y+z)$.

19. If $\dfrac{x}{a+b-c} = \dfrac{y}{b+c-a} = \dfrac{z}{c+a-b}$, then will $(a-b)x + (b-c)y + (c-a)z = 0$.

20. If $\dfrac{a}{b} = \dfrac{c}{d} = \dfrac{e}{f}$ then $\left(\dfrac{a^2+c^2+e^2}{b^2+d^2+f^2}\right)^2 = \dfrac{a^4+c^4+e^4}{b^4+d^4+f^4}$.

21. If $\dfrac{bx+ay}{a-b} = \dfrac{cy+bz}{b-c} = \dfrac{az+cx}{c-a}$, shew that $(a+b+c)(x+y+z) = ax+by+cz$.

22. If $\dfrac{x^3 - 5x^2a - a^3 + 5xa^2}{x^5 + x^2a + xa^2 + a^3} = \dfrac{x-a}{x+a}$, shew that each of these expressions $= 1$.

23. If $\dfrac{1}{6}\left(\dfrac{a-b}{a+b}\right) = \dfrac{1}{5}\left(\dfrac{b-c}{b+c}\right) = \dfrac{1}{10}\left(\dfrac{c-a}{c+a}\right)$, and a, b, c be different, shew that $16a + 11b + 15 = 0$.

24. If $\left(\dfrac{x+yz}{y+zx}\right)^2 = \dfrac{1-y^2}{1-x^2}$, prove that $x^2 + y^2 + z^2 + 2xyz = 1$.

25. If $\dfrac{a}{x-y} = \dfrac{b}{y-z} = \dfrac{c}{z-x}$, shew that $a + b + c = 0$.

26. If $\dfrac{a}{b} = \dfrac{c}{d}$, prove that $\dfrac{a+b}{a-b} = \dfrac{\sqrt{(ac)} + \sqrt{(bd)}}{\sqrt{(ac)} - \sqrt{(bd)}}$.

27. If $\dfrac{a}{b} = \dfrac{c}{d} = \dfrac{e}{f}$, then each is equivalent to $\dfrac{la+mc+ne}{lb+md+nf}$, hence shew that

$$\dfrac{a}{2z+2x-y} = \dfrac{b}{2x+2y-z} = \dfrac{c}{2y+2z-x}, \text{ when}$$

$$\dfrac{x}{2a+2b-c} = \dfrac{y}{2b+2c-a} = \dfrac{z}{2c+2a-b}.$$

28. If $\dfrac{a}{b} = \dfrac{c}{d}$, prove that $\left(\dfrac{a-b}{c-d}\right)^n = \sqrt{\left(\dfrac{a^{2n}+b^{2n}}{c^{2n}+d^{2n}}\right)}$.

29. If $\dfrac{x}{a(y+z)} = \dfrac{y}{b(x+z)} = \dfrac{z}{c(x+y)}$, prove that

$$\dfrac{x}{a}(y-z) + \dfrac{y}{b}(z-x) + \dfrac{z}{c}(x-y) = 0.$$

30. If $\dfrac{a}{lx(ny-mz)} = \dfrac{b}{my(lz-nx)} = \dfrac{c}{nz(mx-ly)}$, then will

$$\dfrac{a}{lx}(l-x) + \dfrac{b}{my}(m-y) + \dfrac{c}{nz}(n-z) = 0.$$

31. If $z = \dfrac{\sqrt{(ay^2-a^2)}}{y}$, and $y = \dfrac{\sqrt{(ax^2-a^2)}}{x}$, shew that

$$x = \dfrac{\sqrt{(az^2-a^2)}}{z}.$$

32. If $\dfrac{x^2-yz}{a^2} = \dfrac{y^2-xz}{b^2} = \dfrac{z^2-xy}{c^2} = 1$, shew that

$$x+y+z = \dfrac{a^2x+b^2y+c^2z}{a^2+b^2+c^2}.$$

33. If $\dfrac{m}{x} = \dfrac{n}{y} = \dfrac{r}{z}$, and $\dfrac{x^2}{a^2} = \dfrac{y^2}{b^2} = \dfrac{z^2}{c^2} = 1$,

prove that $\dfrac{m^2}{a^2} + \dfrac{n^2}{b^2} + \dfrac{r^2}{c^2} = 3\dfrac{m^2+n^2+r^2}{x^2+y^2+z^2}.$

34. If $\dfrac{a}{b} = \dfrac{c}{d} = \dfrac{e}{f} = \&c.$, then

$$\dfrac{a^{3n}-c^{3n}}{b^{3n}-d^{3n}} = \dfrac{a^n c^n e^n - (a^n - c^n + e^n)^3}{b^n d^n f^n - (b^n - d^n + f^n)^3}$$

35. If $\dfrac{a_1}{b_1} = \dfrac{a_2}{b_2} = \dfrac{a_3}{b_3} = \ldots = \dfrac{a_n}{b_n}$, then

$$\dfrac{a_1 a_2 - a_2 a_3 + \ldots (-1)^{n-1} a_{n-1} a_n}{b_1 b_2 - b_2 b_3 + \ldots (-1)^{n-1} b_{n-1} b_n} = \dfrac{a_1 \sqrt{a_2 a_3} + a_2 \sqrt{a_3 a_4} + \&c}{b_1 \sqrt{b_2 b_3} + b_2 \sqrt{b_3 b_4} + \&c}$$

36. If $\dfrac{A+B+C}{abc} = \dfrac{A}{a} + \dfrac{B}{b} + \dfrac{C}{c}$,

and $(A+B+C)(a+b+c) = Aa + Bb + Cc$,

then will $\dfrac{A}{1+a^2} + \dfrac{B}{1+b^2} + \dfrac{C}{1+c^2} = 0$.

and also $\dfrac{A}{a+\dfrac{1}{a}} + \dfrac{B}{b+\dfrac{1}{b}} + \dfrac{C}{c+\dfrac{1}{c}} = 0$.

37. If $\dfrac{xh}{a^2} = \dfrac{yk}{b^2} = \dfrac{zl}{c^2}$, and $\dfrac{x^2}{a^2} = \dfrac{y^2}{b^2} = \dfrac{z^2}{c^2} = 1$,

then will $\left(\dfrac{x}{h} + \dfrac{y}{k} + \dfrac{z}{c}\right)^2 = \dfrac{a^2}{l h^2} + \dfrac{b^2}{k^2} + \dfrac{c^2}{l^2}$.

SECTION IV.—COMPLETE SQUARES, &c.

1. What quantity must be added to $x^2 + px$ to make it a complete square?

Let r be the quantity.
Then $x^2 + px + r =$ complete square $= (x + \sqrt{r})^2$
$= x^2 + 2x\sqrt{r} + r$.
Equating coefficients we have
$$2\sqrt{r} = p$$
$$\therefore r = \dfrac{p^2}{4} = \left(\dfrac{p}{2}\right)^2$$

Or thus: Since $(a+x)^2 = a^2 + 2ax + x^2$; we observe, (See Art. XII), that *four times the product of the extremes is equal to the square of the mean*,
$$\therefore 4x^2 r = p^2 x^2;$$
$$\therefore r = \left(\dfrac{p}{2}\right)^2, \text{ as before.}$$

COMPLETE SQUARES.

Or we may extract the square root and equate the remainder to zero. thus

$$\begin{array}{r|l} & x^2 + px + r(x + \frac{p}{2} \\ & \underline{x^2} \\ 2x + \frac{p}{2} & px + r \\ & px + \frac{p^2}{4} \\ \hline & r - \frac{p^2}{4}. \end{array}$$

Now, if the expression be a complete square, this remainder must vanish; hence we have

$$r = \frac{p^2}{4} = \left(\frac{p}{2}\right)^2$$

2. Find the relation connecting a, b, c, if $ax^2 + bx + c$ is a complete square.

Assume $ax^2 + bx + c = (\sqrt{a}.x + \sqrt{c})^2 = ax^2 + 2\sqrt{(ac)}.x + c$.

Now, since this holds for all values of x, we have $2\sqrt{ac} = b$, or $b^2 = 4ac$, the relation required.

3. Determine the relation amongst a, b, c, in order that
$a^2x^2 + bx + bc + b^2$ may be a perfect square.

As in Ex. 1, we have $4a^2x^2(bc + b^2) = b^2x^2$;

$$\therefore \frac{1}{4a^2} - \frac{c}{b} = 1.$$

Or thus:

Assume $a^2x^2 + bx + bc + b^2 = (ax + \sqrt{bc + b^2})^2$
$= a^2x^2 + 2a\sqrt{bc + b^2} + bc + b^2$.

Equating coefficients, we have $b = 2a\sqrt{bc + b^2}$;

$$\therefore \frac{1}{4a^2} - \frac{c}{b} = 1, \text{ as before.}$$

The same result may also be obtained by extracting the square root and equating the remainder to zero.

4. Show that if $x^4+ax^3+bx^2+cx+d$ be a complete square, the coefficients satisfy the equation $c^2-a^2d=0$.

Is it necessary that the coefficients satisfy any other equation?

Extracting the square root of $x^4+ax^3+bx^2+cx+d$ in the usual manner, we have for the final remainder

$$\left\{c-\frac{a}{2}\left(b-\frac{a^2}{4}\right)\right\}x+d-\frac{1}{4}\left(b-\frac{a^2}{4}\right)^2.$$

Now, if the expression be a complete square, this remainder must vanish; and, that it may vanish for general values of x, we must have

$$c-\frac{a}{2}\left(b-\frac{a^2}{4}\right)=0 \quad \ldots \ldots \ldots \quad (1),$$

$$d-\frac{1}{4}\left(b-\frac{a^2}{4}\right)^2=0 \quad \ldots \ldots \ldots \quad (2);$$

Eliminating $b-\dfrac{a^2}{4}$, we have $c^2-a^2d=0$ . . . (3).

The coefficients must satisfy the equations (1) and (2), and therefore either of these equations, together with the equation (3), which results from them.

The same result may be obtained by assuming
$$x^4+ax^3+bx^2+cx+d=(x^2+\tfrac{1}{2}ax+\sqrt{d})^2$$
$$=x^4+ax^3+2x^2\sqrt{d}$$
$$+\tfrac{1}{4}a^2x^2+ax\sqrt{d}+d.$$

Equating coefficients, we have $2\sqrt{d}+\tfrac{1}{4}a^2=b$ . . . (1)
and $a\sqrt{d}=c$ . . . (2).

From (2) we have $c^2-a^2d=0$, as before.

5. What must be the value of m and n if
$4x^4-12x^3+25x^2-4mx+8n$ is a perfect square?

Assume the expression $=\{(2x^2-3x+\sqrt{(8n)})\}^2$
$$=4x^4-12x^3+4x^2\sqrt{(8n)}+9x^2-6x\sqrt{(8n)}+8n.$$

Equating coefficients, we have $6\sqrt{(8n)}=4m$ . . . (1),
and $4\sqrt{(8n)}+9=25$ . . . . (2);
$$\therefore n=2,$$
$$m=6.$$

Or thus: Extracting the square root in the ordinary way, the remainder is found to be $(-4m+24)x+8n-16$; $\therefore$ we must have
$$4m+24=0, \text{ or } m=6,$$
$$\text{and } 8n-16=0, \text{ or } n=2.$$

6. If ax^3+bx^2+cx+d be a complete cube, shew that $ac^3 = db^3$, and $b^2 = 3ac$.

Assume $ax^3+bx^2+cx+d = (x+d^{\frac{1}{3}})^3$
$$= a x^3 + 3a^{\frac{2}{3}}d^{\frac{1}{3}}x^2 + 3a^{\frac{1}{3}}d^{\frac{2}{3}}x+d$$

Equating coefficients,
$$b = 3a^{\frac{2}{3}}d^{\frac{1}{3}} \quad \ldots \ldots \ldots \ldots (1)$$
$$c = 3a^{\frac{1}{3}}d^{\frac{2}{3}} \quad \ldots \ldots \ldots \ldots (2);$$

dividing (1) by (2), $\dfrac{b}{c} = \dfrac{a^{\frac{1}{3}}}{d^{\frac{1}{3}}}$;
$$ac^3 = db^3.$$

Also, $\quad b^2 = 9a^{\frac{2}{3}}d^{\frac{2}{3}} \quad \ldots \ldots \ldots \ldots (3);$

dividing (3) by (2), $\dfrac{b^2}{c} = 3a$;
$$\therefore b^2 = 3ac.$$

7. Find the relations subsisting between a, b, c, d, e, when $ax^4+bx^3+cx^2+dx+e$ is a complete *fourth* power.

Assume $ax^4+bx^3+cx^2+dx+e = (a^{\frac{1}{4}}x+e^{\frac{1}{4}})^4$
$$= ax^4+4a^{\frac{3}{4}}e^{\frac{1}{4}}x^3+6a^{\frac{1}{2}}e^{\frac{1}{2}}x^2+4a^{\frac{1}{4}}e^{\frac{3}{4}}x+e.$$

Equating coefficients, we have
$$b = 4a^{\frac{3}{4}}e^{\frac{1}{4}},$$
$$c = 6a^{\frac{1}{2}}e^{\frac{1}{2}},$$
$$d = 4a^{\frac{1}{4}}e^{\frac{3}{4}};$$

whence $bd = 16ae.$ $\quad \ldots \ldots \ldots \ldots (1).$
$$bc = 24a^{\frac{5}{4}}e^{\frac{3}{4}} = 6a.4a^{\frac{1}{4}}e^{\frac{3}{4}} = 6ad. \quad \ldots (2).$$
$$cd = 24a^{\frac{3}{4}}e^{\frac{5}{4}} = 6e.4a^{\frac{3}{4}}e^{\frac{1}{4}} = 6be. \quad \ldots (3).$$

8. Shew that $x^4+px^3+qx^2+rx+s$ can be so resolved into two rational quadratic factors if s be a perfect square, negative, and equal to $\dfrac{r^2}{p^2-4q}$.

Since $-s$ is a perfect square, let it be n^2.
Assume $x^4 + px^3 + qx^2 + rx - n^2$
$$= (x^2 + mx + n)(x^2 + m'x - n)$$
$$= x^4 + (m+m')x^3 + mm'x^2 - n(m-m')x - n^2.$$
Equating coefficients, we have
$$m + m' = p$$
$$mm' = q$$
$$m - m' = \frac{r}{n}$$
$$m^2 + 2mm' + m'^2 = p^2$$
$$4mm' = 4q \ ;$$
$$\therefore (m-m')^2 = p^2 - 4q = \frac{r^2}{n^2}$$
$$\therefore \frac{r^2}{p^2 - 4q} = n^2 = -s.$$

Exercise xlvii.

1. What is the condition that $(a-x)(b-x) - c^2$, may be a perfect square.

2. Find the value of n which will make $2x^2 + 8x + n$, a perfect square.

3. Find a value of x which will make $x^4 + 6x^3 + 11x^2 + 3x + 31$, a perfect square.

4. Extract the square root of
$$(a-b)^4 - 2(a^2 + b^2)(a-b)^2 + 2(a^4 + b^4)$$

5. Find the values of m and n which will make $4x^4 - 4x^3 + 5x^2 - mx + n$, a perfect square.

6. What must be added to $x^4 - \sqrt{(4x^4 - 16x^2 + 16)} - 4x^2$ in order to make it a complete square?

7. The expression $x^4 + x^3 - 16x^2 - 4x + 48$, is resolvable into two factors of the form $x^2 + mx + 6$, and $x^2 + nx + 8$; determine the factors.

8. Find the value of c which will make $4x^4 - cx^3 + 5x^2 + \frac{cx}{2} + 1$, a complete square.

9. Obtain the square root of
$4\{(a^2 - b^2)cd + ab(c^2 - d^2)\}^2 + \{(a^2 - b^2)(c^2 - d^2) - 4abcd\}^2$.

10. If $(a-b)x^2 + (a+b)^2 x + (a^2 - b^2)(a+b)$, is a complete square, then $a = 3b$, or $b = 3a$.

11. Find the simplest quantity which, subtracted from $a^2 x^2 + 4abx + 4acx + 5bc + b^2 c^2$, will give for remainder an exact square.

12. $x^4 - 4x^3 - x^2 + 16x - 12$ is resolvable into quadratic factors of the form $x^2 + mx + p$, and $x^2 + nx + q$: find them.

13. Find the values of m which will make $x^2 + max + a^2$ a factor of $x^4 - ax^3 + a^2 x^2 - a^3 x + a^4$.

14. Shew that if $x^4 + ax^3 + bx^2 + cx + d$ be a perfect square, the coefficients satisfy the relations
$$8c = a(4b - a^2), \text{ and}$$
$$64d = (4b - a^2)^2.$$

15. Investigate the relations between the coefficients in order that $ax^2 + by^2 + cz^2 + dxy + eyz + fxz$ may be a complete square.

16. If $x^3 + ax^2 + bx + c$ is exactly divisible by $(x+d)^2$, shew that
$$\tfrac{1}{2}(b - d^2) - \frac{c}{d} = d(a - 2d)$$

17. Determine the relations among a, b, c, d, when $ax^3 - bx^2 + cx - d$, is a complete cube.

18. The polynome $ax^3 + 3bx^2 + 3cx + d$ is exactly divisible by $(a-x)^2$; shew that $(ad - bc)^2 = 4(ac - b^2)(bd - c^2)$.

19. Find the relation between p and q, when $x^3 + px^2 + q$, is exactly divisible by $(x-a)^2$.

20. If $x^2 + nax + a^2$ is a factor of $x^4 + ax^3 + a^2 x^2 + a^3 x + a^4$, shew that $n^2 - n - 1 = 0$.

21. If $x^4 + ax^3 + bx^2 + cx + d$, be the product of two complete squares, shew that
$$(4b - a^2)^2 = 64d, \ (4b - a^2)a = 8c, \ a\sqrt{(3a^2 - 2b)} = 3b.$$

22. Prove that $x^4+px^3+qx^2+rx+s$ is a perfect square, if $p^2s=r^{\bar{2}}$, and $q=\dfrac{p^2}{4}+2\sqrt{s}$.

23. If $ax^3+3bx^2+3cx+d$ contain $ax^2+2bx+c$ as a factor, the former will be a complete cube, and the latter a complete square.

24. If $m^2x^2+px+pq+q^2$ be a perfect square, find p in terms of m, q, and x.

25. Find the relation between p and q in order that x^3+px^2+qx+r may contain $(x+2)^2$ as a factor.

26. If x^3+px^2+qx+r be algebraically divisible by $3x^2+2px+q$, shew that the quotient is $x+\dfrac{p}{3}$.

Relation in Involution.

Art. XXXVIII. If $aa'=bb'=cc'$, prove that

1. $(a+b')(b+c')(c+a')=(a'+b)(b'+c)(c'+a)$
$(a+b')\times a'=aa'+b'a'=bb'+b'a'=(b+a')\times b'$
$(b+c')\times b'=bb'+c'b'=cc'+c'b'=(c+b')\times c'$
$(c+a')\times c'=cc'+a'c'=aa'+a'c'=(a+c')\times a'$
$\therefore (a+b')(b+c')(c+a')\times a'b'c'=$
$\qquad (a'+b)(b'+c)(c'+a)\times b'c'a'$
$\therefore (a+b')(b+c')(c+a')=(a'+b)(b'+c)(c'+a)$.

2. $(a+b)(a+b')(a'-c)(a'-c')=(a'+b)(a'+b')(a-c)(a-c')$.
$(a+b)\times a'=aa'+a'b=bb'+a'b=(b'+a')\times b$
$(a+b')\times a'=aa'+a'b'=bb'+a'b'=(b+a')\times b'$
$(a'-c)\times a=aa'-ac=cc'-ac=(c'-a)\times c$
$(a'-c')\times a=aa'-ac'=cc'-ac'=(c-a)\times c'$
$\therefore (a+b)(a+b')(a'-c)(a'-c')\times (aa')^2=$
$\qquad (b'+a')(b+a')(c'-a)(c-a)\times bb'.cc'$
But $bb'.cc'=(aa')^2$,
and $(c'-a)(c-a)=(a-c)(a-c')$
$\therefore (a+b)(a+b')(a'-c)(a'-c')=(a'+b)(a'+b')(a-c)(a-c')$.

Exercise xlviii.

If $aa' = bb' = cc'$ prove that

1. $(a-b')(b-c)(c'-a') = (b-a')(a-c)(c'-b')$.
2. $(b-c')(c-a)(a'-b') = (c-b')(b-a)(a'-c')$.
3. $(c-a')(a-b)(b'-c') = (a-c')(c-b)(b'-a')$.
4. $(a-b')(b-c')(c-a') = (a-c')(b-a')(c-b')$.
5. $\dfrac{(a-b)(a-b')}{(a'-b)(a'-b')} = \dfrac{(a-c)(a-c')}{(a'-c)(a'-c')}$,
6. $\dfrac{(b-c)(b-c')}{(b'-c)(b'-c')} = \dfrac{(b-a)(b-a')}{(b'-a)(b'-a')}$,
7. $\dfrac{(c-a)(c-a')}{(c'-a)(c'-a')} = \dfrac{(c-b)(c-b')}{(c'-b)(c'-b')}$.

8. Shew that the seven preceding relations may be derived from the single relation

$$(a+a')(bb'-cc') + (b+b')(cc'-aa') + (c+c')(aa'-bb') = 0.$$

CHAPTER V.

SIMPLE EQUATIONS OF ONE UNKNOWN QUANTITY.

Art. XXXIX. Preliminary Equations. Although the following exercise belongs in theory to this chapter, in practice the *numerical* examples should immediately follow Exercise I., and the literal examples Exercise III. Like those exercises, this one is merely a specimen of what the teacher should give till his pupils have thoroughly mastered this preliminary work. But few numerical examples are given, it being left to the teacher to supply these.

Exercise xlix.

What values must x have that the following equations may be true?

1. $x-5=0.\quad x-3\frac{1}{4}=0.\quad x-a=0.\quad x+3=0.$
2. $x+4\frac{1}{2}=0.\quad x+a=0.\quad x+3=5.\quad x-4=6.$
3. $x-a=b.\quad x+a=c.\quad x-b=-c.\quad 6-x=3.$
4. $8-x=10.\quad 5+x=11.\quad 9+x=4.\quad 7-x=-5.$
5. $8+x=-6.\quad a-x=3b.\quad 2a=x+3b.\quad 3a=5b-x.$
6. $2x-6=8.\quad 3x+8=20.\quad ax=a^2.\quad mx=bm.$
7. $3x=c.\quad ax=5.\quad ax=0.\quad (a+b)x=b+a.$
8. $(a-b)x=b-a.\quad (a+bx)=(a+b)^2.\quad (a-b)x=a^2-b^2.$
9. $(a+b)x=b^2-a^2.\quad (a^2-ab+b^2)x=a^3+b^3.$
10. $(a^2-b^2)x=a-b.\quad (a^2-b^2)x=a+b.\quad (a^2+b^2)x=1.$
11. $(a+x-b)=(a+b).\quad x-a+b=b-x+a.$
12. $2a-x=x-2b.\quad ax+b\acute{x}=c.\quad ax-b=cx.$
13. $ax-b=bx-c.\quad ax-ab=ac.$
14. $ax-a^2=bx-b^2.\quad ax-a^3=bx-b^3.$

SIMPLE EQUATIONS.

15. $ax - a^3 = b^3 - bx$; $\quad ax + b + c = a + bx + cx$.
16. $a - bx - c = b - ax + cx$; $\quad a + bx + cx^2 = ax - b + cx^2$.
17. $bx - cx^2 + e = ex - b - cx^2$; $\quad 3x = \frac{2}{5}$; $\quad 4x = \frac{4}{7}$.
18. $10x = \frac{1}{6} - 1$; $\quad ax = \frac{b}{c}$; $\quad ax = \frac{a^2}{b}$.
19. $abx = \frac{a}{b} + \frac{b}{a}$; $\quad bcx = \frac{ac^2}{b} + \frac{ab^2}{c}$.
20. $\frac{1}{2}x = 5$; $\quad \frac{2}{3}x = 8$; $\quad \cdot 5x = 2$; $\quad \cdot 3x = \cdot 06$.
21. $\cdot 02x = 20$; $\quad \cdot \dot{3}x = \cdot 2$; $\quad \cdot \dot{4}x = \cdot \dot{6}$.
22. $\cdot 1\dot{8}x = 1 \cdot 8$; $\quad \frac{x}{a} = b$; $\quad \frac{ax}{b} = c$.
23. $\frac{ax}{b} = \frac{b}{c}$; $\quad \frac{x}{a+b} = c$; $\quad \frac{ax}{a+b} = b$.
24. $\frac{a+b}{a-b}x = \frac{a}{b}$; $\quad \frac{a-b}{a+b}x = \frac{a+b}{b-a}$.
25. $\frac{a}{b-a}x = \frac{a}{a-b}$; $\quad \frac{b-a}{a+b}x = \frac{a-b}{b+a}$.
26. $\frac{a+b}{a+c}x = \frac{a-c}{a+b}$; $\quad \frac{1}{x} = \frac{1}{2}$; $\quad \frac{2}{x} = \frac{3}{5}$.
27. $\frac{1}{x} = \frac{1}{ab}$; $\quad \frac{1}{x} = \frac{a}{b}$; $\quad \frac{a}{x} = \frac{b}{c}$; $\quad \frac{7}{x} = \frac{1}{3} + \frac{1}{4}$.
28. $\frac{3}{20} + \frac{4}{5x} = \frac{33}{5x} - \frac{1}{3}$; $\quad \frac{a}{x} + \frac{b}{c} = 0$.
29. $\frac{5}{x-7} = 6 - \frac{7}{x-7}$; $\quad \frac{5}{3x-4} = 7 + \frac{9}{4-3x}$.
30. $(x-4) - (x+5) + x = 3$; $\quad 2x - (x-5) - (4-3x) = 5$.
31. $2(3-x) + 3(x-8) = 0$; $\quad 2(3x-4) - 3(3-4x) + 9(2-x) = 10$.
32. $a(1-2x) - (2x-a) = 1$; $\quad x - 5(a-x) = bx - 5a$.
33. $mx(3a-4) + 3mx - 3a + 1 = 0$.
34. $a(bx-c) + b(cx-a) + c(ax-b) = 0$.
35. $a(ax-b) + b(cx-c) + c(cx-a) = 0$.

36. $a(bx-a)+b(cx-b)+c(ax-c)=0$.
37. $a(x-2b)+b(x-2c)+c(x-2a)=a^2+b^2+c^2$.
38. $3(3\{3(3x-2)-2\}-2)-2=1$.
39. $9(7\{5(3x-2)-4\}-6)-8=1$.
40. $\frac{1}{3}\{\frac{1}{3}(\frac{1}{3}\{\frac{1}{3}(x+2)+2\}+2)+2\}=1$.
41. $\frac{1}{9}\{\frac{1}{7}(\frac{1}{5}\{\frac{1}{3}(x+2)+4\}+6)+8\}=1$.
42. $\frac{1}{2}(\frac{1}{2}\{\frac{1}{2}(\frac{1}{2}x-\frac{1}{2})-\frac{1}{2}\}-\frac{1}{2})-\frac{1}{2}=0$.
43. $\frac{2}{3}(\frac{2}{3}\{\frac{2}{3}(\frac{2}{3}x-1\frac{1}{3})-1\frac{1}{3}\}-1\frac{1}{3})-1\frac{1}{3}=0$.
44. $\frac{13}{15}\{\frac{9}{11}(\frac{5}{7}\{\frac{3}{4}(\frac{2}{3}x+4)+8\}+12)+20\}+32=58$.
45. $\frac{4}{5}\{\frac{3}{4}(\frac{2}{3}\{\frac{1}{2}(x+7)-3\}+6)-1\}=4$.
46. $r\{q(p\{n(mx-a)-b\}-c)-d\}-e=0$.
47. $(1+6x)^2+(2+8x)^2=(1+10x)^2$.
48. $9(2x-7)^2+(4x-27)^2=13(4x+15)(x+6)$.
49. $(3-4x)^2+(4-4x)^2=2(5+4x)^2$.
50. $(9-4x)(9-5x)+4(5-x)(5-4x)=36(2-x)^2$.

Art. XL. In order that the product of two or more factors may vanish, it is necessary, and it is sufficient, that one of the factors should vanish. Thus, in order that $(x-a)(x-b)$ may $=0$, either $x-a$ must $=0$, or $x-b$ must $=0$, and it is sufficient that one of them should do so.

Hence the single equation $(x-a)(x-b)=0$ is really equivalent to the two disjunctive equations, either $x-a=0$ or $x-b=0$, for either of these will fulfil the condition of the given equation, and that is all that is required.

Similarly, were it required to find what values of x would make the product $(x-a)(x-b)(x-c)$ vanish, they would be given by

$x-a=0$, or $x-b=0$, or $x-c=0$ ∴ $x=a$ or b or c.

Hence the single equation

$$(x-a)(x-b)(x-c)=0$$

is equivalent to the three disjunctive equations

$x-a=0$, or $x-b=0$, or $x-c=0$.

SIMPLE EQUATIONS.

EXAMPLES.

1. Solve $x^2 - x - 20 = 0$.

The expression $= (x-5)(x+4)$, which will vanish if either of its factors does, that is, if $x - 5 = 0$, or $x + 4 = 0$,

$$\therefore x = 5, \quad \text{or } x = -4$$

2. Solve $x^4 - x^3 - x^2 + x = 0$.

This gives $x^3(x-1) - x(x-1) = x(x-1)(x^2-1)$

$= x(x-1)(x+1)(x-1)$, which vanishes for

$x = 0, x = 1, x = -1$.

3. Solve $x^3 + a^2 x^2 - ax - a^3 = 0$.

This $= x(x^2 - a) + a^2(x^2 - a)$

$= (x + a^2)(x^2 - a)$, which vanishes for

$x + a^2 = 0$, and $x^2 - a = 0$, or

$x = -a^2$, and $x^2 = a$.

4. Solve $x^2(a-b) + a^2(b-x) + b^2(x-a) = 0$.

The factors of the expression are (Ex. 2, page 79)

$x - a, x - b, a - b$; hence the expression vanishes if

$x - a = 0$, or $x - b = 0$.

5. Solve $221x^2 - 5x - 6 = 0$.

Here we have the factors $17x - 3$ and $13x + 2$;

$\therefore$ the equation is satisfied by $17x - 3 = 0$, or $x = \frac{3}{17}$,

and $13x + 2 = 0$, or $x = -\frac{2}{13}$.

6. Solve $2x^4 + 2x^3 + 6x - 18 = 0$.

In this case we have $2(x^4 - 9) + 2x(x^2 + 3)$

$= 2(x^2 + 3)\{x^2 - 3 + x\}$, which vanishes for

$x^2 + 3 = 0$, or $x^2 + x - 3 = 0$.

7. Solve $(x-a)^3 + (a-b)^3 + (b-x)^3 = 0$.

The expression is equal to $3(x-a)(a-b)(b-x)$,

and therefore vanishes for $x - a = 0$, or $x = a$;

and for $x - b = 0$, or $x = b$.

Exercise 1.

1. If an equation in x has the factors $2x-4$ and $2x-6$, find the corresponding values of x.

2. If an equation gives the factors $2x-1$ and $3x-1$, what are the corresponding values of x ?

3. If an equation gives the factors $3x^2-12$ and $4x-5$, find the corresponding values of x.

Find the values of x for which the following expressions will vanish;

4. x^2-2x+1; $4x^2-12x+9$.

5. $9x^2-4$; $x^2-(a+b)^2$; $x^2-2ax+a^2$.

6. $x^2-9x+20$; $4x^2-18x+20$.

7. x^2+x-6 : x^2-x-12; $9x^2 \cdot 9x-28$.

8. $6x^2-12x+6$; $6x^2-13x+6$; $6x^2-20x+6$.

9. $6x^2-5x-6$; $6x^2-37x+6$; $6x^2+x-12$.

10. A certain equation of the fourth degree gives the factors x^2-x-2, and $4x^2-2x-2$, find all the values of x.

Find values of x in the following cases :

11. $x^3-2bx^2-3b^2x=0$.

12. $x^3-ax^2+a^2x-a^3=0$.

13. $x^3-2x+1=0$; $x^3-3x+2=0$.

14. $x^4-2ax^3+2a^3x-a^4=0$.

15. $x^3+(b+c)x^2-bcx-b^2c-bc^2=0$.

16. $\dfrac{x-a}{x-b}+\dfrac{x-b}{x-a}-\dfrac{(a-b)^2}{(x-a)(x-b)} = \dfrac{x^2-a^2}{(x-a)(x-b)}$.

17. $x^3-bx^2-a^2x-a^2b=0$.

18. $3x^3+4abx^2-6a^2b^2x-4a^3b^3=0$.

19. $x^3(a-b)+a^3(b-x)+b^3(x-a)=0$.

20. $\dfrac{(x-b)(x-c)}{(a-b)(a-c)}+\dfrac{(x-c)(x-a)}{(b-c)(b-a)}=1$.

21. $x\left(\dfrac{x-2a}{x+a}\right)^3+a\left(\dfrac{2x-a}{x+a}\right)^3=x^2-a^2$.

SIMPLE EQUATIONS. 143

22. $(x+a+b)^3 - x^3 - a^3 - b^3 = (x+a)(a^2-b^2)$.

23. $\dfrac{ab}{(b-a)(x-a)} + \dfrac{bx}{(x-a)(a-b)} + \dfrac{ax}{(a-b)(b-x)} = \dfrac{1}{a-b}$.

24. Form the polynome which will vanish for x equal 5, or -6, or 7.

25. Form the polynome which will vanish for $x=a$, or $4a$, or $3a$, or $-4a$.

26. Form the equation whose roots are 0, 1, -2, and 4.

27. Form the equation whose roots are $1+\sqrt{2}$, $1-\sqrt{2}$, $1-\sqrt{3}$, and $1+\sqrt{3}$.

Art. XLI. In solving fractional equations, the principles illustrated in the section on fractions may frequently be applied with advantage, as in the following cases.

When an equation involves several fractions, we may *take two or more of them together*.

EXAMPLES.

1. Solve $\dfrac{8x+5}{14} + \dfrac{7x-3}{6x+2} = \dfrac{4x+6}{7}$.

Here, instead of multiplying through by the L. C. M. of the denominators, we combine the first fraction with the last, getting at once

$$\dfrac{7x-3}{6x+2} = \dfrac{7}{14} = \dfrac{1}{2} \quad \therefore\ 7x-3 = 3x+1,\ \text{and}\ x=1.$$

2. $\dfrac{2x+8\frac{1}{2}}{9} - \dfrac{13x-2}{17x-32} + \dfrac{x}{3} = \dfrac{7x}{12} - \dfrac{x+16}{36}$.

In this case, taking together all the fractions having only numerical denominators, we get

$$\dfrac{8x+34+12x-21x+x+16}{36} = \dfrac{13x-2}{17x-32};\ \text{or}$$

$$\dfrac{25}{18} = \dfrac{13x-2}{17x-32};$$

$\therefore\ 425x - 800 = 234x - 36$, hence $x = 4$.

It is often advantageous to *complete* the *divisions* represented by the fractions.

3. $\dfrac{4x-17}{9} - \dfrac{3\frac{2}{3} - 22x}{33} = x - \dfrac{6}{x}\left(1 - \dfrac{x^2}{54}\right).$

Here, completing the divisions, we have

$\dfrac{4x}{9} - \dfrac{17}{9} - \dfrac{1}{9} + \dfrac{2x}{3} = x - \dfrac{6}{x} + \dfrac{x}{9},$

$\dfrac{10x}{9} - 2 = x + \dfrac{x}{9} - \dfrac{6}{x} \quad \therefore \quad -2 = -\dfrac{6}{x},$ or $x = 3.$

4. $\dfrac{ax+b}{x-m} + \dfrac{cx+d}{x-n} = a+c$

$\because \quad a + \dfrac{am+b}{x-m} + c + \dfrac{cn+d}{x-n} = a+c$

$\therefore \quad (am+b)(x-n) + (cn+d)(x-m) = 0$

$\therefore \quad (am+b+cn+d)x = (a+c)mn + bn + dm.$

5. Similarly may be solved

$\dfrac{ax+b}{x-m} + \dfrac{cx+d}{x-n} + \dfrac{ex^2+fx-g}{(x-m)(x-n)} = a+c+e$

$\therefore \quad \dfrac{am+b}{x-m} + \dfrac{cn+d}{x-n} + \dfrac{\{e(m+n)+f\}x - emn - g}{(x-m)(x-n)} = 0.$

$\therefore \quad (am+b)(x-n) + (cn+d)(x-m) + \{e(m+n)+f\}x - emn - g = 0.$

$\therefore \quad \{(a+c)m+b+(c+e)n+d+f\}x = (a+b+e)mn + bn + dm + g.$

6. $\dfrac{132x+1}{3x+1} + \dfrac{8x+5}{x-1} = 52;$

$\therefore \quad 44 - \dfrac{43}{3x+1} + 8 + \dfrac{13}{x-1} = 52,$ or

$\dfrac{13}{x-1} = \dfrac{43}{3x+1}; \quad \therefore \quad 39x + 13 = 43x - 43,$ and $x = 14.$

7. $\dfrac{25 - \frac{1}{3}x}{x+1} + \dfrac{16x + 4\frac{1}{3}}{3x+2} = 5 + \dfrac{23}{x+1}.$

Taking the last fraction with the first, and multiplying the resulting equation by 15, we have

$$\frac{240x+63}{3x+2} = 75 + \frac{5x-30}{x+1};$$

$$\therefore 80 - \frac{97}{3x+2} = 75 + 5 - \frac{35}{x+1}, \text{ or}$$

$$\frac{97}{3x+2} = \frac{35}{x+1}; \therefore 8x = 27, \text{ and } x = 3\tfrac{3}{8}.$$

8. $\dfrac{x-a}{b+c} + \dfrac{x-b}{a+c} + \dfrac{x-c}{b+c} = 3.$

$$\therefore \frac{x-a}{b+c} - 1 + \frac{x-b}{a+c} - 1 + \frac{x-c}{b+a} - 1 = 0;$$

$$\therefore \frac{x-(a+b+c)}{b+c} + \frac{x-(a+b+c)}{a+c} + \frac{x-(a+b+c)}{b+a} = 0,$$

which is satisfied by $x-(a+b+c)=0$; $\therefore x = a+b+c.$

9. $\dfrac{m}{x+a} + \dfrac{n}{x-b} = \dfrac{m+n}{x-c};$

$$\therefore \frac{m(x-c)}{x-a} + \frac{n(x-c)}{x-b} = m+n,$$

which may be solved as in Ex. 1.

10. $\dfrac{3x+5}{x+1} - \dfrac{4x+9}{2x+4} = \dfrac{15x+7}{3x+1} - \dfrac{12x+17}{3x+4}.$

$$\therefore 3 + \frac{2}{x+1} - 2 - \frac{1}{2x+4} = 5 + \frac{2}{3x+1} - 4 - \frac{1}{3x+4}, \text{ or}$$

$$\frac{2}{x+1} - \frac{1}{2x+4} = \frac{2}{3x+1} - \frac{1}{3x+4};$$

$$\therefore \frac{3x+7}{2x^2+6x+4} = \frac{3x+7}{9x^2+15x+4}.$$

This can be divided by $3x+7$, giving $3x+7=0$, or $x = -\tfrac{7}{3}$. The result of the division is

$$\frac{1}{2x^2+6x+4} = \frac{1}{9x^2+15x+4}, \text{ or}$$

$9x^2+15x+4 = 2x^2+6x+4$, or $7x^2 = -9x$, which we can divide by x, $\therefore x = 0$; the result of the division is $7x = -9$, or $x = -\tfrac{9}{7}$.

Exercise li.

1. $\dfrac{10x+17}{18} - \dfrac{12x+2}{18x-16} = \dfrac{5x-4}{9}.$

2. $\dfrac{6x+18}{15} - \dfrac{9x+15}{5x-25} + 3 = \dfrac{2x+15}{5}.$

3. $\dfrac{7x+1}{x-1} = \dfrac{35}{9} \times \dfrac{x+4}{x+2} + 3\tfrac{1}{5}.$

4. $\dfrac{4x-7}{2x-9} + \dfrac{2-14x}{7} + \dfrac{3\tfrac{1}{3}+x}{14} = \dfrac{10-3\tfrac{6}{7}x}{2} - \dfrac{19}{21}.$

5. $\dfrac{2x+a}{3(x-a)} + \dfrac{3x-a}{2(x+a)} = 2\tfrac{1}{6}.$

6. $\dfrac{x-4}{6x+5} + \dfrac{3x-13}{18x-6} = \dfrac{1}{3}.$

7. $\dfrac{3x+1}{2x-15} - \dfrac{x-11}{2x-10} = 1;\quad \dfrac{x-9}{x-5} + \dfrac{x-5}{x-8} = 2.$

8. $\dfrac{x-12}{x-7} + \dfrac{x-4}{x-12} = 2 + \dfrac{7}{x-7};\quad \dfrac{3x-19}{x-13} + \dfrac{3x-11}{x+7} = 6.$

9. $\dfrac{x-2}{2x+1} + \dfrac{x-1}{3(x-3)} = \dfrac{5}{6};\quad \dfrac{x+1}{4(x+2)} + \dfrac{x+4}{5x+13} = \dfrac{9}{20}.$

10. $\dfrac{5(2x^2+3)}{2x+1} + \dfrac{5-7x}{2x-5} = 5x-6;\quad \dfrac{3}{x-7} + \dfrac{1}{x-9} = \dfrac{4}{x-8}.$

11. $\dfrac{7x+55}{2x+5} - \dfrac{3x}{2} = 9 - \dfrac{3x^2+8}{2x-4};\quad \dfrac{17}{x-16} + \dfrac{15}{x-18} = \dfrac{32}{x-17}.$

12. $\dfrac{1-25x}{15} - \dfrac{3-2\tfrac{1}{2}x}{14(x-1)} = \dfrac{28-5x}{3} - \dfrac{10x-11}{30} + \dfrac{x}{3}.$

13. $\dfrac{1}{x-2} - \dfrac{2+2\tfrac{1}{2}x^2 - \tfrac{1}{2}x^3}{6-5x+x^2} - \dfrac{1}{2}x = \dfrac{5}{x-5}.$

14. $\dfrac{30+6x}{x+1} + \dfrac{60+8x}{x+3} = \dfrac{48}{x+1} + 14.$

15. $\dfrac{5x^3+x-3}{5x-4} = \dfrac{7x^2-3x-9}{7x-10}.$

16. $\dfrac{x}{x-2} + \dfrac{x-9}{x-7} = \dfrac{x+1}{x-1} + \dfrac{x-8}{x-6}.$

17. $\dfrac{x^2-3x-9}{x-5} + \dfrac{x^2-7x-17}{x-9} = \dfrac{x^2-6x-15}{x-8} \cdot 2$

18. $\dfrac{4x+7}{4x+5} + \dfrac{4x+9}{4x+7} = \dfrac{4x+6}{4x+4} + \dfrac{4x+10}{4x+8}$.

19. $\dfrac{2x-3}{2x-4} - \dfrac{2x-4}{2x-5} = \dfrac{2x-7}{2x-8} - \dfrac{2x-8}{2x-9}$.

20. $\dfrac{7x+6}{28} - \dfrac{2x+4\frac{2}{7}}{28x-6} + \dfrac{x}{4} = \dfrac{11x}{21} - \dfrac{x-3}{42}$.

21. $\dfrac{x^2-5}{x^2-6} + \dfrac{x^2-11}{x^2-12} = \dfrac{x^2-7}{x^2-8} + \dfrac{x^2-9}{x^2-10}$.

22. $\dfrac{x-1\frac{25}{27}}{2} - \dfrac{2-6x}{13} = x - \dfrac{5x-\frac{1}{4}(10-3x)}{39}$.

23. $\dfrac{1-2x}{3(x^2-x+1)} + \dfrac{1+x}{2(x^2+1)} + \dfrac{1}{6(x+1)} = \dfrac{1}{9(x^2+1)}$.

24. $\dfrac{2x^2+x-30}{2x-7} + \dfrac{x^2+4x-4}{x-1} = \dfrac{x^2-17}{x-4} + \dfrac{2x^2+7x-18}{2x-3}$.

25. $\dfrac{x-a}{x-b} + \dfrac{x-b}{x-a} - \dfrac{(a-b)^2}{(x-a)(x-b)} = \dfrac{2(a-x)}{a+x}$.

26. $\dfrac{12x+10a}{3x+a} + \dfrac{28x+117a}{2x+9a} = 18$.

27. $\dfrac{13\frac{1}{2}x-5}{13\frac{1}{2}x-6} + \dfrac{13\frac{1}{2}x-11}{13\frac{1}{2}x-12} = \dfrac{13\frac{1}{2}x-7}{13\frac{1}{2}x-8} + \dfrac{13\frac{1}{2}x-9}{13\frac{1}{2}x-10}$.

28. $\dfrac{1}{2(x-1)^2} + \dfrac{1}{2(x-1)} - \dfrac{x}{2(x^2+1)} = \dfrac{16x}{(x-1)(x^2+1)}$.

29. $\frac{1}{2}(\frac{2}{3}x+4) - \dfrac{7\frac{1}{2}-x}{3} = \dfrac{x}{2}\left(\dfrac{6}{x}-1\right)$.

30. $\dfrac{3x}{2} - \dfrac{81x^2-9}{(3x-1)(x+3)} = 3x - \dfrac{3}{2} \cdot \dfrac{2x^2-1}{x+3} - \dfrac{57-8x}{2}$.

31. $1 + \dfrac{2x+1}{2(x-1)} - \dfrac{4x+5}{2(x+1)} = \dfrac{x^2+3x+2}{x^2-2x+1} - 1$.

148   SIMPLE EQUATIONS.

32. $\dfrac{7x-30}{10\frac{1}{2}} - \dfrac{5x-7}{\frac{1}{2}x-3} - \dfrac{2-21x}{21} =$

$\dfrac{42x-171}{63} - 10 + \dfrac{2x-9}{63-14x} - \dfrac{1}{7}(4-$

33. $\dfrac{18x-22}{13-2x} + 6x + \dfrac{1+6x}{8} = 13\frac{1}{4} - \dfrac{101-64x}{8}.$

34. $\dfrac{4-9x}{1-3x} - \dfrac{5-12x}{7-4x} = 2 - \dfrac{24x^2-5}{7-25x+12x^2}.$

35. $\dfrac{8x+25}{2x+5} + \dfrac{16x+93}{2x+11} = \dfrac{18x+86}{2x+9} + \dfrac{6x+26}{2x+7}.$

36. $\dfrac{1}{x+a+b} + \dfrac{1}{x-a+b} + \dfrac{1}{x+a-b} + \dfrac{1}{x-a-b} = 0.$

Art. XLII. The results deduced in Section III., Chapter IV., may often be applied with advantage.

EXAMPLES.

1. $\dfrac{ax+b}{cx+d} = \dfrac{m}{n}$

$\therefore \dfrac{(ax+b)d - (cx+d)b}{(cx+d)a - (ax+b)c} = \dfrac{md-nb}{na-mc}$ (page 123).

$\therefore x = \dfrac{md-nb}{na-mc}.$

2. $\dfrac{ax^2+bx+c}{mx^2+nx+p} = \dfrac{a}{m}$

$\therefore \dfrac{(ax^2+bx+c) - ax^2}{(mx^2+nx+p) - mx^2} = \dfrac{a}{m}$ (page 122).

$\therefore \dfrac{bx+c}{nx+p} = \dfrac{a}{m}$ &c.

3. $\dfrac{3x+7}{x+4} = \dfrac{3x-13}{x-4}.$

By (5) each of these fractions =
$\dfrac{\text{difference of numerators}}{\text{difference of denominators}}$, $\therefore \dfrac{20}{8} = \dfrac{3x+7}{x+4} = 8 - \dfrac{5}{x+4}$,

or $\dfrac{1}{2} = \dfrac{5}{x+4}$, $\therefore x = 6$.

4. $\dfrac{mx+a+b}{nx-c-d} = \dfrac{mx+a+c}{nx-b-d}$,

$\therefore \dfrac{mx+a+b}{mx+a+c} = \dfrac{nx-c-d}{nx-b-d}$; or By 4, page 122,

$\dfrac{mx+a+b}{b-c} = \dfrac{nx-c-d}{b-c}$; or $(n-m)x$

$= a+b+c+d$, $\therefore x = $ &c.

5. $\dfrac{\sqrt{(a+x)}+\sqrt{(a-x)}}{\sqrt{(a+x)}-\sqrt{(a-x)}} = a$.

Here by (6), page 122, we have

$\dfrac{2\sqrt{(a+x)}}{2\sqrt{(a-x)}} = \dfrac{a+1}{a-1}$; or, cancelling the 2 in left hand member, and squaring,

$\dfrac{a+x}{a-x} = \dfrac{(a+1)^2}{(a-1)^2}$, whence, again by (6),

$\dfrac{2x}{2a} = \dfrac{(a+1)^2-(a-1)^2}{(a+1)^2+(a-1)^2} = \dfrac{4a}{2(a^2+1)} = \dfrac{2a}{a^2+1}$;

$\therefore x = \dfrac{2a^2}{a^2+1}$.

6. $\dfrac{\sqrt{(x-a+b)}-\sqrt{(x+a-b)}}{\sqrt{(x-a+b)}+\sqrt{(x+a-b)}} = \dfrac{a-b}{a+b}$.

$\therefore \dfrac{\sqrt{(x-a+b)}}{\sqrt{(x+a-b)}} = \dfrac{a}{b}$;

squaring and again applying (6),

$\therefore -\dfrac{2x}{2(a-b)} = \dfrac{a^2+b^2}{a^2-b^2}$, and $x = -\dfrac{a^2+b^2}{a+b}$.

SIMPLE EQUATIONS.

Exercise lii.

1. $\dfrac{1+x}{1-x} = \dfrac{1}{a}$; $\dfrac{x+a}{x-a} = m$; $\dfrac{ax+b}{ax-b} = \dfrac{m}{n}$.

2. $\dfrac{a+x}{b+2x} = 1$; $\dfrac{a(b+x)}{a-x} = b$; $\dfrac{a}{a-x} = \dfrac{b}{b-x}$.

3. $\dfrac{a+x}{a-x} = \dfrac{a+b}{a-b}$; $\dfrac{x+m}{x-m} = \dfrac{a+b}{a-b}$; $\dfrac{a+b}{1+cx} = \dfrac{a-b}{1-cx}$.

4. $\dfrac{a+bx}{a+b} = \dfrac{c+dx}{c+d}$; $\dfrac{a+bx}{a-b} = \dfrac{c+dx}{c-d}$; $\dfrac{a-x}{b-x} = \dfrac{a+a}{b+x}$.

5. $\dfrac{2x^2-5x+6}{2x^2-7x+3} = \dfrac{x^2-7x+5}{x^2-9x+2}$.

6. $\dfrac{ax+b-c}{ax-b+c} = \dfrac{(b-c)^2}{(b+c)^2}$.

7. If $\dfrac{\sqrt{(x+y)}+\sqrt{(x-y)}}{\sqrt{(x+y)}-\sqrt{(x-y)}} = \dfrac{x}{y}$, shew that $\dfrac{x+y}{x-y} = 1$.

8. $\dfrac{2x-7}{2x-3} = \dfrac{x+7}{x+11}$; $\dfrac{4x-5}{2x+10} = \dfrac{10x-32}{5x-8\frac{1}{2}}$.

9. $\dfrac{57x-43}{19x+13} = \dfrac{39x-7}{13x+25}$; $\dfrac{23x+5\frac{4}{5}}{115x-29} = \dfrac{36x-7}{180x+23}$.

10. $\dfrac{210x-73}{310x-66} = \dfrac{21x+7\cdot 3}{31x+8}$; $\dfrac{mx-a-b}{nx-c-d} = \dfrac{mx-a-c}{nx-b-d}$.

11. $\dfrac{3x+\sqrt{(4x-x^2)}}{3x-\sqrt{(4x-x^2)}} = 2$; $\dfrac{\sqrt{(12x+1)}+\sqrt{(12x)}}{\sqrt{(12x+1)}-\sqrt{(12x)}} = 18$.

12. $\dfrac{x^3+ax^2-bx+c}{x^3-ax^2+bx+c} = \dfrac{x^2+ax-b}{x^2-ax+b}$.

13. $\dfrac{\sqrt{(2a^2-x^2)}+b\sqrt{(2a-x)}}{\sqrt{(2a^2-x^2)}-b\sqrt{(2a-x)}} = \dfrac{\sqrt{a+b}}{\sqrt{a-b}}$.

14. $\dfrac{\sqrt{(x^2+a^2)}+\sqrt{(x^2-a^2)}}{\sqrt{(x^2+a^2)}-\sqrt{(x^2-a^2)}} = a^2$.

15. $\dfrac{8x^3+12x^2-8x+5}{8x^3-12x^2+8x+5} = \dfrac{4x^2+6x-4}{4x^2-6x+4}$.

SIMPLE EQUATIONS.

16. $\dfrac{\sqrt[3]{(x+1)} - \sqrt[3]{(x-1)}}{\sqrt[3]{(x+1)} + \sqrt[3]{(x-1)}} = \dfrac{1}{2}$

17. $\dfrac{28 + \sqrt{x}}{28 - \sqrt{x}} = \dfrac{9 + 3\sqrt{x}}{9 + 2\sqrt{x}}$

18. $\dfrac{a^3x^3 + a^2bx^2 - acx + d}{a^3x^3 - a^2bx^2 + acx + d} = \dfrac{a^2x^2 + abx - c}{a^2x^2 - abx + c}$

19. $\dfrac{5\sqrt{(2x-1)} + 2\sqrt{(3x-3)}}{4\sqrt{(2x-1)} - 2\sqrt{(3x-3)}} = 2\tfrac{11}{13}$

20. $\dfrac{\sqrt{2x} + \sqrt{(3-2x)}}{\sqrt{2x} - \sqrt{(3-2x)}} = \dfrac{3}{2}$

21. $\dfrac{2\sqrt[3]{(3x+3)} + \sqrt[3]{(7x+8)}}{2\sqrt[3]{(3x+3)} - \sqrt[3]{(7x+8)}} = 5$

22. $33\{13 - 2\sqrt{(x-5)}\} = 3\{13 + 2\sqrt{(x-5)}\}$

23. $(\sqrt{n}+1)\{\sqrt{(nx+1)} - \sqrt{nx}\} = (\sqrt{n}-1)\{\sqrt{(nx+1)} + \sqrt{nx}\}$

24. $\dfrac{\sqrt{(x+c)} + \sqrt{b}}{\sqrt{(x+c)} - \sqrt{b}} = \dfrac{\sqrt{x} + \sqrt{a}}{\sqrt{x} - \sqrt{a}}$

25. $\dfrac{\sqrt{x+28}}{\sqrt{x+4}} = \dfrac{\sqrt{x+38}}{\sqrt{x+6}}; \quad \dfrac{\sqrt[3]{2x+17}}{\sqrt[3]{2x+9}} = \dfrac{\sqrt[3]{2x+27}}{\sqrt[3]{2x+15}}$

26. $\dfrac{\sqrt{x+2a}}{\sqrt{x+b}} = \dfrac{\sqrt{x+4a}}{\sqrt{x+3b}}; \quad \dfrac{3x-1}{\sqrt{3x+1}} = \dfrac{1+\sqrt{3x}}{2}$

27. $\dfrac{\sqrt{a} - \sqrt{(a-x)}}{\sqrt{a} + \sqrt{(a-x)}} = a; \quad \dfrac{\sqrt{x}+\sqrt{b}}{\sqrt{x}-\sqrt{b}} = \dfrac{a}{b}$

28. $\dfrac{ax + 1 + \sqrt{(a^2x^2 - 1)}}{ax + 1 - \sqrt{(a^2x^2 - 1)}} = b$

29. $\dfrac{1 - \sqrt{\{1 - \sqrt{(1-x)}\}}}{1 + \sqrt{\{1 - \sqrt{(1-x)}\}}} = a$

30. $\dfrac{a+x}{\sqrt{(2ax+x^2)}} = \dfrac{b+1}{b-1}; \quad \dfrac{1+x+x^2}{1-x+x^2} = \dfrac{62}{63} \times \dfrac{1+x}{1-x}$

31. $\dfrac{5x^4 + 10x^2 + 1}{x^5 + 10x^3 + 5x} = \dfrac{a^5 + 10a^2 + 1}{5a^4 + 10a^2 + 1}$

Art. XLIII. Various other artifices may be employed to simplify the solution of equations.

Examples.

1. Solve $2+\sqrt{(4x^2-9x+8)}-2x=0$: here there is but one surd, and it is convenient to make that surd one side of the equation and transpose all the rational terms to the other; this gives $\sqrt{(4x^2-9x+8)}=2x-2$; squaring both sides,
$4x^2-9x+8=4x^2-8x+4$, $\therefore x=4$.

2. $\sqrt{(a+x)}+\sqrt{(a-x)}=2\sqrt{x}$. We might square this as it stands, but the work will be simplified if we first transpose, thus $\sqrt{(a+x)}=2\sqrt{x}-\sqrt{(a-x)}$; now squaring,
$a+x=4x+a-x-4\sqrt{(ax-x^2)}$, or
$x=2\sqrt{(ax-x^2)}$. Again squaring,
$x^2=4ax-4x^2$, whence $x=0$, or $\dfrac{4a}{5}$.

3. Clear of radicals
$\sqrt[3]{x}+\sqrt[3]{y}+\sqrt[3]{z}=0$. Transposing,
$\sqrt[3]{x}+\sqrt[3]{y}=-\sqrt[3]{z}$; cube by formula [6],
$x+y+3\sqrt[3]{xy}(\sqrt[3]{x}+\sqrt[3]{y})=-z$; and substituting for $\sqrt[3]{x}+\sqrt[3]{y}$ its value $-\sqrt[3]{z}$, this becomes
$x+y-3\sqrt[3]{xyz}=-z$, or
$x+y+z=3\sqrt[3]{xyz}$; $\therefore$ cubing again,
$(x+y+z)^3=27xyz$.

4. $\dfrac{a+x+\sqrt{(2ax+x^2)}}{a+x}=b$.

Dividing and transposing, we have

$\dfrac{\sqrt{(2ax+x^2)}}{a+x}=b-1$, $\therefore \dfrac{2ax+x^2}{a^2+2ax+x^2}=(b-1)^2$; again by division in left-hand member,

$-\dfrac{a^2}{(a+x)^2}+1=(b-1)^2 \therefore \dfrac{a}{a+x}=\sqrt{\{1-(b-1)^2\}}$, or

$\dfrac{x+a}{a}=1 \div \sqrt{\{1+(b-1)^2\}}$, or

$\dfrac{x}{a}+1=$ &c.

SIMPLE EQUATIONS.   153

5. Solve $\sqrt{(4x^2+19)} + \sqrt{(4x^2-19)} = \sqrt{47} + 3$.
We have the identity
$$(4x^2+19)-(4x^2-19) = 38 = 47-9.$$
Now dividing the members of this identity by those of the given equation, we have
$\sqrt{(4x^2+19)} - \sqrt{(4x^2-19)} = \sqrt{47} - 3$. Adding this to the given equation, then
$2\sqrt{(4x^2+19)} = 2\sqrt{47}$, $\therefore$ $4x^2+19 = 47$, and $x = \pm\sqrt{7}$.

6. $\sqrt[3]{(25+x)} + \sqrt[3]{(25-x)} = 2$.
Cubing by formula [6], (See Ex. 8), we have
$$25+x+25-x+6\sqrt[3]{(25^2-x^2)} = 8, \text{ or}$$
$$\sqrt[3]{(625-x^2)} = -7, \text{ or } (625-x^2) = -343;$$
$\therefore$ $x^2 - 525 + 453 = 968$, and $x = \pm 22\sqrt{2}$.

Exercise lili.

1. $\sqrt{(x+4)} + \sqrt{(x-3)} = 7$.
2. $\sqrt{(3x+1)} + \sqrt{(4x+4)} = 1$.
3. $\sqrt{(2x+10)} + \sqrt{(2x-2)} = 6$.
4. $\sqrt{(mx)} - \sqrt{(nx)} = m - n$.
5. $\sqrt{(bx)} + \sqrt{(ab+bx)} = \sqrt{x}$.
6. $\sqrt{x} + \sqrt{(x+3)} = \dfrac{5}{\sqrt{(x+3)}}$.
7. $\sqrt{(ax+x^2)} = (1+x)$.
8. $\sqrt[3]{(17x-26)} = \dfrac{2}{9}$.
9. $\sqrt{x} - \sqrt{(a+x)} = \sqrt{\dfrac{a}{x}}$.
10. $b+x-\sqrt{(b^2+x^2)} = c^2$.
11. $\sqrt{(8+x)} - \sqrt{x} = 2\sqrt{(1+x)}$.
12. $\sqrt{(2x-27a)} = 9\sqrt{a} - \sqrt{(2x)}$.

SIMPLE EQUATIONS.

13. $\sqrt[3]{(1-x)} + \sqrt[3]{(1+x)} = \sqrt[3]{3}$.
14. $\sqrt[3]{(3+x)} + \sqrt[3]{(3-x)} = \sqrt[3]{7}$.
15. $\sqrt[3]{(x+1)} - \sqrt[3]{(x-1)} = \sqrt[3]{11}$.
16. $\sqrt[3]{(a+x)} + \sqrt[3]{(a-x)} = \sqrt[3]{b}$.
17. $\sqrt[3]{(1+\sqrt{x})} + \sqrt[3]{(1-\sqrt{x})} = 2$.
18. $\sqrt{x} - \sqrt{\{a - \sqrt{(ax+x^2)}\}} = \tfrac{1}{2}\sqrt{a}$.
19. Clear of radicals $\sqrt[3]{a} + \sqrt[3]{b} - \sqrt[3]{c}$.
20. Solve $x + \sqrt{(a^2+x^2)} = \dfrac{na^2}{\sqrt{(a^2+x^2)}}$.
21. Clear of radicals $\sqrt{x} + \sqrt{y} + \sqrt{z} - \sqrt{n}$.

Solve the following equations:

22. $\sqrt{(1+x)} + \sqrt{\{1+x+\sqrt{(1-x)}\}} = \sqrt{(1-x)}$.
23. $\sqrt{(x+\sqrt{x})} - \sqrt{(x-\sqrt{x})} = a\sqrt{\dfrac{x}{x+\sqrt{x}}}$.
24. $\sqrt{(1+x+x^2)} + \sqrt{(1-x+x^2)} = mx$.
25. $\sqrt{(a^2-x^2)} + x\sqrt{(a^2-1)} = a^2(1-x^2)$.
26. $\dfrac{bx-c^2}{\sqrt{(bx)}+c} = \dfrac{\sqrt{(bx)}+c}{n} - a$.
27. $\sqrt{(2x^2+5)} + \sqrt{(2x^2-5)} = \sqrt{15} + \sqrt{5}$.
28. $\sqrt{(3x^2+10)} + \sqrt{(3x^2-10)} = \sqrt{17} + \sqrt{3}$.
29. $\sqrt{(3x^2+9)} - \sqrt{(3x^2-9)} = \sqrt{34} + 4$.
30. $\sqrt{(3a-3b+x^2)} + \sqrt{(2a-2b+x^2)} = \sqrt{a} + \sqrt{x}$.
31. $\sqrt{(4a^2-3b^2-2x^2)} + \sqrt{(3a^2-3b^2-x^2)} = a+x$.
32. Clear of radicals, $\sqrt[3]{(2x)} - \sqrt[3]{(2y)} - \sqrt[3]{(2z)}$.
33. $\sqrt{(a+x)} + \sqrt{(a-x)} = 2x \div \sqrt{\{a + \sqrt{(a^2+x^2)}\}}$.
34. $\sqrt{(x^2+2ax)} + \sqrt{(x^2-2ax)} = \dfrac{nax}{\sqrt{(x^2+2ax)}}$.
35. $\sqrt{\left(\dfrac{\sqrt{x}+a}{\sqrt{x}-a}\right)} - \sqrt{\left(\dfrac{\sqrt{x}-a}{\sqrt{x}+a}\right)} = \sqrt{(x-a^2)}$.
36. $\sqrt{\{(2a+x)^2+b^2\}} + \sqrt{\{(2a-x)^2+b^2\}} = 2a$.

SIMPLE EQUATIONS.   155

Art. XLIV. Sometimes a factor can be discovered, and the principle of Art. XL. applied.

EXAMPLES.

1. $\dfrac{x^4+a^2x^2+a^4}{x-a} = x^3+(a-b)x^2+(a^2-ab)x-a^2b.$

Factoring we have

$\dfrac{(x^2+ax+a^2)(x^2-ax+a^2)}{x-a} = (x-b)(x^2+ax+a^2),$

or $x^2-ax+a^2=(x-a)(x-b)$;

$\therefore (a+b-a)x = ab-a^2,$ and $x=a-\dfrac{a^2}{b}.$

2. $\dfrac{3abc}{a+b} - \dfrac{bx}{a} + \dfrac{a^2b^2}{(a+b)^3} = 3cx - \dfrac{b^2x}{a} \cdot \dfrac{2a+b}{(a+b)^2}.$

Transpose $\dfrac{bx}{a}$ and factor, then

$\dfrac{ab}{a+b}\left\{3c + \dfrac{ab}{(a+b)^2}\right\} = x\left\{3c + \dfrac{b}{a}\left(1 - \dfrac{2ab+b^2}{(a+b)^2}\right)\right\}$

$= x\left\{3c + \dfrac{b}{a} \cdot \dfrac{a^2}{(a+b)^2}\right\}$

$= x\left\{3c + \dfrac{ab}{(a+b)^2}\right\}$

$\therefore \dfrac{ab}{a+b} = x.$

3. $\dfrac{x+a}{(a-b)(c-a)} - \dfrac{x-b}{(a-b)(b-c)} - \dfrac{x-c}{(b-c)(c-a)} = \dfrac{b+c}{(a-b)(b-c)(c-a)}.$

Add term by term the identity (Th. iii., page 54).

$\dfrac{x-a}{(a-b)(c-a)} + \dfrac{x-b}{(a-b)(b-c)} + \dfrac{x-c}{(b-c)(c-a)} = 0.$

$\therefore \dfrac{2x}{(a-b)(c-a)} = \dfrac{b+c}{(a-b)(b-c)(c-a)}.$

$\therefore x = \dfrac{1}{2} \cdot \dfrac{b+c}{b-c}.$

4. $(x+a+b)^3+(a+b)^3-(x+b)^3-(x+a)^3+x^3+a^3+b^3=abc.$

The left hand member vanishes for $x=0$, and ∴ by symmetry for $a=0$ and $b=0$; ∴ it is of the form $mabx$ in which m is numerical.

Put $x=a=b$, and m is found to be 6,

∴ the equation reduces to

$$6abx = abc, \quad \therefore \text{ and } x=\tfrac{1}{6}c.$$

5. $\left(\dfrac{x-a}{x-b}\right)^3 = \dfrac{x-2a+b}{x-2b+a}$; let $x-b=m$, $x-a=n$, and ∴ $m-n=a-b$, then we have

$$\dfrac{m^3}{n^3} = \dfrac{n-(m-n)}{m+(m-n)} = \dfrac{2n-m}{2m-n}$$

∴ $2m^4 - nm^3 = 2n^4 - n^3m$, and

$2(m^4 - n^4) - mn(m^2 - n^2) = 0$, which is divisible by $m^2 - n^2$,

∴ $m^2 - n^2 = 0$, or $m+n=0$;

But $m+n = 2x-a-b = 0$, ∴ $x = \tfrac{1}{2}(a+b)$.

6. $\dfrac{1}{3}\cdot\dfrac{x^2-4x+2}{x^2-4x-1} + \dfrac{1}{6}\cdot\dfrac{x^2-4x+3}{x^2-4x-3} - \dfrac{2}{9}\cdot\dfrac{x^2-4x+3}{x^2-4x-6} = \dfrac{5}{18}.$

Let $y = x^2 - 4x$, then this equation becomes

$$\dfrac{1}{3}\cdot\dfrac{y+2}{y-1} + \dfrac{1}{6}\cdot\dfrac{y+3}{y-3} - \dfrac{2}{9}\cdot\dfrac{y+3}{y-6} = \dfrac{5}{18}, \text{ or by division,}$$

$$\dfrac{1}{3} + \dfrac{1}{y-1} + \dfrac{1}{6} + \dfrac{1}{y-3} - \dfrac{2}{9} - \dfrac{2}{y-6} = \dfrac{5}{18}, \text{ or}$$

$$\dfrac{1}{y-1} + \dfrac{1}{y-3} - \dfrac{2}{y-6} = 0; \text{ this may be written}$$

$$\dfrac{1}{y-1} - \dfrac{1}{y-6} + \dfrac{1}{y-3} - \dfrac{1}{y-6} = 0, \text{ or}$$

$$\dfrac{5}{y-1} + \dfrac{3}{y-3} = 0, \therefore 5y-15+3y-3=0, \text{ or}$$

$y = 2\tfrac{1}{4}$ ∴ $x^2 - 4x = 2\tfrac{1}{4}$, or $x^2 - 4x + 4 = 4 + 2\tfrac{1}{4}$,

and $x - 2 = \pm\tfrac{5}{2}$. We might assume $(x-2)^2 = y$, when the given equation would take the form

$$\frac{1}{3} \cdot \frac{y-2}{y-5} + \frac{1}{6} \cdot \frac{y-1}{y-7} - \frac{2}{9} \cdot \frac{y-1}{y-10} = \frac{5}{18}.$$

And reducing as before, we should find

$y = 6\frac{1}{4} = (x-2)^2$, $\therefore x-2 = \pm\frac{5}{2}$, as before.

Exercise liv.

1. $\dfrac{x^4 + a^2x^2 + a^4}{x - b} = x^3 - (a-b)x^2 + (a^2 - ab)x + a^2b.$

2. $\dfrac{x^4 + 4a^4}{x + b} = x^3 + 2a(a-b)x + (2a-b)x^2 - 2a^2b.$

3. $\dfrac{a^2}{a^2 + ab + b^2} - \dfrac{a^2 \cdot c}{a^3 - b^3} = \dfrac{2c}{a-b} - 2cx.$

4. $\dfrac{1}{a+b+x} - \dfrac{1}{a} - \dfrac{1}{b} - \dfrac{1}{x} = 2ab(x+b)x^2.$

5. $\dfrac{1}{(x-b)(x-c)} + \dfrac{1}{(a+c)(a+b)} =$
$\dfrac{1}{(a+c)(x-c)} + \dfrac{1}{(a+b)(x-b)}.$

6. $\dfrac{bx}{a} - \dfrac{3ab}{a-b} + \dfrac{a^2b^2}{(a-b)^3} = 3x - \dfrac{b^2x}{a} \cdot \dfrac{2a-b}{(a-b)^2}.$

7. $\dfrac{x^3 + 2ax}{x^4 - 11x^2a^2 + a^4} = \dfrac{x-a}{x^2 - 3ax - a^2}.$

8. $\dfrac{1}{2}\left(\dfrac{x-a}{x+a}\right)^2 - \dfrac{1}{2}(-x+a) = \dfrac{x-a}{x+a}.$

9. $x^3 - (a+b+c)x^2 + (a^2+b^2+c^2)x - \frac{1}{3}(a^3+b^3+c^3)$
$= (x-a)(x-b)(x-c).$

10. $\dfrac{1}{ax} + \dfrac{1}{bx} + \dfrac{1}{cx} = \frac{1}{2}(a+b+c)^2 - \dfrac{1}{2}\left(\dfrac{a}{bcx} + \dfrac{b}{acx} + \dfrac{c}{abx}\right).$

11. $\dfrac{1-ax}{bc} + \dfrac{1-bx}{ac} + \dfrac{1-cx}{ab} = \left(\dfrac{2}{a} + \dfrac{2}{b} + \dfrac{2}{c}\right)x.$

12. $\dfrac{(a-b)^2}{abc} - 1 + \dfrac{a}{b} = \dfrac{a^2-b^2}{abc} + \left(1 + \dfrac{a}{b}\right)x.$

13. $x^3 + (b+c)^3 + 3b(b+c)x = b^3.$

SIMPLE EQUATIONS.

14. $x - a - 8\sqrt[3]{(abx)} = b$.

15. $11x^4 + 10x^3 - 40x = 176$.

16. $\dfrac{x}{(a+b)^2} + \dfrac{ac}{(a-b)^3} - \dfrac{c}{(a^2-b^2)(a+b)} = \dfrac{ax}{(a-b)^2}$.

17. $x^3 - \dfrac{a-b}{a+b} \cdot x^2 + \dfrac{2cx^2}{1+cx} = x - \dfrac{a-b}{a+b} \cdot \dfrac{1-cx}{1+cx}$.

18. $\dfrac{4x^4 + 4a^4 - 33x^2a^2}{2x+a} = \frac{1}{2}(4x^3 - 8x^2a - 9xa^2 - 2a^3)$.

19. $\dfrac{7}{x^2-11x+28} + \dfrac{7}{x^2-17x+70} = \dfrac{3\frac{1}{2}x^2}{x^2-14x+40}$.

20. $\dfrac{8}{x^2-6x+5} + \dfrac{8}{x^2-14x+45} = \dfrac{x^4}{x^2-10x+9}$.

21. $\dfrac{x+a}{(a-b)(c-a)} - \dfrac{x-b}{(a-b)(b-c)} + \dfrac{x+c}{(b-c)(c-a)}$
$= \dfrac{a+c}{(a-b)(b-c)(c-a)}$.

22. $(x-a)^3 + (a-b)^3 + (b-x)^3 = x^2 - a^2$.

23. $x\left(\dfrac{x+2a}{x-a}\right)^3 + a\left(\dfrac{a+2x}{a-x}\right)^3 = 2a$.

24. $(x+a)^3 - (a+b)^3 + (b-x)^3 = (x+a)(x+b)(a+b)$.

25. $x^3 - (x-b)^3 - (x-a+b)^3 - a^3 + (x-a)^3 + (a-b)^3 + b^3$
$= (a-b)c^2$

26. $(x+a)^3 - (x+b)^3 - (x-b)^3 - (2a)^3 + (x-a)^3 + (a+b)^3 + (a-b)^3 = (a^2 - b^2)c$.

27. $\dfrac{x+a}{x^2+ax+a^2} - \dfrac{x-a}{x^2-ax+a^2} = \dfrac{a^4}{x(x^4+a^2x^2+a^4)}$.

28. $(x+a+b)^4 - (x+a)^4 - (x+b)^4 + x^4 - (a+b)^4 + a^4 + b^4$
$= 12ab\{x^2 + (a+b)^2\}$.

29. $\dfrac{a-x}{a^2-bc} + \dfrac{b-x}{b^2-ca} + \dfrac{c-x}{c^2-ab} = \dfrac{3x}{ab+bc+ca}$.

30. $x^3(b-a^2) + a^3(x-b^2) + b^3(a-x^2) + abx(abx-1)$
$= (a-x^2)(b^3-a^4)$.

SIMPLE EQUATIONS. 159

31. $(1+x+x^2)^2 = \dfrac{ax+1}{ax-1} \cdot (1+x^3+x^4)$.

32. $\sqrt{\left(\dfrac{x+a}{x+b}\right)} - \dfrac{a-b}{2(x+c)} = 1$.

33. $\dfrac{x+a}{x+b} = \left(\dfrac{2x+a+c}{2x+b+c}\right)^2$.

34. $\sqrt{(x^2+27x+180)} - \sqrt{(x^2+26x+168)} = \sqrt{\left(\dfrac{x+15}{x+12}\right)}$

35. $\{(x+a+\sqrt{(x^2+2ax+b^2)})\}^3 + \{x+a-\sqrt{(x^2+2ax+b^2)}\}^3$
$= 14(x+a)^3$. (See page 17, Ex. 1).

36. $\{x+a+\sqrt{(x^2-2ax-2b^2)}\}^2 + \{x+a-\sqrt{(x^2-2ax-2b^2)}\}^2$
$= {}^2-b^2 - 2a(a-b)$.

37. $\left(\dfrac{x+a}{-b}\right)^3 = \dfrac{x+2a+b}{x-a-2b}$.

38. $(5x-7)^3 - (2x-4)^3 = 27(x^3-1)$.

39. $\dfrac{1}{3} \cdot \dfrac{x^2-6x-1}{x^2-6x-4} + \dfrac{1}{5} \cdot \dfrac{x^2-6x-4}{x^2-6x-9} - \dfrac{2}{9} \cdot \dfrac{x^2-6x-7}{x^2-6x-16}$
$= \dfrac{14}{45} + \dfrac{4}{x^2-6x-9}$.

40. $\dfrac{1}{5} \cdot \dfrac{x^2-2x-3}{x^2-2x-8} + \dfrac{1}{9} \cdot \dfrac{x^2-2x-15}{x^2-2x-24} - \dfrac{2}{13} \cdot \dfrac{x^2-2x-35}{x^2-2x-48}$
$= \dfrac{2}{585}$.

41. $\{x+a-b+\sqrt{(x^2+a^2-b^2)}\}^3 +$
$\{x+a-b-\sqrt{(x^2+a^2-b^2)}\}^3 = 8(x+a-b)^3$.

42. $\dfrac{1}{(x+a)^2-b^2} + \dfrac{1}{(x+b)^2-a^2} =$
$\dfrac{1}{x^2-(a+b)^2} + \dfrac{1}{x^2-(a-b)^2}$.

43. $41\left(\dfrac{6}{x+1} + \dfrac{7x+67}{4}\right) + 130 =$
$89\left(\dfrac{8x+57}{x+2} + \dfrac{9x+68}{x+3}\right)$.

44. $51\left(\dfrac{16x+45}{x-1} - \dfrac{9x+25}{x-4}\right) + 863 =$

$ 61\left(\dfrac{27x-3}{x-2} - \dfrac{7x+30}{x-3}\right).$

45. $(x+a)(x+3a)(x+4a)(x+6a) = x^4 + 6a^2(x^2 + 7ax + 6a^2).$

46. $\dfrac{1}{x+6a} + \dfrac{2}{x-3a} + \dfrac{3}{x+2a} = \dfrac{6}{x+a}.$

Exercise lv.

1. $a(b-x) + b(c-x) = b(a-x) + cx.$
2. $(a+bx)(a-b) - (ax-b) = ab(x+1).$
3. $(a-b)(x-c) + (a+b)(x+c) = 2(bx+ad).$
4. $(a-b)(x-c) - (a+b)(x+c) + 2a(b+c) = 0.$
5. $(a-b)(a-c)(a+x) + (a+b)(a+c)(a-x) = 0.$
6. $(\,-b)(a-c+x) + (a+b)(a+c-x) = 2a^2.$
   (solve in $\{x-c\}$).
7. $(m+a)(a+b-x) + (a-m)(b-x) = a(m+b).$
8. $m(a+b-x) = n(x-a-b).$
9. $(m+n)(m-n-x) + m(x-n) - n(x-m) = m^2 - n^2.$
10. $\dfrac{m-x}{m} + \dfrac{n-x}{n} + \dfrac{p-x}{p} = 3.$
11. $\dfrac{a^2b-x}{a} + \dfrac{b^2c-x}{b} + \dfrac{c^2a-x}{c} = 0.$
12. $\dfrac{a-x}{bc} + \dfrac{b-x}{ca} + \dfrac{c-x}{ab} = 0.$
13. $\dfrac{1-ax}{bc} + \dfrac{1-bx}{ca} + \dfrac{1-cx}{ab} = 0.$

(Deduce the solution from that of No. 12).

14. $\dfrac{a-bx}{bc} + \dfrac{b-cx}{ca} + \dfrac{c-ax}{ab} = 0.$
15. $(a+b+c)x - \dfrac{a^2+b^2}{a-b} = \dfrac{2abx}{a+b} + \dfrac{a+b}{a-b}.\,c.$

16. $\dfrac{abc}{a+b} + \dfrac{a^2b^2}{(a+b)^3} + \dfrac{(2a+b)b^2x}{a(a+b)^2} = \dfrac{(b+3ac)x}{a}$.

17. $\dfrac{10}{x} + \dfrac{4}{9} = \dfrac{9}{x} + \dfrac{2}{3}$. $\left(\text{Solve in } \dfrac{1}{x}\right)$.

18. $\dfrac{7}{x} + \dfrac{1}{3} = \dfrac{23-x}{8x} + \dfrac{7}{12} - \dfrac{1}{4x}$.

19. $\dfrac{7}{3} + \dfrac{13}{5x} = \dfrac{2(5x-12)}{3x} - \dfrac{17}{20} + \dfrac{10}{x}$.

20. $\dfrac{10-x}{3} + \dfrac{13+x}{7} = \dfrac{7x+266}{x+21} - \dfrac{4x+17}{21}$.

21. $\dfrac{5}{x+3} + \dfrac{3}{2(x+3)} = \dfrac{1}{2} - \dfrac{7}{2(x+3)}$.

22. $\dfrac{6x+5}{8x-15} - \dfrac{1+8x}{15} = \dfrac{1-x}{3} - \dfrac{x-3}{5}$.

23. $\dfrac{a - \dfrac{1}{x}}{a + \dfrac{1}{x}} - \dfrac{1}{x} = \dfrac{x - \dfrac{1}{a}}{x + \dfrac{1}{a}} - \dfrac{1}{a}$

24. $\dfrac{a^2}{b - \dfrac{c^2}{d - \dfrac{e^2}{x}}} = 1$.

25. $(x-1)(x-2) - (x-3)(x-4) = 3$;
    $(x-3)(x-4) = (x-2)(x-6)$.

26. $2(x-4)(3x+4) + (2x-3)(3x+2) - 6(x-2)(2x-8) = 0$.

27. $(a-x)(b-x) = x^2$; $(a-x)(x-b) = x^2 - c^2$.

28. $(a-x)(b+x) = b^2 - x^2$; $(x-a)(x-b) = x^2 - a^2$.

29. $(a+x)(b+x) = (a-x)(b-x)$;
    $(ax+b)(bx+a) = (b-ax)(a-bx)$.

30. $(a-x)(b-x) + (a-c-x)(x-b+c) = 0$.

SIMPLE EQUATIONS.

31. $(a-x)(b-x)-(c-x)(d-x)=(c+d)x-cd.$
32. $(x-a)(x-b)-(x-c)(x-d)=(d-a)(d-b).$
33. $\{(a^2-b^2)x-ab\}\{a-(a+b)x\}+2ab^2x=$
    $\{(a+b)^2x+ab\}\{b-(a-b)x\}.$
34. $(x+1)(x+2)(x+3)=(x-3)(x+4)(x+5).$
35. $(x+1)(x+2)(x+3)=(x-1)(x-2)(x-3)+3(x+1)(4x+1).$
36. $(x+1)(x+4)(x+7)=(x+2)(x+5)^2.$
37. $(x+2)(x+5)^2=(x+3)^2(x+6).$
38. $(x-1)(x-4)(x-6)-x(x-2)(x-9)=136.$
39. $(a+x)(b+x)(c+x)-(a-x)(b-x)(c-x)=2(x^3+abc).$
40. $\dfrac{(x-a)(x-b)(x-c)-(d-a)(d-b)(d-c)}{x-d}=(x-d)^2.$
41. $x(x-a)^2-(x-a+b)(x-a+c)(x-b-c)=(a^2+bc)(b+c).$
42. $(x-a+b)(x-b+c)(x-c+d)-x^2(x-a+d)=bc(d-a).$
43. $(x-a+b)(x-b+c)(x-c+d)-x(x-a+c)(x-c+d)$
    $=bc(d-a).$
44. $(x-2a)(x-2b)(x-2c)-(x-a-b)(x-b-c)(x-c-a)$
    $=(a+b+c)(a^2+b^2+c^2)-9abc.$
45. $x^3-(x-a+b)(x-b+c)(x-c+a)$
    $=(a+b+c)(a^2+b^2+c^2)-2(a^2b+b^2c+c^2a)-3abc.$
46. $x\left(a-\dfrac{1}{x}\right)\left(b-\dfrac{1}{x}\right)\left(c-\dfrac{1}{x}\right)+\dfrac{1}{x^2}=\dfrac{a+b+c}{x}.$
47. $(x+a)(x+b)+(x+c)(x+a)=(x+b)(x+d)+(x+d)(x+c).$
48. $(ax+b)(ax-c)-a(b-x)(ax+b)=a^2(x-c)(x-b)-$
    $a(ax-c)(c-x).$
49. $\dfrac{2x-3}{x-4}+\dfrac{3x-2}{x-3}=\dfrac{5x^2-29x-4}{x^2-12x+32}.$
50. $\dfrac{5x-1}{3(x+1)}-\dfrac{3x+2}{2(x-1)}=\dfrac{x^2-30x+2}{6x^2-6}.$

SIMPLE EQUATIONS.

51. $\dfrac{3x-7}{2x-9} - \dfrac{3(x+1)}{2(x+3)} = \dfrac{11x+3}{2x^2-3x-27}.$

52. $\dfrac{7x-5}{3x-2} + \dfrac{8x-7}{3x-1} + \dfrac{10x+7}{9x^2-9x+2} = 5.$

53. $\dfrac{2x+7}{3x-7} + \dfrac{3x-6}{2x-5} + \dfrac{5(x-1)}{9x-25} = \dfrac{3x-2}{2x-5} + \dfrac{5x-8}{9x-25} + \dfrac{2x+2}{3x-7}.$

54. $\dfrac{4x^2-3x}{1+x} - \dfrac{3x}{1-x} = \dfrac{4x^3+2x}{x^2-1}\,;\ \dfrac{x-a}{x-m} - \dfrac{x-b}{x-n} = 0.$

55. $\dfrac{\frac{1}{4}-x}{\frac{1}{4}+x} + \dfrac{1}{4} = \dfrac{x}{\frac{1}{4}-2x} - \dfrac{1}{4}\,;\ \dfrac{a}{c} + \dfrac{cx}{ax-b} = \dfrac{c}{a} + \dfrac{ax}{cx-b}.$

56. $\dfrac{\frac{3}{2}-\frac{1}{x}}{\frac{3}{2}+\frac{1}{x}} - \dfrac{\frac{2}{3}-\frac{1}{x}}{\frac{2}{3}+\frac{1}{x}} = \dfrac{\frac{3}{2}-\frac{2}{3}}{\frac{2}{3}\cdot\frac{1}{x}} + 1$

57. $\dfrac{2(x-1)}{x-7} + \dfrac{x+8}{x-4} = \dfrac{3(5x+16)}{5x-28}.$

58. $\dfrac{ax}{mx-p} + \dfrac{cx}{nx-q} = \dfrac{a}{m} + \dfrac{c}{n}\,;$

   $\dfrac{ax+b}{mx-p} + \dfrac{cx+d}{nx-q} = \dfrac{a}{m} + \dfrac{c}{n}.$

59. $\dfrac{b-x}{a+x} + \dfrac{c-x}{a-x} = \dfrac{a(c-2x)}{a^2-x^2}\,;$

   $\dfrac{a+b}{x-a} + \dfrac{b+c}{x-b} = \dfrac{a+c+2b}{x-c}.$

60. $\dfrac{ax+b}{ax-b} - \dfrac{bx}{ax+b} = \dfrac{ax}{ax-b} - \dfrac{(ax^2-2b)b}{a^2x^2-b^2}.$

61. $\dfrac{ax-b}{mx-p} + \dfrac{cx-d}{nx-q} + \dfrac{(bn+dm)x-(bq+dp)}{(mx-p)(nx-q)} = \dfrac{a}{m} + \dfrac{c}{n}.$

62. $\dfrac{m}{x-a} + \dfrac{n}{x-b} + \dfrac{p}{x-c} = \dfrac{m}{x-c} + \dfrac{n}{x-a} + \dfrac{p}{x-b}.$

SIMPLE EQUATIONS.

63. $\dfrac{ax-2a}{ax-2b} = \dfrac{ax-2b}{ax+2a}$; $\dfrac{\dfrac{1}{a}-\dfrac{1}{x}}{\dfrac{1}{a}+\dfrac{1}{x}} = \dfrac{a-\dfrac{1}{x}}{a+\dfrac{1}{x}}$;

$\dfrac{2x^2-3x+5}{7x^2-4x+2} = \dfrac{2}{7}$.

64. $\dfrac{ax^2-bx+c}{mx^2-nx+p} = \dfrac{a}{m}$; $\dfrac{ax^3-bx^2+ax-d}{mx^3-nx^2+mx-q} = \dfrac{ax-b}{mx-n}$.

65. $\dfrac{\frac{1}{4}-x}{\frac{1}{4}+x} + \dfrac{1}{4} = \dfrac{x}{\frac{1}{4}+x} - \dfrac{1}{4}$;

$\dfrac{\frac{2}{3}x-\frac{2}{3}}{\frac{2}{3}-x} - \dfrac{2}{3} = \dfrac{2}{3} + \dfrac{\frac{2}{3}x+\frac{2}{3}}{x-\frac{2}{3}}$.

66. $\dfrac{21}{x-98} - \dfrac{71}{x-94} = \dfrac{21}{x+44} - \dfrac{71}{x-52}$.

67. $\dfrac{7}{x-6} + \dfrac{8}{x-11} = \dfrac{9}{x-7} + \dfrac{1}{x-12}$;

$\dfrac{9}{x-51} - \dfrac{9}{x-15} = \dfrac{2}{x-81} - \dfrac{2}{x+81}$.

68. $\dfrac{5}{x-6} + \dfrac{4}{x-9} = \dfrac{8}{x-7} + \dfrac{1}{x-10}$;

$\dfrac{1}{x-6} + \dfrac{8}{x-3} = \dfrac{5}{x-2} + \dfrac{4}{x-5}$.

69. $\dfrac{m-n}{x-a} - \dfrac{a-b}{x-m} = \dfrac{m-n}{x-b} - \dfrac{a-b}{x-n}$.

70. $\dfrac{a+b}{x-b} - \dfrac{a+c}{x-c} = \dfrac{b+d}{x-(a+b+2c+d)} - \dfrac{c+d}{x-(a+2b+c+c)}$.

71. $(x-a+b)^3 - (x-a)^3 + (x-b)^3 - x^3 + a^3 - (a-b)^3 - b^3$
$= (a-b)c^2$.

72. $(x+a+b)^5 - (a+b)^5 - (x+b)^5 - (x+a)^5 + x^5 + a^5 + b^5$
$= 10abx(2x+a+b)(x+a+b)$.

73. $\dfrac{(m-n)(x-a)}{b+c} + \dfrac{(n-p)(x-b)}{c+a} + \dfrac{(p-m)(x-c)}{a+b} = 0$.

SIMPLE EQUATIONS. 165

74. $\dfrac{ax-1}{a^2(c+b)} + \dfrac{bx-1}{b^2(c+a)} + \dfrac{cx-1}{c^2(a+b)} = \dfrac{3x}{ab+bc+ca}.$

75. $\dfrac{x-2a}{b+c-a} + \dfrac{x-2b}{c+a-b} + \dfrac{x-2c}{a+b-c} = 3.$

76. $\dfrac{x-2a}{b+c-a} + \dfrac{x-2b}{c+a-b} + \dfrac{x-2c}{a+b-c} = \dfrac{3x}{a+b+c}.$

77. $\dfrac{a-x}{a^2-bc} + \dfrac{b-x}{b^2-ac} + \dfrac{c-x}{c^2-ab} = \dfrac{3}{a+b+c}.$

78. $\dfrac{x+2ab}{a+b+c} + \dfrac{2ab-x}{b+c-a} = \dfrac{x-2ab}{a-b+c} + \dfrac{x+2ab}{a+b-c}.$

79. $\dfrac{a}{x+b-c} + \dfrac{b}{x+a-c} = \dfrac{a-c}{x+b} - \dfrac{b+c}{x+a}.$

80. $\dfrac{m^2(a-b)}{x-m} + \dfrac{n^2(b-c)}{x-n} + \dfrac{p^2(c-d)}{x-p} +$
$\dfrac{q\{pd+(n-p)c+(m-n)b-ma\}}{x-q} = 0.$

81. $\dfrac{(x-2)(x-5)(x-6)(x-9)+(a+2)(a-4)(a-5)(a-11)}{x}$
$+ \dfrac{(b+1)(b+5)(b+8)(b+12)}{x} = (x-4)(x-7)(x-11) +$
$\dfrac{(a^2-1)(a-8)(a-10)+(b+2)(b+3)(b+10)(b+11)}{x}.$

Art XLV. Employing the language of algebra, the principle illustrated in Art. XL. may be stated as follows:

DEFINITION.—Any quantity which substituted for x makes the expression $f(x)$ vanish, is said to be a root of the *equation* $f(x) = 0$. Thus, if a is a root of the equation $f(x) = 0$, then $f(a) = 0$.

By Th. I., if $x-a$ is a factor of the *polynome* $f(x)^n$, then $f(a)^n = 0$, and a must be a root of the *equation* $f(x)^n = 0$; hence in solving the equation we are merely finding a value, or values, of x which will make the corresponding polynome vanish. Suppose $f(x)^n = (x-a)\varphi(x)^{n-1} = 0$, we are required to find a value, or values, of x which will make $(x-a)\varphi(x)^{n-1}$ vanish. The polynome will certainly vanish if *one* of its factors vanishes, whether the other does or not, and will not vanish unless at least *one* of its factors vanishes. Hence $(x-a)\varphi(x)^{n-1}$ will vanish if $x-a=0$, quite irrespective of the value of $\varphi(x)^{n-1}$. Also, if $\varphi(x)^{n-1}=0$, the polynome will vanish, irrespective of the value of $x-a$. It follows, therefore, that if $f(x)^n$ can be resolved into two or more factors, each of these factors equated to zero will give one or more roots of the equation $f(x)^n = 0$.

When there can be found two or more values of x which satisfy the conditions of given equations, they are sometimes distinguished thus: x_1, x_2, x_3, &c., to be read "one value of x," "a second value of x," "a third value of x," &c. Thus, if

$$(x-a)(x-b)(x-c) = 0,$$
$$\therefore x_1 = a,\ x_2 = b,\ x_3 = c.$$

EXAMPLES.

1. Solve $2x^3 - 13x^2 + 27x - 18 = 0$.

    Factoring,
    $$(x-2)(x-3)(2x-3) = 0,$$
    $$\therefore x_1 = 2,\ x_2 = 3,\ x_3 = 1\tfrac{1}{2}.$$

2. $x^2 - (a+b)x + (a+c)b = (a+c)c,$
    $\therefore x^2 - (a+b)x + (a+c)(b-c) = 0,$
    $\therefore x^2 - \{(a+c)+(b-c)\}x + (a+c)(b-c) = 0,$
    $\therefore \{x - (a+c)\}\{x - (b-c)\} = 0,$
    $\therefore x_1 = a+c,\ x_2 = b-c.$

3. $x^2(a-b) + a^2(b-x) + b^2(x-a) = 0.$
    $\therefore x^2(a-b) - x(a^2 - b^2) + ab(a-b) = 0,$
    $\therefore (x-a)(x-b)(a-b) = 0.$

If $a - b = 0$, the given equation holds irrespective of the values of $x - a$ and $x - b$, and therefore of the values of x; but i. $a - b$ is not zero, $x_1 = a,\ x_2 = b$.

4. $x = \dfrac{(a^2+b^2)x-(a^2-b^2)}{(a^2-b^2)x-(a^2+b^2)}$;

$\therefore \dfrac{x+1}{x-1} = \dfrac{a^2(x-1)}{b^2(x+1)}$ $\therefore \left(\dfrac{x+1}{x-1}\right)^2 - \dfrac{a^2}{b^2} = 0$,

$\therefore \dfrac{x_1+1}{x_1-1} - \dfrac{a}{b} = 0$ $\therefore x_1 = \dfrac{a+b}{a-b}$,

$\dfrac{x_2+1}{x_2-1} + \dfrac{a}{b} = 0$ $\therefore x_2 = \dfrac{a-b}{a+b}$.

5. $\dfrac{(a-x)^2+(b-x)^2}{(a-x)^2+(a-x)(b-x)+(b-x)^2} = \dfrac{34}{49}$.

$\therefore \dfrac{(a-x)^2+2(a-x)(b-x)+(b-x)^2}{(a-x)^2-2(a-x)(b-x)+(b-x)^2} = \dfrac{2(49)-34}{3(34)-2(49)} = 16$,

$\therefore \left\{\dfrac{(a-x)+(b-x)}{(a-x)-(b-x)}\right\}^2 - 4^2 = 0$,

$\therefore \dfrac{(a-x_1)+(b-x_1)}{a-b} - 4 = 0$, $\therefore x_1 = \frac{1}{2}(5b-3a)$;

$\dfrac{(a-x_2)+(b-x_2)}{a-b} + 4 = 0$, $\therefore x_2 = \frac{1}{2}(5a-3b)$.

6. $\dfrac{(x-a)(x-b)}{(c-a)(c-b)} + \dfrac{(x-b)(x-c)}{(a-b)(a-c)} = 1$.

Subtract term by term from the identity (See page 53),

$\dfrac{(x-a)(x-b)}{(c-a)(c-b)} + \dfrac{(x-b)(x-c)}{(a-b)(a-c)} + \dfrac{(x-c)(x-a)}{(b-c)(b-a)} = 1$

$\therefore (x-c)(x-a) = 0$, $\therefore x_1 = c$, $x_2 = a$.

7. Find the *rational* roots of $x^4 - 12x^3 + 51x^2 - 90x + 56 = 0$.
Factoring the left-hand member by the method of Art. xxviii.,

$(x-2)(x-4)(x^2-6x+7) = 0$

$\therefore x_1 = 2$, $x_2 = 4$, or $x^2 - 6x + 7 = 0$.

Since $x^2 - 6x + 7$ cannot be resolved into rational factors we know that it will not give rational roots, $\therefore x_1 = 2$, $x_2 = 4$ are the only values that meet the condition of the problem.

Any literal equation of the second, third, or fourth degree, and many equations of the higher degree can be resolved into a series of disjunctive equations. A full analysis for the first four degrees will be given in Part II., meanwhile the following special forms of the Theorem in Art. XLV., will enable the student to solve nearly all the equations commonly proposed.

(A). In order that two expressions having a common factor may be equal, it is necessary either that the common factor should vanish, or else that the product of the remaining factors of one of the expressions should be equal to the product of the remaining factors of the other expression, and it is sufficient if one of these conditions be fulfilled. In symbols this is

If $(x-a)f(x) = (x-a)\varphi(x)$, $\therefore$ $x_1 = a$ or $f(x) = \varphi(x)$.

(B). If an equation reduces to the form $(mx+n)^2 = c^2$

$\therefore (mx+n)^2 - c^2 = 0$,

$\therefore (mx_1+n) - c = 0$ and $\therefore x_1 = \dfrac{c-n}{m}$,

or $(mx_2+n) + c = 0$ and $\therefore x_2 = \dfrac{-c-n}{m}$.

(C). If an equation reduces to the form

$$\left\{\dfrac{mx+n}{px+q}\right\}^2 = \dfrac{a^2}{b^2},$$

then $x_1 = \dfrac{qa-nb}{mb-pa}$, $x_2 = \dfrac{-qa-nb}{mb+pa}$. (See Exs. 4 and 5 above).

(D). If an equation appears under the form

$(a-x)(x-b) = c$, \hfill (1)

then $x_1 = \tfrac{1}{2}(a+b+r)$, $x_2 = \tfrac{1}{2}(a+b-r)$,

in which $r^2 = (a-b)^2 - 4c$.

From the identity $(a-x)+(x-b) = a-b$

we get $(a-x)^2 + 2(a-x)(x-b) + (x-b)^2 = (a-b)^2$ \hfill (2)

$(2)-4(1)$ $\therefore (a-x)^2 - 2(a-x)(x-b) + (x-b)^2$

$= (a-b)^2 - 4c = r^2$ say

$\therefore \{(a-x) - (x-b)\}^2 - r^2 = 0$,

$\therefore \{(a-x_1) - (x_1-b)\} + r = 0$, and $\therefore x_1 = \tfrac{1}{2}(a+b+r)$;

or $\{(a-x_2) - (x_2-b)\} - r = 0$, and $\therefore x_2 = \tfrac{1}{2}(a+b-r)$.

SIMPLE EQUATIONS.   169

8. $x + \dfrac{1}{x} = a + \dfrac{1}{a}$. $\therefore x - a = \dfrac{1}{a} - \dfrac{1}{x}$,

$\therefore \dfrac{x-a}{1} = \dfrac{x-a}{ax}$. Applying (A),

$\therefore x - a = 0$, or $ax = 1$,

$\therefore x_1 = a$, $x_2 = \dfrac{1}{a}$.

9. $(x+a+b)(x+b+c) = (x-3a+b)(2x-3a+2b-c)$;

$\therefore \dfrac{x+a+b}{x-3a+b} = \dfrac{2x-3a+2b-c}{x+b+c}$

$\phantom{\therefore \dfrac{x+a+b}{x-3a+b}} = \dfrac{x-4a+b-c}{3a+c}$ Page 122. (5).

$\therefore \dfrac{2(x-a+b)}{x-3a+b} = \dfrac{x-a+b}{3a+c}$,

$\therefore (A)\quad x_1 = a - b$

$\tfrac{1}{2}(x_2 - 3a + b) = 3a + c$ $\therefore x_2 = 9a - b + 2c$.

10. $\dfrac{(x+2)^2}{x^2 - 2x} = \dfrac{a}{b}$, $\therefore \dfrac{(x+2)^2}{m(x+2)^2 + n(x^2 - 2x)} = \dfrac{a}{ma+nb}$ (1)

But (C) can be applied if m and n are so determined that $m(x+2)^2 + n(x^2 - 2x)$ is a square.

This requires that $4m(m+n) = (2m-n)^2$,

$\therefore 4m^2 + 4mn = 4m^2 - 4mn + n^2$,

$\therefore 8m = n$.

Assume $m = 1$, then $n = 8$, and (1) becomes, on substitution and reduction,

$\dfrac{(x+2)^2}{(3x-2)^2} = \dfrac{a}{a+8b} = r^2$, say

$\therefore x_1 = \dfrac{2(1+r)}{3r-1}$, $x_2 = \dfrac{2(r-1)}{1+3r}$.

11. $\dfrac{(x+1)^4}{(x^2+1)(x-1)^2} = \dfrac{a}{b}$. $\therefore \dfrac{(x^2+2x+1)^2}{(x^2+1)(x^2-2x+1)} = \dfrac{a}{b}$.

For $x^2 + 1$ write xz

$\therefore \dfrac{(xz+2x)^2}{xz(xz-2x)} = \dfrac{a}{b}$ $\therefore \dfrac{(z+2)^2}{z(z-2)} = \dfrac{a}{b}$.

This equation was solved in Ex. 10, hence z may be treated as known.

But $\dfrac{x^2+1}{x} = z$, $\therefore \dfrac{x^2+2x+1}{x^2-2x+1} = \dfrac{z+2}{z-2}$.

$\therefore \left(\dfrac{x+1}{x-1}\right)^2 = \dfrac{z+2}{z-2}$, a formed solved in (C).

12. $(a-x)^4 + (b-x)^4 = c$.

In the identity
$$(u+v)^4 = u^4 + v^4 + 4(u+v)^2 uv - 2u^2 v^2,$$
Let $u = a-x$, $v = x-b$, $\therefore u+v = a-b$ and $u^4+v^4 = c$,

$\therefore (a-b)^4 = c + 4(a-b)^2(a-x)(x-b) - 2(a-x)^2(x-b)^2$

Write z for $(a-x)(x-b)$

$\therefore z^2 - 2(a-b)^2 z + (a-b)^4 = \tfrac{1}{2}\{c+(a-b)^4\} = t^2$, say,

$\therefore \{z-(a-b)^2\}^2 = t^2$

$\therefore$ by (B) $z_1 = (a-b)^2 - t$; $z_2 = (a-b)^2 + t$, $\therefore z$ is known;

But $(a-x)(x-b) = z$

$\therefore$ by (D) $x_1 = \tfrac{1}{2}(a+b+r)$; $x_2 = \tfrac{1}{2}(a+b-r)$ \hfill (1).

in which $r^2 = (a-b)^2 - 4z$,

$$\left. \begin{array}{l} = (a-b)^2 - 4\{(a-b)^2 - t\} = 4t - 3(a-b)^2 \\ \text{or}\quad (a-b)^2 - 4\{(a-b)^2 + t\} = -4t - 3(a-b)^2 \end{array} \right\} (2)$$

and $t^2 = \tfrac{1}{2}\{c + (a-b)^4\}$. \hfill (3)

Hence x is expressed in terms of a, b, and r,

r is expressed in terms of a, b, and t,

t is expressed in terms of a, b, and c,

and the expressions for r and t are cases of (B).

13. $(a-x)(b+x)^4 + (a-x)^4(b+x) = ab(a^3+b^3)$

Let $a-x = n-z$ and $b+x = n+z$ $\therefore n = \tfrac{1}{2}(a+b)$ \hfill (1).

The equation reduces to

$(n^2 - z^2)\{(n+z)^3 + (n-z)^3\} = ab(a^3+b^3)$

$\therefore (n^2 - z^2)(2n^3 + 6nz^2) = ab(a^3+b^3)$

$\therefore (n^2 - z^2)(n^2 + 3z^2) = ab(a^2 - ab + b^2)$

SIMPLE EQUATIONS. 171

z^2 may now be found by (D), and from the result z may be found by (B), and from (1) $x = \frac{1}{2}(a-b)+z$;
$$3z^2 = \tfrac{3}{4}(a-b)^2 \text{ or } \tfrac{1}{4}(10ab - a^2 - b^2)$$
$\therefore x = 0$, or $a-b$, or $\tfrac{1}{2}(a-b) + \tfrac{1}{6}\sqrt{(30ab - 3a^2 - 3b^2)}$.

14. $\{\sqrt[4]{(a+x)} + \sqrt[4]{(a-x)}\}^2 \{\sqrt{(a+x)} + \sqrt{(a-x)}\} = 2cx$.

Divide the terms of the identity
$$\sqrt[4]{(a+x)^4} - \sqrt[4]{(a-x)^4} = 2x$$
by the corresponding terms of the equation,

$$\therefore \sqrt[4]{\left(\frac{a+x}{a-x}\right)} = \frac{c+1}{c-1}, \quad \therefore \frac{a+x}{a-x} = \left(\frac{c+1}{c-1}\right)^4;$$

$$\therefore x = a \cdot \frac{(c+1)^4 - (c-1)^4}{(c+1)^4 + (c-1)^4}.$$

15. $\sqrt[3]{(a-x)^2} + \sqrt[3]{\{(a-x)(b-x)\}} + \sqrt[3]{(b-x)^2} = \sqrt[3]{(a^2 + ab + b^2)}$

Divide the terms of the identity
$$\sqrt[3]{(a-x)^3} - \sqrt[3]{(b-x)^3} = a - b$$
by the corresponding terms of the equation.

$$\therefore \sqrt[3]{(a-x)} - \sqrt[3]{(b-x)} = \frac{a-b}{\sqrt[3]{(a^2+ab+b^2)}}.$$

Cube, using the form $(u-v)^3 = u^3 - v^3 - 3uv(u-v)$.

$$(a-x) - (b-x) - 3\sqrt[3]{\{(a-x)(b-x)\}} \cdot \frac{a-b}{\sqrt[3]{(a^2+ab+b^2)}}$$

$$= \frac{(a-b)^3}{a^2+ab+b^2} = a - b - \frac{3ab(a-b)}{a^2+ab+b^2},$$

$$\therefore \sqrt[3]{\{(a-x)(b-x)\}} = \frac{ab}{\sqrt[3]{(a^2+ab+b^2)^2}}$$

$$\therefore (a-x)(b-x) = \frac{a^3b^3}{(a^2+ab+b^2)^{\frac{3}{2}}}$$

a form solved in (D).

16. $\dfrac{\{\sqrt{(a-x)} + \sqrt{(x-b)}\}^2}{\sqrt{(a-x)} - \sqrt{(x-b)}} = \sqrt{c}$

Assume $\sqrt{(a-x)} = z \sqrt{(x-b)}$
$$\therefore (a-x) + (x-b) = (z^2+1)(x-b).$$
$$\therefore x - b = \frac{a-b}{z^2+1}$$

172   SIMPLE EQUATIONS.

The proposed equation now becomes
$$\frac{\sqrt{(x-b)(z+1)^2}}{z-1} = \sqrt{c}$$
$$\therefore \frac{(x-b)(z-1)^4}{(z-1)^2} = c.$$
$$\therefore \frac{(z+1)^4}{(z^2+1)(z-1)^2} = \frac{c}{a-b}, \text{ a form solved in Ex. 11.}$$

17. $(x-2)(x-5)(x-6)(x-9)+(y+2)(y-4)(y-5)(y-11)+$
$(z+1)(z+5)(z+8)(z+12) = x(x-4)(x-7)(x-11) +$
$(y+1)(y-1)(y-8)(y-10) + (z+2)(z+3)(z+10)(z+11).$
Let $x' = x^2 - 11x$, $y' = y^2 - 9y$ and $z' = z^2 + 13z$,
$\therefore (x'+18)(x'+30)+(y'-22)(y'+20)+(z'+12)(z'+40) =$
$x'(x'+28) + (y'-10)(y'+8)+(z'+22)(z'+30)$
$\therefore x'^2 + 48x' + 540 + y'^2 - 2y' - 440 + z'^2 + 52z' + 480 =$
$x'^2 + 28x' \qquad + y'^2 - 2y' - 80 + z'^2 + 52z' + 660,$
$\therefore 20x' = 0,\ \therefore x^2 - 11x = 0,\ \therefore x_1 = 0, x_2 = 11.$

Exercise lvi.

What can you deduce from the following statements?
1. $A \cdot B = 0$.   2. $A \cdot B \cdot C = 0$.   3. $(a-b)x = 0$.   4. $12xy = 0$.
5. What is the difference between the equation
$$(x-5y)(x-4y+3) = 0$$
and the simultaneous equations
$$x - 5y = 0 \text{ and } x - 4y + 3 = 0.$$
What values of x will satisfy the following equations?
6. $x(x-a) = 0$.   7. $ax(x+b) = 0$.   8. $(x-a)(bx-c) = 0$.
9. $ax^2 = 3ax$.   10. $x^2 = (a+b)x$.   11. $x(x^2 - a^2) = 0$.
12. $a^2x^3 = b^2x$.   13. $x^2 + (a-x)^2 = a^2$.
14. $x^2 + (a-x)^2 = (a-2x)^2$.   15. $(a-x)^2 + (x-b)^2 = a^2 + b^2$.
16. $(a-x)(x-b) + ab = 0$.
17. $(a-x)^2 - (a-x)(x-b)+(x-b)^2 = a^2 + ab + b^2$.
18. $x^2 - (a-b)x - ab = 0$.
19. $x^3 - (a+b+c)x^2 + (ab+bc+ca)x - abc = 0$.

SIMPLE EQUATIONS. 173

If x must be positive, what value or values of x will satisfy the following equations?

20. $(x-5)(x+4)=0.$ 21. $x^2+29x-30=0.$
22. $x^2-17x-84=0.$ 23. $3x^2+10x+3=0.$
24. $x^4-13x^2+36=0.$ 25. $x^3-2x^2-5x+6=0.$

Solve the following equations:

26. $(a-x)^2+(x-b)^2=(a-b)^2.$
27. $(a-x)^2-(a-x)(x-b)+(x-b)^2=(a-b)^2.$
28. $a^2(a-x)^2=b^2(b-x)^2.$ 29. $a^2(b-x)^2=b^2(a-x)^2.$
30. $(x-a)^3+(a-b)^3+(b-x)^3=0.$ 31. $(x-1)^2=a(x^2-1).$
32. $\dfrac{a-x}{x-b} = \dfrac{x-a}{c+x}.$ 33. $\dfrac{a+b-x}{a-c-x} = \dfrac{a-c+x}{a+c-x}.$
34. $(x-a+b)(x-a+c)=(a-b)^2-x^2.$
35. $(x-a)^2-b^2+(a+b-x)(b+c-x)=0.$
36. $(a+b+c)x^2-(2a+b+c)x+a=0.$
37. $\dfrac{a+b-x}{c} = \dfrac{a+b-c}{x}.$
38. $(a-x)^3+(a-b)^3=(a+b-2x)^3.$
39. $x(a+b-x)+(a+b+c)c=0.$
40. $(n-p)x^2+(p-m)x+m-n=0.$
41. $\dfrac{ax^2-bx+c}{mx^2-nx+p} = \dfrac{c}{p}.$ 42. $\dfrac{ax^2-bx+c}{mx^2-nx+p} = \dfrac{a-b+c}{m-n+p}.$
43. $4x^2+a^2-b^2-2(a+b)x=(a-x)(b+x)-(a+x)(b-x).$
44. $(2a-b-x)^2+9(a-b)^2=(a+b-2x)^2.$
45. $(2a+2c-x)^2=(2b+x)(3a-b+3c-2x).$
46. $(3a-5b+x)(5a-3b-x)=(7a-b-3x)^2.$
47. $(3a-b+x)(3a+b-x)=(5a+3b-3x)^2.$
48. $a(a-b)-b(a-c)x+c(b-c)x^2=0.$

SIMPLE EQUATIONS.

49. $(ab+bc+ca)(x^2+x+1)+(a-b)^2 = (2ac+b^2)(x^2+x+1) + (a-c)^2 x.$

50. $(x+1)(x+3)(x-4)(x-7)+(x-1)(x-3)(x+4)(x+7) = 96.$

51. $(x-1)(x+3)(x-5)(x+9)+(x+1)(x-3)(x+5)(x-9)+18 = 0.$

52. $x + \dfrac{1}{x} = 3\tfrac{1}{3}.$

53. $x + \dfrac{1}{x} = \dfrac{a+b}{a-b} + \dfrac{a-b}{a+b}.$

54. $x - \dfrac{1}{x} = \dfrac{a}{b} - \dfrac{b}{a}.$

55. $\dfrac{a+x}{b+x} + \dfrac{b+x}{a+x} = 2\tfrac{1}{2}.$

56. $\dfrac{a-x}{x-b} + \dfrac{x-b}{a-x} = \dfrac{13}{6}.$

57. $\dfrac{a-x}{b+x} - \dfrac{b+x}{a-x} = \dfrac{m}{n} - \dfrac{n}{m}.$

58. $\dfrac{a}{x} + \dfrac{x}{a} = \dfrac{m}{n}.$

59. $\dfrac{x^2+ax+a^2}{x^2-ax+a^2} = c.$

60. $\dfrac{x^2+a^2}{x^2-ax+a^2} = c.$

61. $\dfrac{x^2+a^2}{(x+a)^2} = c.$

62. $\dfrac{(a-x)^2+(x-b)^2}{(a-x)(x-b)} = \dfrac{5}{2}.$

63. $\dfrac{a-x}{x-b} + \dfrac{x-b}{a-x} = \dfrac{m}{n}.$

64. $\dfrac{(x+a)^2+(x-b)^2}{(x+a)^2-(x-b)^2} = \dfrac{a^2+b^2}{2ab}.$

65. $\dfrac{(a-x)^2-(x-b)^2}{(a-x)(x-b)} = \dfrac{4ab}{(a^2-b^2)}.$

66. $\dfrac{(a-x)^2+(a-x)(x-b)+(x-b)^2}{(a-x)^2-(a-x)(x-b)+(x-b)^2} = \dfrac{49}{19}.$

67. $\dfrac{2a^2+a(a-x)+(a+x)^2}{2a^2+a(a+x)+(a-x)^2} = \dfrac{c+1}{c-1}.$ (Also for $c = 5$).

68. $(5-x)^4+(2-x)^4 = 17.$

69. $x^4+(a-x)^4 = c;\ x^4+(x-4)^4 = 82.$

70. $(a-x)^4+(x-b)^4 = (a-b)^4.$ 71. $(a-x)^5+(x-b)^5 = c.$

72. $x^5+(a-x)^5 = a^5;\ x^5+(6-x)^5 = 1056.$

73. $(a-x)^3(x-b)^2+(a-x)^2(x-b)^3 = a^2b^2(a-b).$

SIMPLE EQUATIONS.

74. $(a-x)(b+x)^4 + (a-x)^2(b+x)^3 + (a-x)^3(b+x)^2 + (a-x)^4(b+x) = (a+b)c$.

75. $\dfrac{(a-x)^4 + (x-b)^4}{(a-x)^2 + (x-b)^2} = \dfrac{41}{20}(a-b)^2$.

76. $\dfrac{(a-x)^5 + (x-b)^5}{(a-x)^4 + (x-b)^4} = \dfrac{211}{97}(a-b)$.

77. $\dfrac{(a-x)^4 + (x-b)^4}{(a-x)^2 + (x-b)^2} = \dfrac{a^4 + b^4}{a^2 + b^2}$.

78. $\dfrac{(a-x)^4 + (x-b)^4}{(a-x)^3 + (x-b)^3} = \dfrac{a^4 + b^4}{a^3 - b^3}$.

79. $\dfrac{(a-x)^5 + (x-b)^5}{(a-x)^3 + (x-b)^3} = \dfrac{a^5 - b}{a^3 - b^3}$.

80. $\dfrac{(a-x)^3}{b-x} + \dfrac{(b-x)^3}{a-x} = \dfrac{a^3}{b} + \dfrac{b^3}{a}$.

81. $\dfrac{a-x}{(x-b)^2} + \dfrac{x-b}{(a-x)^2} = \dfrac{a}{b^2} - \dfrac{b}{a^2}$.

82. $\dfrac{(a-x)^4 + (x-b)^4}{(a+b-2x)^2} = \dfrac{a^4 + b^4}{(a+b)^2}$.

83. $\dfrac{(a-x)^5 + (x-b)^5}{(a+b-2x)^2} = \dfrac{a^5 - b^5}{(a+b)^2}$.

84. $\dfrac{(a-x)^5 + (x-b)^5}{(a-x)^2 + (x-b)^2} = (a-b)^3$.

85. $\dfrac{(a-x)^4 - (x-b)^4}{(a-x) - (x-b)} = \dfrac{(a-b)c}{(a-x)(x-b)}$.

86. $\dfrac{(a-x)^5 + (x-b)^5}{(a-x)^2 + (x-b)^2} = c(a-x)(x-b)$.

87. $\dfrac{(a-x)^3 + (x-b)^3}{(a-x)^4 + (x-b)^4} = \dfrac{c}{(a-x)(x-b)}$.

88. $(1+x^2)^3 = (x^3 - 8)^2$.

SIMPLE EQUATIONS.

89. $\dfrac{x^4+1}{2x(x^2+1)} = \dfrac{a}{b}.$

90. $\dfrac{(x+1)^2(x^2+1)}{(x-1)^2(x^2-x+1)} = \dfrac{a}{b}.$

91. $\dfrac{(x-1)^2 x}{(x^2-x+1)^2} = \dfrac{a}{b}.$

92. $\dfrac{(x^2+x+1)^2}{(x+1)^2(x^2+1)} = \dfrac{a}{b}.$

93. $\dfrac{(x^2+1)^2}{x(x+1)^2} = \dfrac{a}{b}.$

94. $\dfrac{(x+1)^4}{x(x^2+1)} = \dfrac{a}{b}.$

95. $\dfrac{x(x+1)^2}{(x-1)^4} = \dfrac{a}{b}.$

96. $\dfrac{x^2+x+1}{(x+1)^2} \cdot \dfrac{x^2+x-1}{(x-1)^2} = \dfrac{a}{b}.$

97. $\dfrac{x^4-x^2+1}{(x^2-1)^2} = \dfrac{a}{b}.$

98. $\dfrac{x(x^2+1)}{(x^2-1)^2} = \dfrac{a}{b}.$

99. $\dfrac{(x+1)(x^3+1)}{(x-1)(x^3-1)} = \dfrac{a}{b}.$

100. $\dfrac{(x+1)(x^5-1)}{(x-1)(x^5+1)} = \dfrac{a}{b}.$

101. $\dfrac{(x+1)^4}{x^4+1} = \dfrac{a}{b}.$

102. $\dfrac{(x+1)^5}{x^5+1} = \dfrac{a}{b}.$

103. $2(a-x)^4 - 9(a-x)^3(x-b) + 14(a-x)^2(x-b)^2 - 9(a-x)(x-b)^3 + 2(x-b)^4 = 0.$

104. $4(a-x)^4 - 17(a-x)^2(x-b)^2 + 4(x-b)^4 = 0.$

Find the rational roots in the following equations:

105. $x^4 - 12x^3 + 49x^2 - 78x + 40 = 0.$ [Let $z = x^2 - 6x$].

106. $x^4 - 6x^3 + 7x^2 + 6x - 8.$

107. $x^4 - 10x^3 + 35x^2 - 50x + 24 = 0.$

108. $32x^4 - 48x^3 - 10x^2 + 21x + 5 = 0.$

109. $x^3 - 6x^2 + 5x + 12 = 0.$

110. $\dfrac{5}{x} - \dfrac{4}{x-a} - \dfrac{9}{x-2a} - \dfrac{4}{x-3a} + \dfrac{5}{x-4a} = 0.$

111. $\dfrac{14}{x+20} + \dfrac{5}{x+5} - \dfrac{4}{x-4} = \dfrac{14}{x-55} + \dfrac{5}{x-40} - \dfrac{4}{x-25}.$

112. $\dfrac{2x+5a}{x} - \dfrac{x+8a}{x-a} + \dfrac{x}{x-2a} = \dfrac{x-a}{x-3a} - \dfrac{x+5a}{x-4a} + \dfrac{2x-5a}{x-5a}$

SIMPLE EQUATIONS.

113. $\dfrac{x+4}{x+2} + \dfrac{x+2}{x} + \dfrac{x+4}{x-1} = \dfrac{x+3}{x-2} + \dfrac{x-1}{x-3} + \dfrac{x-3}{x-5}$.

114. $\dfrac{7}{x} - \dfrac{31}{x-1} + \dfrac{20}{x-2} + \dfrac{8}{x-3} + \dfrac{20}{x-4} - \dfrac{31}{x-5} + \dfrac{7}{x-6} = 0$.

115. $\sqrt{(x^2-a^2-b^2)} + \sqrt{(x^2-b^2-c^2)} - \sqrt{(x^2-c^2-a^2)} = x$.

116. $\dfrac{\sqrt{(a^2+2x)}+\sqrt{(a^2-2x)}}{\sqrt{(a^2+2x)}-\sqrt{(a^2-2x)}} = \dfrac{m^2x^2}{a^2} \cdot \dfrac{\sqrt{(m^2x+2)}+\sqrt{(m^2x-2)}}{\sqrt{(m^2x+2)}-\sqrt{(m^2x-2)}}$

117. $\sqrt{(x^2-a^2)} + \sqrt{(x^2-b^2)} + \sqrt{(x^2-c^2)} = x$.

118. $\{\sqrt{(a-x)}+\sqrt{(b-x)}\}\{\sqrt{(a-x)}-\sqrt{(b-x)}\} = c$.

119. $\dfrac{\sqrt[3]{(a-x)}-\sqrt[3]{(x-b)}}{\sqrt[3]{(a-x)}+\sqrt[3]{(x-b)}} = \dfrac{a+b-2x}{a-b}$.

120. $\sqrt[3]{(a+x)} + \sqrt[3]{(a-x)} = \sqrt[3]{(2a)}$.

121. $\dfrac{\{\sqrt[3]{(a-x)^2}+\sqrt[3]{(x-b)^2}\}^2}{\sqrt[3]{(a-x)}+\sqrt[3]{(x-b)}} = a-b$.

[Write u for $\sqrt[3]{(a-x)}$, and v for $\sqrt[3]{(x-b)}$].

CHAPTER VI.

Simultaneous Equations.

Art. XLVI. There are three general methods of resolving simultaneous linear equations, 1° by substitution, 2° by comparison, 3° by elimination. The last is often subdivided into the method by cross-multipliers, and the method by arbitrary multipliers.

In applying the elimination-method the work should be done with detached coefficients, each equation should be numbered, and a register of the operations performed should be kept.

Ex. Resolve $u+v+x+y+z = 15.$
$u+2v+4x+8y+16z = 57.$
$u+3v+9x+27y+81z = 179.$
$u+4v+16x+64y+256z = 453.$
$u+5v+25x+125y+625z = 975.$

Register

	u	v	x	y	z		
	1	1	1	1	1	= 15	(1)
	1	2	4	8	16	57	(2)
	1	3	9	27	81	179	(3)
	1	4	16	64	256	453	(4)
	1	5	25	125	625	975	(5)
$(2)-(1).$		1	3	7	15	42	(6)
$(3)-(2).$		1	5	19	65	122	(7)
$(4)-(3).$		1	7	37	175	274	(8)
$(5)-(4).$		1	9	61	369	522	(9)
$(7)-(6).$			2	12	50	80	(10)
$(8)-(7).$			2	18	110	152	(11)
$(9)-(8).$			2	24	194	248	(12)
$(11)-(10).$				6	60	72	(13)
$(12)-(11).$				6	84	96	(14)
$(14)-(13).$					24	24	(15)
$(15)\div 24.$					1	1	(16)
$\frac{1}{6}\{(13)-60(16)\}.$				1		2	(17)
$\frac{1}{2}[(10)-\{12(17)+50(16)\}].$			1			3	(18)
$(6)-\{3(18)+7(17)+15(16)\}.$		1				4	(19)
$(1)-\{(19)+(18)+(17)+(16)\}.$	1					5	(20)

SIMULTANEOUS EQUATIONS.

An examination of the Register will show how easy it would have been to shorten the process, thus (10) is $(7)-(6)$ which is $(3)+(1)-2(2)$; similarly (11) is $(4)+(2)-2(3)$; $\therefore$ (13) is $(4)+3(2)-3(3)-(1)$, &c.

A general systematic arrangement of the elimination-method will be given in Part II. For two or three simultaneous equations it may be stated as follows.

$$a_1x+b_1y+c_1 = 0$$
$$a_2x+b_2y+c_2 = 0.$$

Arrange the coefficients thus—

$$\begin{array}{cccc} a_1 & b_1 & c_1 & a_1 \\ a_2 & b_2 & c_2 & a_2 \end{array}$$

Form their products diagonally from left to right downwards, thus— $\quad a_1b_2 \quad b_1c_2 \quad c_1a_2.$

Form their products diagonally from right to left downwards, thus— $\quad b_1a_2 \quad c_1b_2 \quad a_1c_2.$

Subtract the latter products in order from the former, thus—
$$a_1b_2-b_1a_2, \quad b_1c_2-c_1b_2, \quad c_1a_2-a_1c_2.$$

Divide the 2° and 3° remainders by the 1° remainder, the first quotient will be the value of x, the second quotient will be the value of y.

[Writing R_1, R_2, R_3 for the three 'remainders' respectively, the general result is $(mx+ny)R_1 = mR_2 + nR_3$].

Ex. 1. Solve $\quad 11x+5y-68=0$
$\qquad\qquad\qquad 6x-7y+31=0$

$$\begin{array}{rrrr} 11 & 5 & -68 & 11 \\ 6 & -7 & 31 & 6 \\ \hline -77 & 155 & -408 & \\ 30 & 476 & 341 & \\ \hline -107) & -321 & -749 & \\ & 3 & 7 & \\ & \| & \| & \\ & x & y. & \end{array}$$

Ex. 2. $\dfrac{12}{x} - \dfrac{25}{y} = 1.$

$\dfrac{22}{x} + \dfrac{30}{y} = 17.$

```
 12    -25     - 1     12
 22     30     -17     22
────   ────   ────
 360    425   - 22
-550    -30   -204
────   ────   ────
 910)   455    182
         1       1
        ───     ───
         2       5
         ||      ||
         1       1
        ───     ───
         x       y
```

$\therefore x = 2$ and $y = 5$.

2° Let the equations be

$$a_1x + b_1y + c_1z + d_1 = 0$$
$$a_2x + b_2y + c_2z + d_2 = 0$$
$$a_3x + b_3y + c_3z + d_3 = 0.$$

Arrange the coefficients thus

$$\begin{array}{cccccc}
a_1 & b_1 & c_1 & -d_1 & -a_1 & -b_1 \\
a_2 & b_2 & c_2 & -d_2 & -a_2 & -b_2 \\
a_3 & b_3 & c_3 & -d_3 & -a_3 & -b_3 \\
a_1 & b_1 & c_1 & -d_1 & -a_1 & -b_1 \\
a_2 & b_2 & c_2 & -d_2 & -a_2 & -b_2
\end{array}$$

Selecting the first three columns form the diagonal products from left to right downwards, thus:

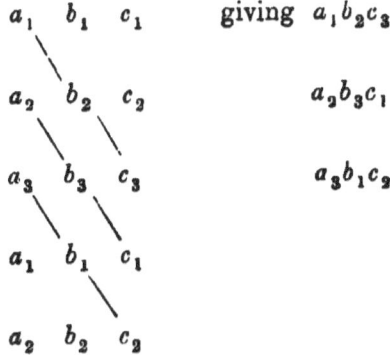

giving $a_1 b_2 c_3$

$a_2 b_3 c_1$

$a_3 b_1 c_2$

Form the diagonal products from right to left downwards, thus:

giving $c_1 b_2 a_3$

$c_2 b_3 a_1$

$c_3 b_1 a_2$

From the sum of the former products take the sum of the latter products obtaining a remainder, which call R_1.
Similarly form a 2° remainder, R_2 from the 2°, 3° and 4° columns
        a 3°   " R_3 "   3°, 4° and 5°   "
        a 4°   " R_4 "   4°, 5° and 6°   "

Then $x = R_2 \div R_1$, $y = R_3 \div R_1$, $z = R_4 \div R_1$, and generally

$$(mx + ny + pz)R_1 = mR_2 + nR_3 + pR_4.$$

Ex. 3. $3x + 2y - 4z + 20 = 0$
          $5x - 7y - 6z - 1 = 0$
          $7x + 5y + 5z - 24 = 0.$

```
      3     2    -4   -20   -3   -2
      5    -7   -6     1   -5    7
      7     5    5    24   -7   -5
      3     2   -4   -20   -3   -2
      5    -7   -6     1   -5    7
```

```
   -105  -288    28  -500      (3 × -7 × 5 = -105
   -100   700   432    14       5 × 5 × -4 = -100
   - 84  - 20   500  -504       7 × 2 × -6 = - 84, &c)
   -----------------------
   -289   392   960  -990
```

```
    196   600  - 15   240      (-4 × -7 × 7 = 196
    -90    10   480   980       -6 × 5 × 3  =  -90
     50   672  -840    15        5 × 2 × 5 = 50, &c.)
   -----------------------
    156  1282  -375  1235
   -445) -890  1335 -2225
   -----------------------
           2    -3     5
           ‖    ‖      ‖
           x    y      z
```

Exercise lvii.

Solve the following systems of equations:

1. $2x+3y=41$
   $3x+2y=39$

2. $5x+7y=17$
   $7x-5y=9.$

3. $11x+12y=100$
   $9x+8y=80.$

4. $18x-35y+13=0$
   $15x+28y-275=0.$

5. $3x+7y=7$
   $5x+3y=-36.$

6. $3x+16y-5=0$
   $28y=5x+19.$

7. $5x+3y+2=0$
   $3x+2y+1=0$

8. $21x+8y+66=0$
   $23y-28x+13=0.$

9. $10x+7y+4=0$
   $6x+5y+2=0.$

10. $23x+15y-4\frac{1}{4}=0$
    $32x+21y-6=0.$

11. $\frac{1}{3}x+\frac{1}{4}y=6.$
    $3x-4y=4.$

12. $\frac{1}{2}x-\frac{1}{3}y=1.$
    $\frac{1}{3}x-\frac{2}{3}y+5=0.$

SIMULTANEOUS EQUATIONS.

13. $\frac{1}{3}y = \frac{1}{2}x - 1.$
    $\frac{1}{4}y = \frac{2}{5}x - 1.$

14. $\frac{2}{3}x + \frac{3}{5}y = 17.$
    $\frac{3}{4}x + \frac{2}{3}y = 19.$

15. $1{\cdot}5x - 2y = 1.$
    $2{\cdot}5x - 3y = 6.$

16. $7x = 10y + {\cdot}1.$
    $11x = 16y + {\cdot}1.$

17. $5x - 4y + 1 = 0.$
    $1{\cdot}7x - 2{\cdot}2y + 7{\cdot}9 = 0.$

18. ${\cdot}16x - {\cdot}04y = 1.$
    ${\cdot}19x - {\cdot}11y = 1.$

19. $3{\cdot}5x + 2\tfrac{1}{3}y = 13 + 4\tfrac{1}{7}x - 3{\cdot}5y.$
    $2\tfrac{1}{7}x + {\cdot}8y = 22\tfrac{1}{2} + {\cdot}7x - 3\tfrac{1}{3}y.$

20. $\dfrac{1}{x} + \dfrac{1}{y} = \dfrac{5}{6}.$
    $\dfrac{1}{x} - \dfrac{1}{y} = \dfrac{1}{6}.$

21. $\dfrac{{\cdot}3}{x} + \dfrac{8}{y} = 3.$
    $\dfrac{15}{x} - \dfrac{4}{y} = 4.$

22. $\dfrac{1{\cdot}6}{x} = \dfrac{2{\cdot}7}{y} - 1.$
    $\dfrac{{\cdot}8}{x} + \dfrac{3{\cdot}6}{y} = 5.$

23. $17x - \dfrac{{\cdot}3}{y} = 3.$
    $16x - \dfrac{{\cdot}4}{y} = 2.$

24. $\dfrac{x}{3} + \dfrac{5}{y} = 4\tfrac{1}{3}.$
    $\dfrac{x}{6} + \dfrac{10}{y} = 2\tfrac{2}{3}.$

25. $\dfrac{5x}{{\cdot}7} + \dfrac{{\cdot}3}{y} = 6.$
    $\dfrac{10x}{7} + \dfrac{9}{y} = 31.$

26. $\tfrac{3}{4}x - \tfrac{1}{2}(y+1) = 1.$
    $\tfrac{1}{2}(x+1) + \tfrac{3}{4}(y-1) = 9.$

27. $\dfrac{5}{x + 2y} = \dfrac{7}{2x + y}.$
    $\dfrac{7}{3x - 2} = \dfrac{5}{6 - y}.$

28. $\dfrac{1}{3x+1} = \dfrac{2}{5y+4}.$
    $\dfrac{1}{4x-3} = \dfrac{2}{7y-6}.$

29. $\dfrac{x + 3y}{x - y} = 8.$
    $\dfrac{7x - 13}{3y - 5} = 4.$

30. $\dfrac{15x + 1}{45 - y} = 8.$
    $\dfrac{12y + 19}{x - 10} = 25.$

31. $\dfrac{3x + 1}{4 - 2y} = \dfrac{4}{3}.$
    $x + y = 1.$

SIMULTANEOUS EQUATIONS.

32. $\dfrac{7-2x}{5-3y} = \dfrac{3}{2}$.

$\dfrac{7-2y}{5-3x} = \dfrac{2}{3}$.

33. $\dfrac{x+2y+1}{2x-y+1} = 2$.

$\dfrac{3x-y+1}{x-y+3} = 5$.

34. $\dfrac{x+3y+13}{\cdot 4x+\cdot 5y-2\cdot 5} = 30$.

$\dfrac{\cdot 8x+\cdot 1y+\cdot 6}{5x+3y-23} = \dfrac{1}{2}$.

35. $\dfrac{x+1}{3} - \dfrac{y+2}{4} = \dfrac{2(x-y)}{5}$.

$\dfrac{x-3}{4} - \dfrac{y-3}{3} = 2y-x$.

36. $\dfrac{2x-y+3}{3} - \dfrac{x-2y+3}{4} = 4$.

$\dfrac{3x-4y+3}{4} + \dfrac{4x-2y-9}{3} = 4$.

37. $20(x+1) = 15(y+1) = 12(x+y)$.
38. $(x-2) : (y+1) : (x+y-3) :: 3 : 4 : 5$.
39. $(x-5) : (y+9) : (x+y+4) :: 1 : 2 : 3$.
40. $\dfrac{x+3}{x+1} = \dfrac{y+8}{y+5}$.

$\dfrac{2x-3}{2(y+1)} = \dfrac{5x-6}{5y+7}$.

41. $(x-4)(y+7) = (x-3)(y+4)$.

$(x+5)(y-2) = (x+2)(y-1)$.

42. $(x-1)(5y-3) = 3(8x+1)$.

$(x-1)(4y+3) = 3(7x-1)$.

43. $(x+1)(2y+1) = 5x+9y+1$.

$(x+2)(3y+1) = 9x+13y+2$.

44. $(3x-2)(5y+1) = (5x-1)(y+2)$.

$(3x-1)(y+5) = (x+5)(7y-1)$.

45. $x+y = 37$.
$y+z = 25$.
$z+x = 22$.

46. $2x+2y = 7$.
$7x+9z = 29$.
$y+8z = 17$.

47. $1\cdot 3x - 1\cdot 9y = 1$.
$1\cdot 7y - 1\cdot 1z = 2$.
$2\cdot 9z - 2\cdot 1x = 3$.

48. $5x+3y+2z = 217$.
$5x-3y = 39$.
$3y-2z = 20$.

49. $\frac{1}{5}x - \frac{1}{2}y = 0$.
$\frac{1}{3}x - \frac{1}{4}z = 1$.
$\frac{1}{2}z - \frac{1}{3}y = 2$.

50. $1\frac{1}{3}x + 1\frac{1}{2}y = 10$.
$2\frac{2}{3}y + 2\frac{2}{5}z = 20$.
$3\frac{1}{4}y + 3\frac{2}{5}z = 30$.

SIMULTANEOUS EQUATIONS.

51. $x+y-z=17.$
    $y+z-x=13.$
    $z+x-y=7.$

52. $x+y+z=9.$
    $x+2y+4z=15.$
    $x+3y+9z=23.$

53. $x+y+z=8.$
    $2x+4y+8z=13.$
    $3x+9y+27z=34.$

54. $7x+6y+7z=100.$
    $x-2y+z=0.$
    $3x+y-2z=0.$

55. $3x+2y+3z=110.$
    $5x+y-4z=0.$
    $2x-3y+z=0.$

56. $x+y+z=9.$
    $x+2y+3z=14.$
    $x+3y+6z=20.$

57. $x+2y+3z=32.$
    $2x+3y+z=42.$
    $3x+y+2z=40.$

58. $x+y+2z=34.$
    $x+2y+z=33.$
    $2x+y+z=32.$

59. $3x+3y+z=17.$
    $3x+y+3z=15.$
    $x+3y+3z=13.$

60. $x+2y-z=4{\cdot}6.$
    $y+2z-x=10{\cdot}1.$
    $z+2x-y=5{\cdot}7.$

61. $x+2y-7z=21.$
    $3x+\cdot 2y-z=24.$
    $\cdot 9x+7y-2z=27.$

62. $x+y=1\tfrac{1}{2}z+8.$
    $y+z=2\tfrac{2}{3}y-14.$
    $z+x=3\tfrac{3}{4}x-32.$

63. $\tfrac{1}{2}x+\tfrac{1}{3}y+\tfrac{1}{4}z=36\tfrac{1}{2}.$
    $\tfrac{1}{3}x+\tfrac{1}{4}y+\tfrac{1}{5}z=27.$
    $\tfrac{1}{5}x+\tfrac{1}{6}y+\tfrac{1}{7}z=18.$

64. $2\tfrac{1}{2}x+3\tfrac{1}{3}y+4\tfrac{1}{4}z=140.$
    $3\tfrac{1}{3}x+4\tfrac{1}{4}y+5\tfrac{1}{5}z=175.$
    $2\tfrac{2}{3}x+3\tfrac{3}{4}y+4\tfrac{4}{5}z=157.$

65. $\dfrac{x+1}{y+1}=2.$
    $\dfrac{y+2}{z+1}=4.$
    $\dfrac{z+3}{x+1}=\tfrac{1}{2}.$

66. $\dfrac{3x+y}{z+1}=2.$
    $\dfrac{3y+z}{x+1}=2.$
    $\dfrac{3z+x}{y+1}=2.$

67. $\dfrac{x+y}{y-z}=10.$
    $\dfrac{x+z}{x-y}=9.$
    $\dfrac{y+z}{x+5}=1.$

68. $\dfrac{x+3}{y+z}=2.$
    $\dfrac{y+3}{x+z}=1.$
    $\dfrac{z+3}{x+y}=\tfrac{1}{2}.$

69. $\dfrac{4}{x} - \dfrac{3}{y} = 1.$
$\dfrac{2}{x} + \dfrac{3}{z} = 4.$
$\dfrac{3}{y} - \dfrac{1}{z} = 0.$

70. $\dfrac{6}{x} + \dfrac{4}{y} + \dfrac{5}{z} = 4.$
$\dfrac{3}{x} + \dfrac{8}{y} + \dfrac{5}{z} = 4.$
$\dfrac{9}{x} + \dfrac{12}{y} - \dfrac{10}{z} = 4.$

71. $\dfrac{xy}{x+y} = \dfrac{1}{5}$
$\dfrac{yz}{y+z} = \dfrac{1}{6}.$
$\dfrac{zx}{z+x} = \dfrac{1}{7}.$

72. $\dfrac{xy}{4y-3x} = 20.$
$\dfrac{xz}{2x-3z} = 15.$
$\dfrac{yz}{4y-5z} = 12.$

73. $(x+2)(2y+1) = (2x+7)y.$
$(x-2)(3z+1) = (x+3)(3z-1).$
$(y+1)(z+2) = (y+3)(z+1).$

74. $(2x-1)(y+1) = 2(x+1)(y-1).$
$(x+4)(z+1) = (x+2)(z+2).$
$(y-2)(z+3) = (y-1)(z+1).$

75. $(x+1)(5y-3) = (7x+1)(2y-3).$
$(4x-1)(z+1) = (x+1)(2z-1).$
$(y+3)(z+2) = (3y-6)(3z-1).$

76. $21x+31y+42z = 115.$
$6(2x+y) = 3(3x+z) = 2(y+z)$

77. $15(x-2y) = 5(2x-3z) = 3(y+z).$
$21x+31y+41z = 135.$

78. $6x(y+z) = 4y(z+x) = 3z(x+y).$
$\dfrac{1}{x} + \dfrac{1}{y} + \dfrac{1}{z} = 9.$

79. $3x+y+z = 20.$
$3u+x+4y = 30.$
$3u+6x+z = 40.$
$5u+8y+3z = 50.$

80. $x+z+8y = 33.$
$5u+y+z = 11.$
$4u+x+z = 11.$
$3u+x+y = 11.$

SIMULTANEOUS EQUATIONS. 187

81. $u+x+y+z=144.$
    $u+2x+2y+2z=267.$
    $u+2x+3y+3z=359.$
    $u+2x+3y+4z=410.$

82. $u+x+y+z=24.$
    $u+2x+3y-9z=0.$
    $3u-x-5y+z=0.$
    $2u+3x-4y-5z=0.$

83. $u+x+y+z=60.$
    $u+2x+3y+4z=100.$
    $u+3x+6y+10z=150.$
    $u+4x+10y+20z=210.$

84. $u+x+y+z=1.$
    $2u+4x+8y+16z=5.$
    $3u+9x+27y+81z=15.$
    $4u+16x+64y+256z=35.$

85. $\frac{1}{2}x+\frac{1}{3}y-\frac{1}{3}z=1.$
    $\frac{1}{3}x-\frac{1}{4}y-\frac{1}{5}u=1.$
    $\frac{1}{6}x+\frac{3}{5}z-\frac{1}{2}u=1.$
    $\frac{3}{4}y-\frac{1}{3}z-\frac{1}{3}u=0.$

86. $\frac{1}{2}u-\frac{1}{3}x+\frac{1}{5}y-\frac{1}{7}z=47.$
    $\frac{1}{3}u+\frac{1}{5}x+\frac{1}{7}y-\frac{1}{2}z=37.$
    $\frac{1}{5}u-\frac{2}{7}x+\frac{1}{2}y-\frac{1}{3}z=17.$
    $\frac{2}{7}u-\frac{1}{2}x-\frac{1}{3}y+\frac{1}{5}z=17.$

Art. XLVII. The principle of symmetry is often of use in the solution of symmetrical equations. For from one relation which may be found to exist between two or more of the letters involved, other relations may be derived by symmetry; also, when the value of one of the unknown quantities has been determined, the values of the others can be at once written down, &c.

1. $\quad (x+y)(x+z) = a.$
   $\quad (x+y)(y+z) = b.$
   $\quad (x+z)(y+z) = c.$

Multiply the equations together and extract the square root.
$$\therefore (x+y)(y+z)(z+x) = \sqrt{(abc)}.$$
Divide this equation by the third.
$$\therefore x+y = \frac{\sqrt{(abc)}}{c}, \text{ and therefore, by symmetry,}$$
$$\therefore y+z = \frac{\sqrt{(abc)}}{a},$$
$$\therefore z+x = \frac{\sqrt{(abc)}}{b}.$$
Hence we get
$$x = \frac{ab-bc+ca}{2\sqrt{(abc)}},$$
whence y and z may be derived by symmetry.

2. $x+y+z=0$ .... (1).
    $ax+by+cz=0$ .... (2).
    $bcx+cay+abz+(a-b)(b-c)(c-a)=0$ .... (3).
    $c\times(1)-(2)$ gives $(c-a)x+(c-b)y=0$.
    $\therefore y = \dfrac{(c-a)x}{b-c}$, and similarly,
    $$z = \dfrac{(a-b)x}{b-c}.$$
    Substitute in (3) these values of y and z, and reduce,
    $\therefore x(a-b)(c-a) = (a-b)(b-c)(c-a)$,
    $\therefore$ or $x=(b-c)$, $\therefore y=c-a$, $z=a-b$.

3. $a(yz-zx-xy) = b(zx-xy-yz) = c(xy-yz-zx) = xyz$.
    Divide the first and the last equations by $axyz$;
    $\therefore \dfrac{1}{a} = \dfrac{1}{x} - \dfrac{1}{y} - \dfrac{1}{z}$, and hence, by symmetry,
    $$\dfrac{1}{b} = \dfrac{1}{y} - \dfrac{1}{z} - \dfrac{1}{x},$$
    $$\dfrac{1}{c} = \dfrac{1}{z} - \dfrac{1}{x} - \dfrac{1}{y}.$$
    $\therefore \dfrac{1}{b} + \dfrac{1}{c} = -\dfrac{2}{x}$, and by symmetry,
    $$\dfrac{1}{c} + \dfrac{1}{a} = -\dfrac{2}{y},$$
    $$\dfrac{1}{a} + \dfrac{1}{b} = -\dfrac{2}{z}.$$

4. $ax+by+cz=1$ .... (1).
    $a^2x+b^2y+c^2z=1$ .... (2).
    $a^3x+b^3y+c^3z=1$ .... (3).
    $c\times(1)-(2)$ gives $a(c-a)x+b(c-b)y=c-1$ .... (4).
    $c\times(2)-(3)$ " $a^2(c-a)x+b^2(c-b)y=c-1$ .... (5).
    $b\times(4)-(5)$ " $ab(c-a)x-a^2(c-a)x=b(c-1)-(c-1)$,
    or $a(a-b)(a-c)x=(c-1)(b-1)$,
    $\therefore x = \dfrac{(1-b)(1-c)}{a(a-b)(a-c)}$;
    whence y and z may be derived by symmetry.

5. Eliminate x, y, z, u (which are supposed all different) from the following equations:

$$x = by + cz + du.$$
$$y = cz + du + ax.$$
$$z = du + ax + by.$$
$$u = ax + by + cz.$$

Subtracting the second equation from the first,

$$\therefore x - y = by - ax, \text{ or}$$
$$(1+a)x = (1+b)y = (\text{by symmetry}) (1+c)z = (1+d)u.$$

These relations may be also obtained by adding ax to both members of the first equation, by, to both members of the second equation, &c.

Now divide the first equation by these equals.

$$\therefore \frac{1}{1+a} = \frac{b}{1+b} + \frac{c}{1+c} + \frac{d}{1+d}.$$

And since $\frac{1}{1+a} = 1 - \frac{a}{1+a}$, we have

$$1 = \frac{a}{1+a} + \frac{b}{1+b} + \frac{c}{1+c} + \frac{d}{1+d}.$$

Exercise lviii.

1. Given $ax + by = c$ and that $x = \frac{b'c - bc'}{b'a - ba'}$,

   $a'x + b'y = c'$ derive the value of y

2. Given $bx = ay$ and that $x = \frac{a(dm - cn)}{bc - ad}$,

   $dx + md = cy + nc$ derive the value of y.

3. Given $ax + by + cz = d$. and that $x =$

   $a^2x + b^2y + c^2z = d^2$ $\frac{a(d-b)(d-c)}{d(a-b)(a-c)}$, write down

   $a^3x + b^3y + c^3z = d^3$ the values of y and z.

4. There is a set of equations in x, y, z, u, and w, with corresponding coefficients (a to x, &c.), a, b, c, d, and e; one of the equations is

$x = by + cz + du + ew$, write down the others.

Solve the following equations:

5. $\dfrac{x}{m} + \dfrac{y}{n} = a, \dfrac{y}{n} + \dfrac{z}{p} = b, \dfrac{x}{m} + \dfrac{z}{p} = c.$

6. $x + ay + bz = m, \ y + az + bx = n, \ z + ax + by = p.$

7. $x + ay = l, \ y + bz = m, \ z + cu = n, \ u + dw = p, \ w + ex = r.$

8. Eliminate x, y, z, (supposed to be all different) from the following equations:

$x = by + cz, \ y = cz + ax, \ z = ax + by.$

9. Eliminate x, y, z, from

$\dfrac{x}{y+z} = a, \ \dfrac{y}{z+x} = b, \ \dfrac{z}{x+y} = c.$

10. Having given

$x = by + cz + du + ew,$
$y = cz + du + ew + ax,$
$z = du + ew + ax + by,$
$u = ew + ax + by + cz,$
$w = ax + by + cz + du,$

Shew that $\dfrac{a}{1+a} + \dfrac{b}{1+b} + \dfrac{c}{1+c} + \dfrac{d}{1+d} + \dfrac{e}{1+e} = 1.$

Art. XLVIII. Resolution of Particular Systems of Linear Equations.

Ex. 1.

	$x+y+z = a$	(1)
	$y+z+u = b$	(2)
	$z+u+x = c$	(3)
	$u+x+y = d$	(4)
(1)+(2)+(3)+(4)	$3(u+x+y+z) = a+b+c+d$	(5′)
3(1)	$3(x+y+z) = 3a$	(6′)
$\tfrac{1}{3}\{(5′)-(6′)\}$	$u \qquad = \tfrac{1}{3}(-2a+b+c+d.)$	

The values of x, y and z may now be written down by symmetry.

The following is a variation of the above method, applicable to a much more general system.

Assume the auxiliary equation

$$u+x+y+z=s, \qquad (5)$$

$\therefore$ (1) becomes $\qquad s-u=a, \qquad (6)$
(2) " $\qquad s-x=b, \qquad (7)$
(3) " $\qquad s-y=c, \qquad (8)$
(4) " $\qquad s-z=d, \qquad (9)$

$(5)+(6)+(7)+(8)+(9) \qquad 4s=s+a+b+c+d.$
$\qquad \therefore s=\frac{1}{3}(a+b+c+d).$

s is now a known quantity, and may be treated as such,

in (6) giving $\qquad u=s-a$
" (7) " $\qquad x=s-b$
" (8) " $\qquad y=s-c$
" (9) " $\qquad z=s-d.$

Ex. 2. $\qquad yz=a(y+z), \qquad (1)$
$\qquad zx=b(z+x), \qquad (2)$
$\qquad xy=c(x+y), \qquad (3)$

$(1) \div ayz, \qquad \dfrac{1}{y}+\dfrac{1}{z}=\dfrac{1}{a},$

$(2) \div bzx, \qquad \dfrac{1}{z}+\dfrac{1}{x}=\dfrac{1}{b},$

$(3) \div cxy, \qquad \dfrac{1}{x}+\dfrac{1}{y}=\dfrac{1}{c}.$

This may now be solved like Ex. 1, using the reciprocals of a, b, c, x, y and z instead of these quantities themselves.

Ex. 3. $\qquad a_1u+b_1(x+y+z)=c_1 \qquad (1)$
$\qquad a_2x+b_2(y+z+u)=c_2 \qquad (2)$
$\qquad a_3y+b_3(z+u+x)=c_3 \qquad (3)$
$\qquad a_4z+b_4(u+x+y)=c_4 \qquad (4)$

Assume the auxiliary equation

$$u+x+y+z=s. \qquad (5)$$

SIMULTANEOUS EQUATIONS.

(1) becomes $b_1 s - (b_1 - a_1)u = c_1$

$$\therefore \frac{b_1}{b_1 - a_1} s - u = \frac{c_1}{b_1 - a_1} \qquad (6)$$

Similarly from (2) $\quad \dfrac{b_2}{b_2 - a_2} s - x = \dfrac{c_2}{b_2 - a_2} \qquad (7)$

"   "   (3) $\quad \dfrac{b_3}{b_3 - a_3} s - y = \dfrac{c_3}{b_3 - a_3} \qquad (8)$

"   "   (4) $\quad \dfrac{b_4}{b_4 - a_4} s - z = \dfrac{c_4}{b_4 - a_4} \qquad (9)$

$(5)+(6)+(7)+(8)+(9) \quad \left(\dfrac{b_1}{b_1-a_1} + \dfrac{b_2}{b_2-a_2} + \dfrac{b_3}{b_3-a_3} + \dfrac{b_4}{b_4-a_4}\right) s$

$$= s + \frac{c_1}{b_1 - a_1} + \frac{c_2}{b_2 - a_2} + \frac{c_3}{b_3 - a_3} + \frac{c_4}{b_4 - a_4} \qquad (10)$$

From (10) we can at once get the value of s, which may therefore be treated as a known quantity.

in (6) giving $\qquad u = \dfrac{b_1 s - c_1}{b_1 - a_1}$

and the value of x, y, and z may be obtained from (7), (8) and (9), or they may be written down by symmetry.

Ex. 4.
$$ax + b(y+z) = c \qquad (1)$$
$$ay + b(z+u) = d \qquad (2)$$
$$az + b(u+x) = e \qquad (3)$$
$$au + b(x+y) = f \qquad (4)$$

Assume $\quad u + x + y + z = s \qquad (5)$

$(1)+(2)+(3)+(4) \quad (a+2b)s = c+d+e+f \qquad (6)$

Hence s is a known quantity and may be treated as such.

From (1) and (5) $\quad bs - bu + (a-b)x = c,$

$\therefore bu - (a-b)x = bs - c, \qquad (7)$

Similarly from (2) and (5) $\quad bx - (a-b)y = bs - d, \qquad (8)$

"   "   (3) "   " $by - (a-b)z = bs - e, \qquad (9)$

"   "   (4) "   " $bz - (a-b)u = bs - f, \qquad (10)$

$b(7) + (a-b)(8) \quad b^2 u - (a-b)^2 y = abs - bc - (a-b)d, (11)$

$b(9) + (a-b)(10) \quad b^2 y - (a-b)^2 u = abs - be - (a-b)f, (12)$

SIMULTANEOUS EQUATIONS.   193

$$b^2(11)+(a-b)^2(12) \quad \{b^4-(a-b)^4\}u = abs\{b^2+(a-b)^2\}$$
$$-a\{b^2d+(a-b)^2f\}-b\{b^3(c-d)+(a-b)^2(e-f)\} \quad (13)$$

The values of x, y, and z may now be written down by symmetry.

Ex. 5.
$$a^3+a^2x+ay+z=0,$$
$$b^3+b^2x+by+z=0,$$
$$c^3+c^2x+cy+z=0.$$

The polynome t^3+xt^2+yt+z vanishes for $t=a$, $t=b$, $t=c$;

∴ by Th. II., p. 46, for *all* values of t,

$$t^3+xt^2+yt+z = (t-a)(t-b)(t-c)$$
$$= t^3-(a+b+c)t^2+(ab+bc+ca)t-abc.$$

∴ Th. III., p. 53, $x = -(a+b+c),$
$$y = ab+bc+ca,$$
$$z = -abc.$$

Ex. 6.
$$x+y+z+u=1, \quad (1)$$
$$ax+by+cz+du=0, \quad (2)$$
$$a^2x+b^2y+c^2z+d^2u=0, \quad (3)$$
$$a^3x+b^3y+c^3z+d^3u=0. \quad (4)$$

Employing the method of arbitrary multipliers,

$(4)+l(3)+m(2)+n(1)$

$$\left. \begin{array}{l} a^3x+ \\ +la^2 \\ +ma \\ +n \end{array} \right| \begin{array}{l} b^3y+ \\ +lb^2 \\ +mb \\ +n \end{array} \left| \begin{array}{l} c^3z+ \\ +lc^2 \\ +mc \\ +n \end{array} \right| \begin{array}{l} d^3 \\ +ld^2 \\ +md \\ +n \end{array} \right| u = n \quad (5)$$

To determine x assume

$$b^3+lb^2+mb+n=0, \quad (6)$$
$$c^3+lc^2+mc+n=0, \quad (7)$$
$$d^3+ld^2+md+n=0, \quad (8)$$
$$\therefore x = \frac{n}{a^3+la^2+ma+n} \quad (9)$$

But the system (6), (7), (8) has been solved in Ex. 5, from which it is seen that

$$l = -(b+c+d), \; m = bc+cd+db, \; n = -bcd,$$
and $\quad a^3+a^2l+am+n = (a-b)(a-c)(a-d);$

∴ using these values in (9)
$$x = \frac{-bcd}{(a-b)(a-c)(a-d)}.$$
The values of y, z and u may now be written down by symmetry.

Ex. 7. $\quad \dfrac{x}{m-a} + \dfrac{y}{m-b} + \dfrac{z}{m-c} = 1.$ (1)

$\dfrac{x}{n-a} + \dfrac{y}{n-b} + \dfrac{z}{n-c} = 1.$ (2)

$\dfrac{x}{p-a} + \dfrac{y}{p-b} + \dfrac{z}{p-c} = 1.$ (3)

Assume $1 - \dfrac{x}{t-a} - \dfrac{y}{t-b} - \dfrac{z}{t-c} = \dfrac{t^3 + Bt^2 + Ct + D}{(t-a)(t-b)(t-c)}$ (4)

But in virtue of equations (1), (2) and (3), the first member of (4) vanishes for $t=m$, $t=n$, and $t=p$, and ∴ $t^3 + Bt^2 + Ct + D$ vanishes for the same values of t, and ∴ Th. II. p. 46,
$$t^3 + Bt^2 + Ct + D = (t-m)(t-n)(t-p),$$
∴ (4) becomes $1 - \dfrac{x}{t-a} - \dfrac{y}{t-b} - \dfrac{z}{t-c}$
$$= \frac{(t-m)(t-n)(t-p)}{(t-a)(t-b)(t-c)}.$$

To obtain the value of x multiply both sides of this equation by $(t-a)$,
$$t - a - x - \frac{y(t-a)}{t-b} - \frac{z(t-a)}{t-c} = \frac{(t-m)(t-n)(t-p)}{(t-b)(t-c)}$$
Now t may have any value in this equation; let $t=a$.
$$\therefore x = \frac{(a-m)(a-n)(a-p)}{(a-b)(a-c)}.$$
The substitution $(xyz|abc)$ will give the values of y and z.

Ex. 8. $\quad \dfrac{x+a}{p} = \dfrac{y+b}{q} = \dfrac{z+c}{r}$ (1)

$lx + my + nz = s^2$ (2)

SIMULTANEOUS EQUATIONS. 195

By Art. XXXVII.,
$$\frac{x+a}{p} = \frac{y+b}{q} = \frac{z+c}{r} = \frac{lx+my+nz+la+mb+nc}{lp+mq+nr}$$

(2) $$= \frac{s^2+la+mb+nc}{lp+mq+nr} = R, \text{ say}$$

$$\therefore x = pR-a, \ y = qR-b, \ z = rR-c.$$

Ex. 9.
$$yz+zx+xy = (a+b+c)xyz \quad (1)$$

$$\frac{yz+zx}{a} = \frac{zx+xy}{b} = \frac{xy+yz}{c} \quad (2)$$

$(1) \div xyz$
$$\frac{1}{x} + \frac{1}{y} + \frac{1}{z} = a+b+c \quad (3)$$

$(2) \div xyz.$
$$\frac{\frac{1}{x}+\frac{1}{y}}{a} = \frac{\frac{1}{y}+\frac{1}{z}}{b} = \frac{\frac{1}{z}+\frac{1}{x}}{c} \quad (4)$$

Page 122 and (3)
$$= \frac{\frac{2}{x}+\frac{2}{y}+\frac{2}{z}}{a+b+c} = 2 \quad (5)$$

(4) and (5) $$\therefore \frac{1}{x}+\frac{1}{y} = 2a, \ \frac{1}{y}+\frac{1}{z} = 2b,$$

$$\frac{1}{z}+\frac{1}{x} = 2c. \quad (6)$$

$(3)-(6)$ $$\frac{1}{x} = a-b+c, \ \frac{1}{y} = a+b-c, \ \frac{1}{z} = -a+b+c.$$

Ex. 10.
$$\frac{x+c}{a+b} + \frac{y+b}{a+c} = 2. \quad (1)$$

$$\frac{x-b}{a-c} + \frac{y-c}{a-b} = 2. \quad (2)$$

(1) $$\therefore \frac{x+c}{a+b} - 1 = 1 - \frac{y+b}{a+c}$$

$$\therefore \frac{x-a-b+c}{a+b} = \frac{a+c-b-y}{a+c} \quad (3)$$

Similarly from (2) $\dfrac{x-a-b+c}{a-c} = \dfrac{a-b+c-y}{a-b}$ (4)

(3) and (4) $\therefore x-a-b+c = \dfrac{a+b}{a+c}(a-b+c-y)$

$$= \dfrac{a-c}{a-b}(a-b+c-y).$$

But unless $\dfrac{a+b}{a+c} = \dfrac{a-c}{a-b}$, this cannot be the case except for

$$a-b+c-y = 0,$$

in which case $x-a-b+c = 0$ also,

giving $x = a+b-c$ and $y = a-b+c.$ (5)

If $\dfrac{a+b}{a+c} = \dfrac{a-c}{a-b} \therefore a^2-b^2 = a^2-c^2$ (6)

$b^2-c^2 = 0,$ or $(b+c)(b-c) = 0,$

$\therefore b = c,$ or $b = -c.$

But if $b = +c$ or $-c$, (1) and (2) are one and the same equation; hence if (1) and (2) are independent, (6) cannot be true, thus leaving only the alternative (5).

Ex. 11. $\quad 2ax = (b+c-a)(y+z),$ (1)

$2by = (c+a-b)(z+x),$ (2)

$(x+y+z)^2 + x^2 + y^2 + z^2 = 4(a^2+b^2+c^2)$ (3)

(1) and page 122 (5) $\dfrac{x}{b+c-a} = \dfrac{y+z}{2a} = \dfrac{x+y+z}{b+c+a}$ (4)

(2) " " $\dfrac{y}{c+a-b} = \dfrac{x+z}{2b} = \dfrac{x+y+z}{c+a+b}$ (5)

(4), (5) and " $\therefore \dfrac{x+y+z}{a+b+c} = \dfrac{x}{b+c-a} = \dfrac{y}{c+a-b} = \dfrac{z}{a+b-c}.$

$\therefore \dfrac{x^2}{(b+c-a)^2} = \dfrac{(x+y+z)^2 + x^2 + y^2 + z^2}{(a+b+c)^2 + (b+c-a)^2 + (c+a-b)^2 + (a+b-c)^2}$

Reduction and (3) = $\dfrac{(x+y+z)^2 + x^2 + y^2 + z^2}{4(a^2+b^2+c^2)} = 1.$

$\therefore x^2 = (b+c-a)^2.$

Ex. 12. $\quad ax = by = cz = \dfrac{1}{x} + \dfrac{1}{y} + \dfrac{1}{z}$ (1)

$(1) \div xyz \quad \therefore \quad \dfrac{a}{yz} = \dfrac{b}{zx} = \dfrac{c}{xy} \therefore = \dfrac{a+b+c}{xy+yz+zx}$ (2)

Also from $(1) \div xyz \quad \dfrac{a}{yz} = \dfrac{1}{xyz}\left(\dfrac{1}{x}+\dfrac{1}{y}+\dfrac{1}{z}\right) = \dfrac{xy+yz+zx}{x^2y^2z^2}$ (3)

$(2)\times(3) \quad \dfrac{a^2}{y^2z^2} = \dfrac{a+b+c}{x^2y^2z^2}$

$\therefore a^2x^2 = a+b+c.$

Ex. 13. $\quad \dfrac{y+z-x}{a} = \dfrac{z+x-y}{b} = \dfrac{x+y-z}{c}$ (1)

$\quad xyz = m^3$ (2)

(1) $\quad \dfrac{z}{a+b} = \dfrac{x}{b+c} = \dfrac{y}{c+a}, = \dfrac{m}{r}$ suppose (3)

then $\quad \dfrac{xyz}{(a+b)(b+c)(c+a)} = \dfrac{m^3}{r^3}$

$\therefore r^3 = (a+b)(b+c)(c+a)$

Hence the value of r is known and from (3)

$\quad rx = m(b+c).$

Ex. 14. $\quad y+z = 2axyz$ (1)
$\quad z+x = 2bxyz$ (2)
$\quad x+y = 2cxyz$ (3)

$\therefore xyz = \dfrac{y+z}{2a} = \dfrac{z+x}{2b} = \dfrac{x+y}{2c} = \dfrac{x+y+z}{a+b+c}$

$\quad = \dfrac{x}{b+c-a} = \dfrac{y}{c+a-b} = \dfrac{z}{a+b-c}$ (4)

$\therefore x^3y^3z^3 = \dfrac{xyz}{(b+c-a)(c+a-b)(a+b-c)}$

$\therefore x^2y^2z^2 = \dfrac{1}{(b+c-a)(c+a-b)(a+b-c)}.$

Hence the value of $x^2y^2z^2$ is known, call it $\dfrac{1}{r^2}$ and substitute in (4)

$$\frac{1}{r} = \frac{x}{b+c-a}$$

$$\therefore rx = b+c-a,$$

in which $r^2 = (b+c-a)(c+a-b)(a+b-c)$.

Ex. 15.
$$y^2 + z^2 - x(y+z) = a \qquad (1)$$
$$z^2 + x^2 - y(z+x) = b \qquad (2)$$
$$x^2 + y^2 - z(x+y) = c \qquad (3)$$

(1)+(2)+(3) $\quad 2(x^2+y^2+z^2-xy-yz-zx) = a+b+c \quad (4)$

(1) may be written $\quad x^2+y^2+z^2-x(x+y+z) = a \quad (5)$

(2) " " $\quad x^2+y^2+z^2-y(x+y+z) = b \quad (6)$

(3) " " $\quad x^2+y^2+z^2-z(x+y+z) = c \quad (7)$

$$\therefore x+y+z = \frac{a-b}{y-x} = \frac{b-c}{z-y} = \frac{c-a}{x-z}$$

$$\therefore (x+y+z)^2 = \frac{(a-b)^2 + (b-c)^2 + (c-a)^2}{(y-x)^2 + (z-y)^2 + (x-z)^2}$$

$$= \frac{a^2+b^2+c^2-ab-bc-ca}{x^2+y^2+z^2-xy-yz-zx}$$

(4) $\qquad = \dfrac{2(a^2+b^2+c^2-ab-bc-ca)}{a+b+c} \qquad (8)$

$\qquad\qquad = \dfrac{2(a^3+b^3+c^3-3abc)}{(a+b+c)^2} \qquad (9)$

Write r^2 for $2(a^3+b^3+c^3-3abc)$.

(9) $\qquad \therefore x+y+z = \dfrac{r}{a+b+c} \qquad (10)$

Returning to (8) $(x+y+z)^2 = \dfrac{2(a^2+b^2+c^2-ab-bc-ca)}{a+b+c} \qquad (8)$

(4) $\quad 2(x^2+y^2+z^2-xy-yz-zx) = \dfrac{(a+b+c)^2}{a+b+c} \qquad (11)$

SIMULTANEOUS EQUATIONS. 199

$\frac{1}{2}\{(8)+(11)\}$ $\quad x^2+y^2+z^2 = \dfrac{a^2+b^2+c^2}{a+b+c}$ (12)

(5) and (10) $\quad x^2+y^2+z^2 - \dfrac{rx}{a+b+c} = a$

$\therefore rx = (a+b+c)(x^2+y^2+z^2) - a(a+b+c)$

(12) $\quad = a^2+b^2+c^2 - a(a+b+c)$

$\quad = b^2+c^2 - a(b+c).$

(5), (6), (7) are symmetrical with respect to $(xyz|abc)$; (10) shows this substitution does not affect r, and consequently the values of y and z may be written down at once from that of x.

Exercise lix.

1. $ax+by=c,$
   $mx+ny=d.$

2. $ax+by=c,$
   $mx-ny=d.$

3. $ax+by=c,$
   $mx+ny=c.$

4. $\dfrac{x}{a}+\dfrac{y}{b}=1,$
   $x+y=c.$

5. $\dfrac{x}{a}+\dfrac{y}{b}=1,$
   $\dfrac{x}{b}+\dfrac{y}{a}=1,$

6. $\dfrac{x}{a}+\dfrac{y}{b}=1,$
   $\dfrac{x}{b}=\dfrac{y}{a}.$

7. $ax+bc=by+ac.$
   $x+y=c.$

8. $\dfrac{a}{x}+\dfrac{b}{y}=m,$
   $\dfrac{b}{x}+\dfrac{a}{y}=n.$

9. $(a+c)x-(a-c)y=2ab,$
   $(a+b)y-(a-b)x=2ac.$

10. $\dfrac{x-c}{y-c}=\dfrac{a}{b},$
    $x-y=a-b.$

11. $\dfrac{x}{y}=\dfrac{a}{b},$
    $\dfrac{x+m}{y+n}=\dfrac{c}{d}.$

12. $\dfrac{x+y}{y+1}=\dfrac{a+b+c}{a-b+c},$
    $\dfrac{y-1}{x+1}=\dfrac{a-b-c}{a+b-c}.$

13. $\dfrac{x-a+c}{y-a+b} = \dfrac{b}{c}$,

$\dfrac{y+b}{x+c} = \dfrac{c+a}{b+a}$.

14. $\dfrac{x+c}{a+b} + \dfrac{y+b}{a+c} = 2$,

$\dfrac{x-b}{a-c} + \dfrac{y-c}{a-b} = 2$,

15. $\dfrac{x}{m-a} + \dfrac{y}{m-b} = 1$,

$\dfrac{x}{n-a} + \dfrac{y}{n-b} = 1$.

16. $x+y+z=0$,
$(b+c)x+(a+c)y+(a+b)z = 0$,
$bcx+acy+abz=1$.

17. $x+y+z=l$,
$ax+by+cz=m$,
$\dfrac{x}{l-a} + \dfrac{y}{l-b} + \dfrac{z}{l-c} = 1$.

18. $\dfrac{x-a}{p} = \dfrac{y-b}{q} = \dfrac{z-c}{r}$,

$l(x-a)+m(y-b)+n(z-c) = 1$.

19. $\dfrac{x-a}{p} = \dfrac{y-b}{q} = \dfrac{z-c}{r}$,

$lx+my+nz=1$.

20. $a(x-a) = b(y-b) = c(z-c)$,
$ax+by+cz = m^2$.

21. $x+y+z = a+b+c$,
$bx+cy+az = a^2+b^2+c^2$,
$cx+ay+bz = a^2+b^2+c^2$.

22. $x+y+z=0$,
$ax+by+cz = ab+bc+ca$,
$(b-c)x+(c-a)y+(a-b)z = 0$.

23. $x+y+z=m$,
$x:y:z = a:b:c$.

24. $ax+by+cz=r$,
$mx = ny$, $qy = pz$.

25. $xy+yz+zx = 0$, $ayz+bzx+cxy = 0$,
$bcyz+acxz+abxy+(a-b)(b-c)(c-a)xyz = 0$.

26. $(a+b)x+(b+c)y+(c+a)z = ab+bc+ca$,
$(a+c)x+(a+b)y+(b+c)z = ab+ac+bc$,
$(b+c)x+(a+c)y+(a+b)z = a^2+b^2+c^2$.

27. $mx+ny+pz+qu = r$,

$\dfrac{x}{a} = \dfrac{y}{b} = \dfrac{z}{c} = \dfrac{u}{d}$.

SIMULTANEOUS EQUATIONS.   201

28. $\dfrac{x(y+z)}{a} = \dfrac{y(x+z)}{b} = \dfrac{z(x+y)}{c}$, $z = \dfrac{1}{a+b-c}$

$\dfrac{1}{x} + \dfrac{1}{y} + \dfrac{1}{z} = a+b+c.$

29. $(a-b)(x+c) - ay + bz = (c-a)(y+b) - cz + ax = 0,$
    $x+y+z = 2(a+b+c).$

30. $ax + by = 1,$ 31. $ly + mx = n,$
    $by + cz = 1,$ $nx + lz = m,$
    $cz + ax = 1,$ $mz + ny = l.$

32. $x + y = a,$ 33. $y+z-x = \dfrac{mn}{l},$
    $y + z = b,$ $z+x-y = \dfrac{ln}{m},$
    $x + z = c.$ $x+y-z = \dfrac{lm}{n},$

34. $\dfrac{1}{y} + \dfrac{1}{z} = 2a,$ 35. $\dfrac{1}{y} + \dfrac{1}{z} - \dfrac{1}{x} = \dfrac{2}{a},$

    $\dfrac{1}{z} + \dfrac{1}{x} = 2b,$ $\dfrac{1}{z} + \dfrac{1}{x} - \dfrac{1}{y} = \dfrac{2}{b},$

    $\dfrac{1}{x} + \dfrac{1}{y} = 2c.$ $\dfrac{1}{x} + \dfrac{1}{y} - \dfrac{1}{z} = \dfrac{2}{c}.$

36. $(a+b)x + (a-b)z = 2bc,$
    $(b+c)y + (b-c)x = 2ac,$
    $(c+a)z + (c-a)y = 2ab.$

37. $x + \dfrac{y}{b} - \dfrac{z}{c} = a,$ 38. $\dfrac{x}{b+c} + \dfrac{y}{c+a} = b-a,$

    $y + \dfrac{z}{c} - \dfrac{x}{a} = b,$ $\dfrac{x}{b+c} + \dfrac{z}{a+b} = a-c,$

    $z + \dfrac{x}{a} - \dfrac{y}{b} = c.$ $\dfrac{y}{c+a} + \dfrac{z}{a+b} = c-b.$

39. $x+y-z=a,$
$y+z-v=b,$
$z+v-x=c,$
$v+x-y=d.$

40. $u+v-x=a,$
$v+x-y=b,$
$x+y-z=c,$
$y+z-u=d,$
$z+u-v=e.$

Exercise lx.

Resolve

1. $(a+b)x+(a-b)y=2(a^2+b^2)$
$(a-b)x+(a+b)y=2(a^2-b^2)$

2. $x+y=a,$
$x^2-y^2=b.$

3. $2x-3y=m,$
$2x^2-3y^2=n^2+xy.$

4. $(a-b)x+(a+b)y=a+b,$
$\dfrac{x}{a+b}-\dfrac{y}{a-b}=\dfrac{1}{a+b}.$

5. $(a-b)x+y=\dfrac{a+b+1}{a+b},$
$x+(a+b)y=\dfrac{a-b+1}{a-b}.$

6. $(a+b-c)x-(a-b+c)y$
$=4a(b-c),$
$\dfrac{x}{y}=\dfrac{a+b-c}{a-b+c}.$

7. $\dfrac{x+y}{x-y}=\dfrac{a}{b-c},$
$\dfrac{x+c}{a+b}=\dfrac{y+b}{a+c}.$

8. $\dfrac{x-a}{y-a}=\dfrac{a-b}{a+b}.$
$\dfrac{x}{y}=\dfrac{a^3-b^3}{a^3+b^3}.$

9. $\dfrac{x-y+1}{x-y-1}=a,$
$\dfrac{x+y+1}{x+y-1}=b.$

10. $\dfrac{x+y+1}{x-y+1}=\dfrac{a+1}{a-1},$
$\dfrac{x+y+1}{x-y-1}=\dfrac{1+b}{1-b}.$

11. $\dfrac{x-y+1}{x+y-1}=a,$
$\dfrac{x+y+1}{x-y-1}=b.$

12. $\dfrac{x}{a+b}+\dfrac{y}{a-b}=a+b,$
$\dfrac{x}{a}+\dfrac{y}{b}=2a.$

13. $(a+c)x+(a-c)y=2ab,$
$(a+b)y-(a-b)x=2ac.$

14. $a^2+ax+y=0,$
$b^2+bx+y=0.$

SIMULTANEOUS EQUATIONS.

15. $y+z-x=a,$
    $z+x-y=b,$
    $x+y-z=c.$

16. $7x+11y+z=a,$
    $7y+11z+x=b,$
    $7z+11x+y=c.$

17. $\dfrac{a}{x} + \dfrac{b}{y} - \dfrac{c}{z} = 2ab,$

    $\dfrac{b}{y} + \dfrac{c}{z} - \dfrac{a}{x} = 2bc,$

    $\dfrac{c}{z} + \dfrac{a}{x} - \dfrac{b}{y} = 2ca.$

18. $(a-b)(x+c) - ay+bz = 0,$

    $(c-a)(y+b) - cz+ax = 0,$

    $x+y+z = 2(a+b+c).$

19. $\dfrac{x}{b+c} + \dfrac{y}{c-a} = a+b,$

    $\dfrac{y}{c+a} + \dfrac{z}{a-b} = b+c,$

    $\dfrac{z}{a+b} + \dfrac{x}{b-c} = c+a.$

20. $\dfrac{x}{b+c} + \dfrac{y}{c-a} - \dfrac{z}{a-b} = 0,$

    $\dfrac{x}{b-c} - \dfrac{y}{c-a} + \dfrac{z}{a+b} = 0,$

    $\dfrac{x}{b+c} + \dfrac{y}{c-a} + \dfrac{z}{a+b} = 2a.$

21. $\dfrac{x}{a} + \dfrac{y}{a-1} + \dfrac{z}{a-2} = 1,$

    $\dfrac{x}{b} + \dfrac{y}{b-1} + \dfrac{z}{b-2} = 1,$

    $\dfrac{x}{c} + \dfrac{y}{c-1} + \dfrac{z}{c-2} = 1.$

22. $\dfrac{xy}{x+y} = a,$

    $\dfrac{yz}{y+z} = b,$

    $\dfrac{zx}{z+x} = c.$

23. $\dfrac{x}{a} + \dfrac{y}{b} + \dfrac{z}{c} = \dfrac{x}{b} + \dfrac{y}{c} + \dfrac{z}{a} =$

    $\dfrac{x}{c} + \dfrac{y}{a} + \dfrac{z}{b} = \dfrac{1}{a} + \dfrac{1}{b} + \dfrac{1}{c}.$

24. $\dfrac{x}{a} = \dfrac{y}{b} = \dfrac{z}{c} = \dfrac{u}{d},$

    $mx+ny+pz+qu = r.$

25. $ax = by = cz = du,$

    $y^2 - z^2 = x - u.$

26. $y+z = au,$
    $x+z = bu,$
    $x+y = cu,$
    $\dfrac{1-x}{1-y} = \dfrac{a}{b}.$

27. $x+y = m,$
    $y+z = n,$
    $z+u = a,$
    $u-x = b.$

28. $11x+9y+z-u=a,$
    $11y+9z+u-x=b,$
    $11z+9u+x-y=c,$
    $11u+9x+y-z=d.$

29. $x+ay+a^2z+a^3u+a^4=0,$
    $x+by+b^2z+b^3u+b^4=0,$
    $x+cy+c^2z+c^3u+c^4=0,$
    $x+dy+d^2z+d^3u+d^4=0.$

30. $x+y=a,$
    $y+z=b,$
    $z+u=c,$
    $u+v=d,$
    $v+x=e.$

31. $x+ly=a,$
    $y+mz=b,$
    $z+nu=c,$
    $u+pv=d,$
    $v+qx=e.$

32. $x+y+z=a,$
    $y+z+u=b,$
    $z+u+v=c,$
    $u+v+x=d,$
    $v+x+y=e.$

33. $x-y+z=a,$
    $y-z+u=b,$
    $z-u+v=c,$
    $u-v+x=d,$
    $v-x+y=e.$

34. $x+y+z-u=a,$
    $y+z+u-v=b,$
    $z+u+v-x=c,$
    $u+v+x-y=d,$
    $v+x+y-z=e.$

35. $x+y+z-u-v=a,$
    $y+z+u-v-x=b,$
    $z+u+v-x-y=c,$
    $u+v+x-y-z=d,$
    $v+x+y-z-u=e.$

36. $2x-y-z+2u-v=3a,$
    $2y-z-u+2v-x=3b,$
    $2z-u-v+2x-y=3c,$
    $2u-v-x+2y-z=3d,$
    $2v-x-y+2z-u=3e.$

37. $v-2x+3u-2y+z=a,$
    $x-2y+3v-2z+u=b,$
    $y-2z+3x-2u+v=c,$
    $z-2u+3y-2v+x=d,$
    $u-2v+3z-2x+y=e.$

Exercise lxi.

Resolve the following systems of equations:

1. $\dfrac{1+x+x^2}{1+y+y^2} = a,$
   $\dfrac{1+y+x^2}{1+x+y^2} = b.$

2. $\dfrac{x+1}{y+1} = a\left(\dfrac{x-1}{y-1}\right),$
   $\dfrac{x^2+x+1}{y^2+y+1} = b^2\left(\dfrac{x-1}{y-1}\right)^2.$

3. $\dfrac{(1+x)(1+y)}{(1-x)(1-y)} = \dfrac{1+a}{1-a},$

4. $\dfrac{x+y}{1+xy} = \dfrac{a^2-\alpha^2}{a^2+\alpha^2},$

SIMULTANEOUS EQUATIONS. 205

$$\frac{(1+x)(1-y)}{(1-x)(1+y)} = \frac{1+b}{1-b}. \qquad \frac{x-y}{1-xy} = \frac{b^2-\beta^2}{b^2+\beta^2}.$$

5. $\dfrac{x+y}{1+xy} = \dfrac{a}{b+c},$

   $\dfrac{x-y}{1-xy} = \dfrac{b-c}{a},$

6. $\dfrac{x+y}{1-xy} = \dfrac{2a}{1-a^2},$

   $\dfrac{x-y}{1+xy} = \dfrac{2b}{1-b^2}.$

7. $\dfrac{x+y}{1-xy} = \dfrac{2a\alpha}{a^2-\alpha^2},$

   $\dfrac{x-y}{1+xy} = \dfrac{2b\beta}{b^2-\beta^2}.$

8. $\dfrac{1+xy}{x+y} + \dfrac{x+y}{1+xy} = \dfrac{2a}{m},$

   $\dfrac{1-xy}{x-y} + \dfrac{x-y}{1-xy} = \dfrac{2b}{n}.$

9. $\dfrac{y(1+x^2)}{x(1+y^2)} = a,$

   $\dfrac{y(1-x^2)}{x(1-y^2)} = b.$

10. $y+z = 2axyz,$
    $x+z = 2bxyz,$
    $x+y = 2cxyz.$

11. $\dfrac{y+z-x}{a} = \dfrac{z+x-y}{b} = \dfrac{x+y-z}{c},$

    $xyz = m^3.$

12. $ax = by = cz,$

    $= \dfrac{1}{x} + \dfrac{1}{y} + \dfrac{1}{z}.$

13. $y^2+z^2-x(y+z) = a,$
    $x^2+z^2-y(x+z) = b,$
    $x^2+y^2-z(x+y) = c.$

14. $2ax = (b+c-a)(y+z),$
    $2by = (c+a-b)(x+z),$
    $(x+y+z)^2 + x^2+y^2+z^2 = 4(a^2+b^2+c^2).$

15. $\dfrac{x-1}{y-1} = \dfrac{a-1}{b-1},$

    $\dfrac{x^3-1}{y^3-1} = \dfrac{a^3-1}{b^3-1}.$

16. $\dfrac{x^2+xy+y^2}{x^2-xy+y^2} = \dfrac{x^2+y^2}{a} = \dfrac{xy}{b}.$

17. $x^4+x^2y^2+y^4 = a,$

    $x^2+xy+y^2 = b.$

18. $x^3+y^3 = \dfrac{a}{x-y},$

    $x^2y - xy^2 = \dfrac{d}{x+y}.$

19. $xy + \dfrac{x}{y} = a(x^2+y^2)$

20. $x^3 = a(x^2+y^2) - bxy,$

SIMULTANEOUS EQUATIONS.

$$xy - \frac{x}{y} = b(x^2+y^2) \qquad y^3 = b(x^2+y^2)-axy.$$

21. $4c(x^2+1) = (a+b)(x-y)^2,$
    $4c(y^2-1) = (a-b)(x-y)^2.$

22. $x^3 - u^3 = \dfrac{b+c}{2a}(x^2+xy+y^2)(x+y),$

    $y^3 - u^3 = \dfrac{b-c}{2a}(x^2+xy+y^2)(x+y).$

23. $\dfrac{x+x^2}{y+y^2} = a,$ 24. $\dfrac{x^2+y^2}{xy} = a,$

    $\dfrac{y+x^2}{x+y^2} = b.$ $\dfrac{1+x^2y^2}{xy} = b.$

25. $x(y+z) = a,$ 26. $(x+y)(x+z) = a,$
    $y(z+x) = b,$ $(y+z)(y+x) = b,$
    $z(x+y) = c.$ $(z+x)(z+y) = c.$

27. $x(x+y+z) = a-yz,$ 28. $x^2 - (y-z)^2 = a,$
    $y(x+y+z) = b-zx,$ $y^2 - (z-x)^2 = b,$
    $z(x+y+z) = c-xy.$ $z^2 - (x-y)^2 = c.$

29. $x^2+y^2 = az,$

    $x+y = bz,$ 30. $\dfrac{1}{x^2} + \dfrac{1}{y^2} = \dfrac{2a}{z^2},$

    $x-y = cz.$ $\dfrac{1}{x^2} - \dfrac{1}{y^2} = \dfrac{2b}{z^2},$

                  $\dfrac{1}{x} + \dfrac{1}{y} = \dfrac{1}{c}.$

31. $x^2 - y^2 = az,$

           32. $xy = \dfrac{z-1}{z+1},$
    $x+y = bz,$ $(x-y)(z+1) = 2a,$
    $x-y = cz.$ $(x^2-y^2)(z+1)^2 = 4bz.$

CHAPTER VII.

EXAMINATION PAPERS: EDUCATION DEPARTMENT AND UNIVERSITY OF TORONTO.

I.

1. State the rules for the addition and subtraction of Algebraic quantities. Express in the simplest form

$(b+c-a)x+(c+a-b)y+(a+b-c)z$
$(c+a-b)x+(a+b-c)y+(b+c-a)z$
$(a+b-c)x+(b+c-a)y+(c+a-b)z$

2. State and prove the Index Laws. Assuming these to be general, interpret x^{-m}.

Find the products in the following cases:

(1) $(x^3+6x^2y+12xy^2+8y^3)(x^3-6x^2y+12xy^2-8y^3)$.
(2) $(a+b+c)(b+c-a)(c+a-b)(a+b-c)$.

3. Prove the rule of signs in Division.

Divide: [Apply Horner's Method to (1)]

(1) $x^6-22x^4+60x^3-55x^2+12x+4$ by x^2+6x+1.
(2) $x^4+9+81x^{-4}$ by x^2-3+9x^{-2}. (3) $x^{n^2}-1$ by x^n-1.

4. Find the square roots of

(1) $4x^{4m} - \dfrac{4}{3}x^{5m} + \dfrac{1}{9}x^{6m}$

(2) $\dfrac{a^2}{b^2} + \dfrac{b^2}{c^2} + \dfrac{c^2}{a^2} - 2\dfrac{a}{c} - 2\dfrac{c}{b} + 2\dfrac{b}{a}$.

5. Distinguish between an algebraic equation and an identity.

Solve

(1) $\sqrt[3]{(1-2x)} + \sqrt[3]{(1+2x)} = 3$.

(2) $\dfrac{x-2}{x+2} + \dfrac{x+2}{x-2} = 2 \cdot \dfrac{x+3}{x-3}$.

6. A person bought a certain number of oxen for \$320. If he had been able to purchase four more for the same sum, each would have cost him \$4 less. Find the number of oxen. Explain the negative result.

7. (1) If $\dfrac{a}{b} = \dfrac{c}{d}$ shew that $\dfrac{a^2+2ab+3b^2}{c^2+2cd+3d^2} = \dfrac{b(a-3b)}{d(c-3d)}$.

(2) Find the value of $x^6 - 200x^5 + 198x^4 + 200x^3 - 197x^2 - 397x$ when $x = 199$.

8. Three towns, A, B, C, are at the angles of a triangle. From A to C, through B, the distance is 82 miles; from B to A, through C, is 97 miles; and from C to B, through A, is 89 miles. Find the direct distances through the towns.

II.

1. Prove $x^m \div x^n = x^{m-n}$.

Simplify $(a+b+c)^3 - 3(a+b+c)^2 c + 3(a+b+c)c^2 - c^3$.

2. Prove the rule for finding the L. C. M. of two quantities. Find the L. C. M. of

$a^3 + b^3 + c^3 - 3abc$, and $(a+b)^2 + 2(a+b)c + c^2$.

3. Prove $\dfrac{a}{b} \times \dfrac{c}{d} = \dfrac{ac}{bd}$.

Simplify $\left(\dfrac{1-x^2}{1-x^3} + \dfrac{1-x}{1-x+x^2}\right) \div \left(\dfrac{1+x}{1+x+x^2} - \dfrac{1-x^2}{1+x^3}\right)$.

4. Reduce to their lowest terms $\dfrac{a^{3m} + a^{2m} - 2}{a^{2m} + a^m - 2}$, and

$\dfrac{a(a+2b)+b(b+2c)+c(c+2a)}{a^2-b^2-c^2-2bc}$

5. (1.) If $a^3 - pa^2 + qa - r = 0$, then $x^3 - px^2 + qx - r$ is exactly divisible by $x - a$.

(2.) Prove that $(a+b+c)(bc+ca+ab) - (b+c)(c+a)(a+b)$ is divisible by abc. Is there any other divisor?

6. If $x = \left(\dfrac{a+b}{a-b}\right)^{\frac{2mn}{n-m}}$, then $\dfrac{1}{2}\dfrac{a^2-b^2}{a^2+b^2}(\sqrt[m]{x}+\sqrt[n]{x}) = \left(\dfrac{a+b}{a-b}\right)^{\frac{m+n}{n-m}}$

7. Solve the equations—

(1.) $\dfrac{3-2x}{1-2x} - \dfrac{5-2x}{7-2x} = 1 - \dfrac{4x^2-2}{7-16x+4x^2}$.

(3.) $\dfrac{x+3}{x+4} - \dfrac{x+1}{x+2} = \dfrac{4x+9}{2x+7} - \dfrac{12x+17}{6x+16}$.

8. A person going at the rate of p miles an hour, and desiring to reach home by a certain time, finds, when he has still r miles to go, that, if he were continuing to travel at the same rate, he would be q hours too late. How much must he increase his speed to reach home in time?

9. Of the three digits comprising a number, the second is double of the third; the sum of the first and third is 9, and the sum of the three digits is 17. Find the number.

10. A owes B \$$a$ due m months hence, and also \$$b$ due n months hence. Find the equation which determines the time at which both sums could be paid at once, reckoning interest at 5 per cent. per annum.

III.

1. If $x=10$, $y=11$, $z=12$, find the value of

$$\left\{x^2-(y+z)^2\right\} \times \dfrac{x+y-z}{x+y+z};$$ and subtract

$(y-z)a^2+(z-x)ab+(x-y)b^2$ from

$(y-x)a^2-(y-z)ab-(z-x)b^2$.

2. Divide $a+(a+b)x+(a+b+c)x^2+(a+b+c)x^3+(b+c)x^4 +cx^5$ by $1+x+x^2+x^3$; and find the square root of

$9-24x+58x^2-116x^3+129x^4-140x^5+100x^6$.

3. Solve (1) $\dfrac{4x+5}{x+1} + \dfrac{x+5}{x+4} = \dfrac{2x+5}{x+2} - \dfrac{x^2-10}{x+3}+x$.

(2) $\frac{1}{2}x+\frac{1}{3}y+\frac{1}{4}z=9$, $\frac{1}{3}x+\frac{1}{4}y-\frac{1}{2}z=-1\frac{3}{4}$,
$\frac{1}{4}x-\frac{1}{2}y+\frac{1}{3}z=1$.

4. A boy bought a number of oranges at the rate of 45 cents a dozen; if he had received 20 oranges more for the same money the whole would have cost him only 40 cents a dozen. How many did he buy?

5. A farmer took to market two loads of wheat, amounting together to 75 bushels; he sold them at different prices per bushel, but received on the whole the same amount for each load; had he sold the whole quantity at the lower price he would have received $78.75; but had he sold it at the higher price he would have received $90. Find the number of bushels in each load.

6. Show how to find the square root of $a+\sqrt{b}$.

Find the square root of $1+\sqrt{(1-a^2)}$

7. Solve $\dfrac{6x+5}{2x-7} + \dfrac{4x-1}{x-2} = \dfrac{7x+1}{x-3}$; and find the value of a when $ax^2-36x+81=0$, has equal roots.

8. If $\dfrac{a}{b} = \dfrac{c}{d}$, prove that $\dfrac{a+c}{b+d} = \sqrt[3]{\dfrac{a^3+c^3}{b^3+d^3}}$; and that $\dfrac{a+b}{a-b} = \dfrac{\sqrt{(ac)}+\sqrt{(bd)}}{\sqrt{(ac)}-\sqrt{(bd)}}$.

9. Show that $a^3(b-c)+b^3(c-a)+c^3(a-b)$ is exactly divisible by $a+b+c$; and resolve the expression into its factors.

IV.

1. Multiply $a^2+b^2-c^2+2ab$ by $a^2-b^2+c^2+2ac$, and divide the product by $a^2-b^2-c^2+2bc$.

2. Simplify
$$\dfrac{18a^2b^2}{x+y} \div \left\{ \dfrac{3ab(x-y)}{7(c+d)} \div \left(\dfrac{4(c-d)}{21ab^2} \div \dfrac{8(c^2-d^2)}{a(x^2-y^2)} \right) \right\}$$

3. Find the L.C.M. of $4x^2-9y^2$, $4x^2-10xy+6y^2$, and $6x^2-13xy+6y^2$, and the G.C.M. of $1+x^{\frac{1}{4}}+x+x^{\frac{7}{4}}$ and $2x+2x^{\frac{7}{4}}+3x^2+3x^{\frac{5}{4}}$

4. Obtain the square root of $\frac{1}{2}-\frac{2}{3}\sqrt{\frac{1}{2}}$, and find the value of c when $4x^4-12x^3y+cx^2y^2-12xy^3+4y^4$ is a perfect square.

5. Distinguish between an *equation* and an *identity*. Give an example of each. What value of m makes $(x-3)^2-(x-1)(x-5) = m$ an identity? Can any value of m make it an equation?

6. Reduce to its simplest form
$$\frac{\sqrt{(2+x)}-\sqrt{(1+x)}}{\sqrt{(1+x)}-\sqrt{x}} + \frac{1+\sqrt{\{1-1\div(1+x)\}}}{1+\sqrt{\{1+1\div(1+x)\}}}$$

7. Solve the equations

(1) $\dfrac{2x+5}{x+2} + \dfrac{2x-5}{x-2} - \dfrac{4x-5}{x-1} = 0.$

(2) $73y-5x = (x-5y)(x+3y),$
$\dfrac{2}{x-5y} - \dfrac{5}{x+3y} = \dfrac{7}{33}.$

8. A person performed a journey of $22\frac{1}{2}$ miles, partly by carriage, at 10 miles an hour, and partly by train, at 36 miles an hour, and the remainder by walking, at 4 miles an hour. He did the whole in 1 hour 50 minutes. Had he walked the first portion, and performed the last by carriage, it would have taken him 2 hours $30\frac{1}{4}$ minutes. Find the respective distances by carriage, train and walking.

9. Solve
$$\frac{x+3}{x+4} - \frac{x+1}{x+2} = \frac{4x+9}{2x+7} - \frac{12x+17}{6x+16}.$$

10. What value of y will make $2x^2+3xy+6y^2$ exactly divisible by $x-8$?

If a and b are the roots of the equation $x^2+x+1=0$, show that $a^3-b^3=0$.

V.

1. Multiply
$$4x^2 - \tfrac{2}{5}x + \tfrac{1}{25} \text{ by } 2x + \tfrac{1}{5}.$$
Prove that
$(\tfrac{1}{2}x - y)^3 - (x - \tfrac{1}{2}y)^3$ is exactly divisible by $x + y$.

2. Express in words the meaning of the formula
$$(x+a)(x+b) = x^2 + (a+b)x + ab.$$
Retaining the order of the terms, how will the right-hand member of this expression be affected by changing, in the left-hand member (1) the sign of b only, (2) the sign of a only, (3) the signs of both a and b?

3. Simplify $(a+b)^4 + (a-b)^4 - 2(a^2 - b^2)^2$; and show that
$$(a+b+c)(b+c-a)(a+c-b)(a+b-c) = 4a^2 b^2$$
when $a^2 + b^2 = c^2$.

4. Prove that $\dfrac{a}{b} \div \dfrac{c}{d} = \dfrac{ad}{bc}$.

Simplify
$$\left(\frac{a^2+b^2}{2ab} + 1\right)\left(\frac{ab^2}{a^3+b^3}\right) \div \frac{4a(a+b)}{a^2 - ab + b^2}.$$

5. I went from Toronto to Niagara, 35 miles, in the steamer "City of Toronto" and returned in the "Rothsay," making the round trip in 5 hours and 15 minutes; on another occasion I went in the "Rothsay" (whose speed on this occasion was 1 mile an hour less than usual), from Toronto to Lewiston, 42 miles, and returned in the "City of Toronto," making the round trip in 6 hours and 30 minutes; find the usual rates per hour which these steamers make.

6. Solve

(1) $\dfrac{3}{x} - \dfrac{2}{y} = \dfrac{1}{a}, \quad \dfrac{2}{x} - \dfrac{1}{y} = \dfrac{2}{a}$

(2) $x^2 + 5x = 5\sqrt{(x^2 + 5x + 28)} - 4.$

7. Find three consecutive numbers whose product is 48 times the middle number.

8. If m and n are the roots of $ax^2+bx+c=0$, then
$$ax^2+bx+c=a(x-m)(x-n).$$
Show that if $ax^2+bx+c=0$ has equal roots, one of them is given by the equation
$$(2a^2-2ab)x+ab-b^2=0.$$

9. If $\dfrac{m}{x}=\dfrac{n}{y}$ and $\dfrac{x^2}{a^2}+\dfrac{y^2}{b^2}=1$, prove that
$$\dfrac{m^2}{a^2}+\dfrac{n^2}{b^2}=\dfrac{m^2+n^2}{x^2+y^2}.$$

VI.

1. Simplify
$$\left(\dfrac{ax^2-ay^2+2bxy}{x^2+y^2}\right)^2+\left(\dfrac{by^2-bx^2+2axy}{x^2+y^2}\right)^2.$$

2. Divide $a^3-b^3-c^3-3abc$ by $a-b-c$, and show, without expansion, that
$$(1+x+x^2)^3-(1-x+x^2)^3-6x(x^4+x^2+1)-8x^3=0.$$

3. Resolve into factors $x^4-\tfrac{1}{4}x^2y^2+y^4$, and
$7x^9-9y^8 \ xy+19x+38y-36$; and prove that
$b^2(c+a)+c^2(a+b)-a^2(b+c)+abc$ is exactly divisible by $b+c-a$.

4. Apply Horner's method of division to find the value of
$5x^5+197x^4+200x^3+196x^2-218x-2000$ when $x=-99$, and the value of $6x^5+5x^4-17x^3-6x^2+10x-2$ when $2x^2=-3x+1$.

5. Find what
$$\dfrac{\sqrt{(a+x)}+\sqrt{(a-x)}}{\sqrt{(a+x)}-\sqrt{(a-x)}}\ \text{becomes when}\ x=\dfrac{2ab}{1+b^2}.$$

6. If a and b be any positive numbers, prove that
$$\dfrac{1}{a}+\dfrac{a}{1+a}>1,\ \dfrac{a}{b}+\dfrac{b}{a}>2.$$

7. Solve the equations—

(1) $x^{\frac{1}{3}} + y^{\frac{1}{3}} = 5$,

$x^{-\frac{1}{3}} + y^{-\frac{1}{3}} = \frac{5}{6}$.

(2) $x + 2y + 3z = 14$,
$2x + 3y + z = 11$,
$3x + y + 2z = 11$.

(3) $(x+1)(x+3)(x+4)(x+6) = 16$.

8. There are three consecutive numbers such that the sum of their cubes is equal to $16\frac{2}{7}$ times the product of the two higher numbers: find the numbers.

9. (1) Form an equation three of whose roots are 0, $\sqrt{(-3)}$, and $1 - \sqrt{2}$.

(2) If one of the roots of the equation $x^2 + px + q = 0$, is a mean proportional between p and q, prove that $p^3 = q(1+p)^2$.

10. Two trains start at the same instant, the one from B to A, the other from A to B; they meet in $1\frac{1}{4}$ hours; and the train for A reaches its destination $52\frac{1}{2}$ minutes before the other train reaches B: compare the rates of the trains.

VII.

1. Give some application of the "rule of signs" in Algebraic Multiplication and Division.

2. Find the numerical value of the quantity

$$bc(c-a)(a-b) - ca(a-b)(b-c) + ab(b-c)(c-a),$$

when $a = 10$, $b = \cdot 01$, $c = 0$; and prove that if

$$x = \frac{c^2 - ab}{a+b}, \text{ then will } (a+b) \cdot \frac{a^2 + b^2 + c^2 + x^2}{a+b-c+x}$$

$$= b^2 + c^2 + (a+b)(a+c).$$

3. Investigate a method of finding by inspection the remainder after dividing any rational and integral function of x by $x+a$.

Show that the quantity
$$a^2b^2 - ab^2x - (a^2 + 2b^2)x^2 + ax^3 + 2x^4$$
is divisible by each of the quantities $x+a$, $x+b$, $a-2x$, $b-x$.

4. Investigate the rule for finding the H.C.F. of two algebraic quantities, showing under what limitations factors may be introduced or suppressed at any step.

Find the H.C.F. of

(1) $6x^4 - 7x^3 - 13x^2 + 19x - 6$ and $x^3 + 2x^2 - 1$.

(2) $(x+y)(ax^2 - by^2) - xy(a-b)(x+y)$, and $(x-y)(ax^2 - by^2) + xy(a-b)(x-y)$.

5. Prove, by general reasoning, that the value of a fraction is not altered by multiplying or dividing both the numerator and denominator by the same quantity.

Simplify (1) $\dfrac{13}{12(2x-3)} - \dfrac{7}{12(2x+3)} - \dfrac{x-4}{4x^2+9}$.

(2) $\left\{\dfrac{1}{(x+a)(x-b)} + \dfrac{1}{(x-a)(x+b)}\right\} \div \left\{\dfrac{1}{(x+a)(x+b)} + \dfrac{1}{(x-a)(x-b)}\right\}$.

6. Solve, with respect to x, the equations

(1) $\dfrac{x-18}{4} + \dfrac{2x-24}{11} + \dfrac{11x-34}{22} = \dfrac{7}{44}$

(2) $\dfrac{5x^2+x-3}{5x-4} - \dfrac{7x^2-3x-9}{7x-10} = \dfrac{x-3}{35x^2-78x+40}$.

(3) $x^2 = ax + by$, and $y^2 = bx + ay$.

VIII.

1. Define the terms "power," "root," "index," and "coefficient"; explain also the reasoning by which it is shown that
$$a - (b - c) = a - b + c.$$

2. Multiply $(x^2+xy+y^2)^2$ by $(x-y)^3$.

Find the values of a and b which will make
x^2+ax+b divisible by $x+p$, and also by $x+q$.

3. Divide $x^6+y^6+2x^3y^3$ by $(x+y)^2$, and
$a^8+a^4b^4+b^8$ by $(a^2-ab+b^2)(a^2+ab+b^2)$.

4. Investigate a rule for the extraction of the square root of any algebraic quantity, and deduce the rule for the extraction of the square root of a number.

If to any square number be added the square of half the number immediately preceding it, the sum will be a complete square: viz., the square of half the number immediately following it.

5. Find the square root of

(1) $a^2x^6+2abx^4+(b^2+2ac)x^2+c^2x^{-2}+2bc$.

(2) $\frac{1}{4}x^{\frac{3}{2}} - \frac{1}{3}x^{\frac{5}{4}} + \frac{1}{4}x^{\frac{4}{3}} + \frac{1}{9}x^{\frac{4}{3}} - \frac{1}{6}x^{\frac{11}{4}} + \frac{1}{16}x^{\frac{7}{3}}$.

6. If x^2+ax+b and $x^2+a'x-b$ have a common measure, it will be $x+\dfrac{a+a'}{2}$, and the condition that they may have a common measure is $4b = a^2 - a'^2$.

Find the H. C. F. of $x^4+p^2x^2+p^4$ and $x^4+2px^3+p^2x^2-p^4$.

Find the L. C. M. of $2\frac{1}{2}(x^2+x-20)$, $3\frac{1}{3}(x^2-x-30)$, and $4\frac{1}{6}(x^2-10x+24)$.

7. Find values of a and b which will render the fraction
$$\dfrac{3x^2 - (4a+b)x + a + 2b^2}{5x^2 - (8a+b)x - a + 4b^2}$$
the same, for all values of x.

8. Solve the equation $2+\sqrt{(x+1)(x+6)} - \sqrt{(x-1)(x+5)} = 0$, and account for the circumstance, that the values of x, determined from it, apparently do not satisfy the equation.

IX.

1. Prove that $a(2n+1)(a^2+n\cdot\overline{n+1})-n(2a+1)(n^2+a\cdot\overline{a+1})$
$=(a-n)^3$.

2. If a, b, and c are positive quantities, and if $a>b$ and $c>a-b$, prove that
$$c-(a-b)=c-a+b.$$
Assuming this equation to hold good when a, b and c are unrestricted, prove that the expression $-(-a)$, occurring in an algebraic operation, is equivalent to $+a$.

3. If x^3+ax^2+b and x^3+px+q have a common measure of the form of x^2+mx+n, then $a^3bq=(b-q)^3$

4. Find the H. C. F. of
$a^2-b^2-abxy+abx^{-1}y^{-1}$, and $a^2x^3-b^3y^{-1}+a^2bx^2y-b^2xy^{-2}$.

5. A and B are two numbers, each of two digits. The left-hand digit of A exceeds that of B by x; the excess of A above B is y; but the sum of the digits of B exceeds the sum of the digits of A by z. Prove that $y+z=9x$; and give an example of two such numbers as A and B.

6. If $\dfrac{a}{b}=\dfrac{b}{c}=\dfrac{c}{d}$, prove that each of these ratios
$$=\sqrt[3]{\dfrac{a}{d}} \text{ and also } =\dfrac{a+b+c}{b++d}.$$

7. Solve the equations

(1) $\dfrac{x+a}{x-a}-\dfrac{x-a}{x+a}=\dfrac{b+x}{b-x}-\dfrac{b-x}{b+x}$.

(2) $a(x^2+y^2)-b(x^2-y^2)=2a$
$(a^2-b^2)(x^2-y^2)\quad=4ab$.

8. A farmer buys a sheep for $\$P$ and sells b of them at a gain of 5 per cent. ; at what price ought he to sell the remainder to gain 10 per cent. on the whole ?

9. The sum of three numbers is 70 ; and if the second is divided by the first, the quotient is 2, and the remainder 1 ; but if the third is divided by the second, the quotient is 3, and the remainder is 3 ; what are the numbers.

X.

1. Divide $ax^3 + 2cxyz + by^3 + ax^2(y+z) + by^2(z+x) + 2cxy(x+y)$ by $x+y+z$.

2. Prove that if $x^4 + px^2 + qx + a^2$ be divisible by $x^2 - 1$, it is also divisible by $x^2 - a^2$.

3. Explain the reason for introducing or suppressing factors in the process of finding the H.C.F. of two algebraical quantities.

Why is the name "Greatest Common Measure" objectionable?

Find the H.C.F. of $x^4 - x^3 - x^2 - x - 2$ and $3x^3 - 7x^2 + 3x - 2$.

4. A traveller leaves A for B at the same time that another leaves B for A; the former walks at the rate of 3 miles an hour till he has performed half the distance; he then rests for an hour; after which he resumes his journey, walking now at the rate of 4 miles an hour; the second traveller goes at the rate of 4 miles an hour till he has got over one-third of the distance between B and A; he then rests for 40 minutes; after which he resumes his journey, walking now at the rate of 3 miles an hour. The travellers reach A and B respectively at the same time. Find the distance between A and B.

5. Show by examining the square of $a+b$ how the square root of an algebraical quantity may be found.

Find the square roots of

(1) $25x^4 - 30ax^3 + 49a^2x^2 - 24a^3x + 16a^4$, and

(2) $\dfrac{x^2}{y^2} + \dfrac{y^2}{x^2} - \left(\dfrac{x}{y} + \dfrac{y}{x}\right)\sqrt{2} + \dfrac{5}{2}$.

6. Show that $a^{\frac{m}{n}} = \sqrt[n]{a^m}$, when m and n are integers, and m is divisible by n; and state the principle on which you would maintain the truth of the equation for all values of m and n.

7. Solve the equations

(1) $\dfrac{5x^2 + x - 3}{5x - 4} = \dfrac{7x^2 - 3x - 9}{7x - 10}$.

(2) $(3x-1)^2 + (4x-2)^2 = (5x-3)^2$.

EXAMINATION PAPERS.

8. Two regular polygons are so related that the number of their sides is as 2 to 3, and the magnitude of their angles as 3 to 4; find the figures.

XI.

1. State in words the several operations to be performed in order to obtain the result expressed by the following algebraical expression:
$$\sqrt[4]{\frac{ma^2+nb^2}{m+n}}.$$
Also find its value when $a=b=4$.

2. Two men, A and B, dig a trench in $3\frac{2}{7}$ days. If A were to do more work by one-third than he does, and B more work by one-half than he does, they would dig the trench in $2\frac{26}{89}$ days. In what time would each dig it alone, at his present rate of work?

3. Perform the multiplications in

(1)
$$\left(2x^{\frac{1}{2}}+3y^{\frac{1}{2}}\right)\left(2x^{\frac{1}{2}}-2y^{\frac{1}{2}}\right)\left(4x^{\frac{1}{2}}+6x^{\frac{1}{2}}y^{\frac{1}{2}}+9y^{\frac{1}{2}}\right)\left(4x^{\frac{1}{2}}-6x^{\frac{1}{2}}y^{\frac{1}{2}}+9y^{\frac{1}{2}}\right).$$

(2) $(\frac{1}{4}x^2+\frac{1}{3}xy+\frac{2}{9}y^2)(\frac{1}{4}x^2-\frac{1}{3}xy+\frac{2}{9}y^2)$.

4. Divide

(1) $x^4+9+81x^{-4}$ by x^2-3+9x^{-2}.

(2) $x^4-(a+b+p)x^3+(ap+bp-c+q)x^2-(aq+bq-cp)x-qc$ by x^2-px+q.

5. Show that $x^{2m+1}-x^{2n-1}$ is always divisible by $x\pm 1$, m and n being any positive integers.

6. Define a fraction; and from your definition prove a rule for adding together two fractions with different denominators.

Add together the fractions,

$$\frac{a^2-bc}{(a+b)(a+c)}, \quad \frac{b^2-ca}{(b+c)(b+a)}, \quad \frac{c^2-ab}{(c+a)(c+b)}.$$

7. Solve the following equations:

(1) $\dfrac{x^2+2x+2}{x+1} + \dfrac{x^2+8x+20}{x+4} = \dfrac{x^2+4x+6}{x+2} + \dfrac{x^2+6x+12}{x+3}$.

(2) $(x^2+y^2)\dfrac{x}{y} = \dfrac{100}{3}$, $(x^2-y^2)\dfrac{y}{x} = \dfrac{21}{4}$.

XII.

1. When m and n are whole numbers, and m greater than n, show that $\dfrac{a^m}{a^n} = a^{m-n}$ and that $\dfrac{1}{a^n}$ is correctly symbolized by a^{-n}.

2. Multiply $(a-b)(a+b)(a^2+b^2)(a^4+b^4) \ldots$ to $(n+1)$ factors.

3. Divide $1-x$ by $1-2x$, to 5 terms, and write down the $(r+1)$th term, and the remainder after $(r+1)$ terms.

4. If the number three be divided into any two parts, show that the difference of the squares is three times the difference of the numbers.

5. Find the L. C. M. of $1-8x+17x^2+2x^3-24x^4$, and $1-2x-13x^2+38x^3-24x^4$.

6. What relation must there be between the coefficients m, n, p and q, in order that
$$(x^2+mx+n)^2+px^2+qx$$
may be an exact square for all values of x?

7. Solve the following equations:

(1) $\dfrac{1+x^3}{(1+x)^2} + \dfrac{1-x^3}{(1-x)^2} = a$.

(2) $\dfrac{ax-b^2}{\sqrt{(ax)}+b} = \dfrac{\sqrt{(ax)}-b}{n} - c$.

(3) $\dfrac{xy}{x+y} = 1$, $\dfrac{xz}{x+z} = 2$, and $\dfrac{yz}{y+z} = 3$.

8. Given $x+y+z = ax = by$, find $(x+y+z) \div z$.

9. Find a number expressed in the decimal notation by two digits, whose sum is 10; and such, that if 1 be taken form its double, the remainder will be expressed by the same digits in a reversed order.

XIII.

1. Find the value, when $a = 2\frac{1}{2}$, $b = 3\frac{1}{2}$, $c = 4\frac{1}{2}$ of
$$\frac{b^2c^2(c^2-b^2)+c^2a^2(a^2-c^2)+a^2b^2(b^2-a^2)}{(b+c)(c+a)(a+b)}.$$

2. Show that the value of the expression, in the preceding question, is not altered by changing a into $a+x$, b into $b+x$, and c inte $c+x$.

3. Multiply $(1+a_1x)(1+a_2x)(1+a_3x) \ldots (1+a_nx)$ to 8 terms.

4. A speculator borrows a sum of money at the yearly interest of 7 per cent.; part of the amount he invests at $8\frac{1}{2}$ per cent., and the remainder at 9; and, at the end of the year, he finds that he has made a profit of $75; but, had the former part been invested at 9 per cent., and the latter at $8\frac{1}{4}$, his profit at the end of the year would have been only $65. Find the whole sum borrowed.

5. Given $ax+by=c$, $a'x+b'y=c'$, determine the value of $mx+ny$, and find the conditions under which the value becomes indeterminate.

6. If $\dfrac{a_1}{a_2} = \dfrac{a_2}{a_3} = \ldots = \dfrac{a_{n-1}}{a_n}$,

then will $a_1+a_2+a_3+ \ldots + a_n = \dfrac{a_1^2-a_2a_n}{a_1-a_2}$.

7. Eliminate x and y from the equations
$$x^{\frac{2}{3}} + y^{\frac{2}{3}} = a^{\frac{2}{3}}$$
$$\alpha = x+3x^{\frac{1}{3}}y^{\frac{2}{3}}$$
$$\beta = y+3x^{\frac{2}{3}}y^{\frac{1}{3}}.$$

8. If $ax^2+bx+c=0$ and $a_1x^2+b_1x+c_1=0$, then will
$(ab_1-a_1b)(bc_1-b_1c) = (ac_1-a_1c)^2$.

9. Find that number of two figures to which if the number formed by changing the places of the digits be added, the sum is 121; and if the same two numbers be subtracted, the remainder is 9.

EXAMINATION PAPERS.

XIV.

1. Simplify

$a(b+c)^2 + b(c+a)^2 + c(a+b)^2 - \{(a-b)(a-c)(b+c) + (b-c)(b-a)(c+a) + (c-a)(c-b)(a+b)\}$.

2. State the law of Indices, and prove it for positive integral indices; and assuming it to be general, interpret the expressions x^{-m}, $x^{\frac{m}{n}}$, where m and n are positive integers.

3. Having given the equations,
$x+y+z=0,$ $x'+y'+z'=0,$
$a^2 = x^2 + x'^2,$ $b^2 = y^2 + y'^2,$
$c^2 = z^2 + z'^2 ;$
prove that $a^2(yz - y'z') + b^2(zx - z'x') + c^2(xy - x'y') = 0$.

4. A traveller P sets out to walk from A to B, proceeding at the rate of 3 miles an hour; and, 32 minutes afterwards, another traveller Q sets out to walk from B to A, proceeding at a uniform rate. They meet half way betwixt A and B. P then quickens his pace by 1 mile an hour; and Q slackens his 1 mile an hour. Q reaches A at the same time that P reaches B. Find the distance between A and B.

5. How are equations classified?
Solve the equations—
(1) $mnx + amn = n^2x + am^2$.
(2) $x^4 - x^2 + y^4 - y^2 = 84,$
      $x^2 + x^2y^2 + y^2 = 49.$

6. What two numbers are those whose difference, sum and product are to each other as the three numbers 2, 3, 5?

XV.

1. What is the meaning of the symbols a, a^2, a^3 . . ?
Show *a priori* that $a^0 = 1$; how do you know that $ab = ba$?
How is it proved that the multiplication of like signs gives a positive, and that of unlike signs, a negative result.

2. Find the value of
$$(b-c)^3 + 2(c-a)^3 + (a-b)^3 - 3(b-c)(c-a)(a-b)$$
when $a = 1$, $b = -\frac{1}{2}$, $c = \frac{3}{2}$.

3. Simplify the following expression:
$$(ac - b^2)(ce - d^2) + (ae - c^2)(bd - c^2) - (ad - bc)(be - ca)$$

4. P and Q are travelling along the same road in the same direction. At noon P, who goes at the rate of m miles an hour, is at a point A; while Q who goes at the rate of n miles in the hour, is at a point B, two miles in advance of A. When are they together?

Has the answer a meaning, when $m - n$ is negative? Has it a meaning when $m = n$? If so, state what interpretation it must receive in these cases.

5. Show how to find the Least Common Multiply of two or more algebraic quantities.

(1) $x^2 - ax - 2a^2$, $x^3 + ax^2$ and $ax^2 - x^3$.

(2) $x^3 - x^2y - a^2x + a^2y$ and $x^3 + ax^2 - xy^2 - ay^2$.

In what algebraic operations is the Lowest Common Multiple of two or more quantities required?

6. State and prove the principle upon which the rules of Addition and Subtraction of fractions are founded.

Simplify the following expressions:

(1) $\dfrac{(a+b-c)^2 - d^2}{(a+b)^2 - (c+d)^2} + \dfrac{(b+c-a)^2 - d^2}{(b+c)^2 - (a+d)^2} + \dfrac{(c+a-b)^2 - d^2}{(c+a)^2 - (b+d)^2}$.

(2) $\dfrac{x^2 + y^2 - z^2 + 2xy}{x^2 - y^2 - z^2 + 2yz}$, $\dfrac{a^3 + a^2b}{a^2b - b^3} - \dfrac{a(a-b)}{(a+b)b} - \dfrac{2ab}{a^2 - b^2}$.

7. If $ax - by + c(x - y) = (a - b)(a + b - c)$,
$by - cz + a(y - z) = (b - c)(b + c - a)$,
$cz - ax + b(z - x) = (c - a)(c + a - b)$

then will $a^2(b - c) + b^2(c - a) + c^2(a - b) = 0$.

8. P is a number, of two digits, x being the left hand digit, and y the right. By inverting the digits, the number Q is obtained. Prove that $11 (x+y)(P - Q) = 9 (x-y) (P+Q)$.

XVI.

1. Show that
$$\{(ax+by)^2+(ay-bx)^2\}\{(ax+by)^2-(ay+bx)^2\} = (a^4-b^4)(x^4-y^4);$$ and that
$$2(a-b)(a-c)+2(b-c)(b-a)+2(c-b)(c-a)$$
is the sum of three squares.

2. If $s = a+b+c+$ &c. to n terms, then
$$\frac{s-a}{s} + \frac{s-b}{s} + \frac{s-c}{s} + \&c. = n-1.$$

3. Show that $a-b$, $b-c$, and $c-a$ cannot be all three positive or all three negative.

4. Extract the square root of
$$4x^8+9x^6-12x^4+16x^2+9-2x(6x^6-8x^4+9x^2-12).$$

5. Given
$$ab - \tfrac{1}{2}(a+b)(p+q)+pq = 0,$$
$$cd - \tfrac{1}{2}(c+d)(p+q)+pq = 0,$$
find the value of $p-q$, and show that if either a or b is equal to c or d, then p is equal to q, unless $a+b = c+d$.

6. Find the value of $\dfrac{x}{y}$, having given
$$\frac{x^{2n}-ay^{2n}}{x^{2n}+ay^{2n}} = \frac{x^n-b(x-y)^n}{x^n+b(x-y)^n}.$$

7. Prove that $(a-b)(b-c)(c-a)$ is a common measure of the quantities
$$(a^2-b^2)^5+(b^2-c^2)^5+(c^2-a^2)^5,$$
$$c^4(a-b)+a^4(b-c)+b^4(c-a).$$

8. Find the conditions that $a_1x+b_1y = c_1$, $a_2x+b_2y = c_2$, and $a_3x+b_3y = c_3$ may be satisfied by the same values of x and y.

9. Two persons, A and B, start at the same instant from two stations (c) miles apart, and proceed in the same direction along the line joining the stations with velocities (a) and (b) miles per hour. Find the distance (x) from the stations where A overtakes B, and interpret the result when $a \angle b$.

XVII.

1. Express in symbols the result of subtracting from unity the quotient obtained by dividing the sum of a and b by their product.

2. Multiply together $x+\sqrt{a+b}$, $x-\sqrt{a+b}$, $x+\sqrt{a-b}$ and $x-\sqrt{a-b}$; and divide $24a^3 - 22a^2b + 2a^2c - 5ab^2 + 27abc - 34ac^2 + 6b^3 - 22b^2c + 16bc^2 + 8c^3$ by $3a - 2b + 4c$.

3. If $x+a$ be the H. C. F. of x^2+px+q and $x^2+p'x+q'$, their L. C. M. will be $(x+a)(x+p-a)(x+p'-a)$.

Show that the difference between
$$\frac{x}{x-a} + \frac{x}{x-b} + \frac{x}{x-c} \text{ and } \frac{a}{x-a} + \frac{b}{x-b} + \frac{c}{x-c}$$
is the same whatever values be given to x.

4. Prove, if the four fractions
$$\frac{bx+cy+dz}{b+c+d-a}, \quad \frac{cx+dy+az}{c+d+a-b}, \quad \frac{dx+ay+bz}{d+a+b-c}, \quad \frac{ax+by+cz}{a+b+c-d}$$
are equal to one another, their common value will be equal to $\frac{x+y+z}{2}$ as long as $a+b+c+d$ does not vanish.

5. What do you mean by *solving an equation*. Show that 3 is a root of the equation
$$\sqrt{(x^2-3x+4)} = \frac{3+\sqrt[3]{(x-2)}}{x^2-7}.$$

6. Eliminate x between the equations
$$x^3 + \frac{1}{x^3} + 3\left(x + \frac{1}{x}\right) = m, \text{ and}$$
$$x^3 - \frac{1}{x^3} - 3\left(x - \frac{1}{x}\right) = n.$$

7. If $\frac{1}{a} + \frac{1}{b} - \frac{1}{c} = \frac{1}{a+b-c}$, a, b, c are not all different.

8. A cask, A, contains m gallons of wine and n gallons of water; an another cask, B, contains p gallons of wine and q gallons of water, how many gallons must be drawn from each cask so as to produce by their mixture b gallons of wine and c gallons of water?

XVIII.

1. Multiply together the factors
$$1-x,\ 1+x,\ 1+x^2,\ 1+x^4,\text{ and } 1+x^8,$$
and show that if n is any uneven number, the sum of the nth powers of any two numbers is always divisible by the sum of the numbers.

2. Find the numerical value of the expression
$$\frac{c}{b} \cdot \frac{\sqrt{a}+\sqrt{c}}{\sqrt{a}-\sqrt{c}},$$
where a, b, c are connected by the equation $a(b-c)^2 - c(b+c)^2 = 0$.

3. A has a younger brother, B. The difference between their ages is $\frac{3}{8}$ of the sum of their ages. By adding twice B's age to 5 times A's, we obtain the age of the father; and by subtracting twice B's age from 5 times A's, we obtain the age of the mother. Show that the age of the mother is $\frac{9}{11}$ that of the father.

4. Find the H.C.F. of
$$x^3 - (2a+b)x^2 + a(2a+b)x - a^2(a+b),\text{ and}$$
$$x^3 - (2b+a)x^2 + b(2b+a)x - b^2(b+a).$$

5. If $\dfrac{1}{b} + \dfrac{1}{c} = \dfrac{4}{a}$, shew that
$$(a+b-c)^3 + 2(b+c-a)^3 + (c+a-b)^3 = 2(b+c)^3.$$

6. Show fully how the rule for finding the square root of a given number is obtained. If $n+1$ figures of the square root of a number have been obtained, prove that the remaining n may be obtained by division.

Extract the square root of
$$x^2(x^2+y^2+z^2) + y^2z^2 + 2x(y+z)(yz-x^2).$$

7. Find the value of the expression
$$\frac{x-y}{1+xy}\text{ when }x = \frac{a+b}{a-b},\ y = \frac{b}{a}.$$

8. Solve the equations :
   (1) $\frac{1}{2}(x-2a) - \frac{1}{3}(x+3a) + \frac{1}{6}(x-6a) = 0$.
   (2) $\sqrt{(2x^2+1)} + \sqrt{(2x^2+3)} = 2(1-x)$.

9. Divide 21 into two parts, so that ten times one of them may exceed nine times the other by 1.

XIX.

1. Multiply together
$$x^2 - \tfrac{9}{4}ax + \tfrac{1}{2}a^2 + \tfrac{5}{4}x - \tfrac{3}{4}a + \tfrac{1}{4};$$
$$x^2 + \tfrac{15}{4}ax - a^2 - \tfrac{7}{4}x + \tfrac{3}{2}a - \tfrac{1}{2}.$$
Divide this product by
$$\tfrac{1}{4}x^2 + \tfrac{1}{2}ax - 2a^2 - \tfrac{1}{4}x + 2a - \tfrac{1}{2};$$
and extract the square root of the quotient.

2. If $x+y+z = \dfrac{1}{x} + \dfrac{1}{y} + \dfrac{1}{z} = 0$, shew that
$$(x^6 + y^6 + z^6) \div (x^3 + y^3 + z^3) = xyz.$$

3. Find the H. C. D. of $20x^4 + x^2 - 1$ and $75x^4 + 15x^3 - 3x - 3$; also of $(x+y)^7 - x^7 - y^7$ and $(x^3 - y^3)^2$.

4. Given that $ab - (a+b)(x+y) + 4xy = 0$,
$$cd - (c+d)(x-y) + 4xy = 0,$$
find the value of $(x-y)^2$.

5. Having given
$$x^2 = y^2 + z^2 - 2ayz$$
$$y^2 = z^2 + x^2 - 2bzx$$
$$z^2 = x^2 + y^2 - 2cxy,$$
Show that $\dfrac{x^2}{1-a^2} = \dfrac{y^2}{1-b^2} = \dfrac{z^2}{1-c^2}$.

6. $\dfrac{1+x+\sqrt{(2x+x^2)}}{1-x+\sqrt{(2x+x^2)}} = 1 - ax$.

7. Determine x in terms of a and b in order that $x^4 + 2ax^3 + 3b^2x^2 - 4a^3x + 4b^4$ may be a perfect square.

8. A company of 90 persons consists of men, women, and children; the men are 4 in number more than the women, and the children exceed the number of men and women by 10. How many men, women, and children are there in the company.

XX.

1. Divide $(1+m)x^3 - (m+n)xy(x-y) - (n-1)y^3$ by $x^2 - xy + y^2$.

2. If $x^3 + px^2 + qx + r$ is exactly divisible by $x^2 + mx + n$, then $nq - n^2 = rm$.

3. Prove that if m be a common measure of p and q, it will also measure the difference of any multiples of p and q.

Find the G.C.M. of $x^4 - px^3 + (q-1)x^2 + px - q$ and
$x^4 - qx^3 + (p-1)x^2 + qx - p$.

4. Prove the rule for multiplication of fractions.

Simplify $\dfrac{x^2 - (y-z)^2}{(y+z)^2 - x^2} \times \dfrac{y^2 - (z-x)^2}{(z+x)^2 - y^2} \times \dfrac{z^2 - (x-y)^2}{(x+y)^2 - z^2}$;

and $\dfrac{a}{a^2 + b^2} - \dfrac{a}{a^2 - b^2} + \dfrac{a^2}{(a-b)(a^2+b^2)} - \dfrac{2a^3 - b^3 - ab^3}{a^4 - b^4}$.

5. What is the distinction between an *identity* and an *equation*? If $x - a = y + b$, prove $x - b = y + a$.

Solve the equation
$$\dfrac{16x-13}{4x-3} + \dfrac{40x-43}{8x-9} = \dfrac{32x-30}{8x-7} + \dfrac{20x-24}{4x-5}.$$

6. What are *simultaneous equations*? Explain why there must be given as many independent equations as there are unknown quantities involved. If there is a greater number of equations than unknown quantities, what is the inference?

Eliminate x and y from the equations $ax + by = c$, $a'x + b'y = c'$, $a''x + b''y = c''$.

7. Solve the equations—
  (1) $\sqrt[3]{(n+x)} + \sqrt[3]{(n-x)} = m$.
  (2) $3x + y + z = 13$, $3y + z + x = 15$, $3z + x + y = 17$.

8. A person has two kinds of foreign money; it takes a pieces of the first kind to make one £, and b pieces of the second kind: he is offered one £ for c pieces, how many pieces of each kind must he take?

9. A person starts to walk to a railway station four and a-half miles off, intending to arrive at a certain time; but after walking a mile and a half he is detained twenty minutes, in consequence of which he is obliged to walk a mile and a half an hour faster in order to reach the station at the appointed time. Find at what pace he started.

10. (a) If $\dfrac{a}{b} = \dfrac{c}{d}$ then will $\dfrac{a^4+c^4}{b^4+d^4} = \dfrac{a^2c^2}{b^2d^2}$.

(b) Find by Horner's method of division the value of $x^5+290x^4+279x^2-2892x^2-586x-312$ when $x = -289$.

(c) Show without actual multiplication that
$(a+b+c)^3 - (a+b+c)(a^2 - ab+b^2 - bc+c^2 - ac) - 3abc = 3(a+b)(b+c)(c+a)$.

NOTE.—In. Ex. 6, p. 87, after proving that $a+b+c$ is a factor, we may proceed as follows to discover the remaining quadratic factor:

The quadratic factor must be of the form
$$m(a^2+b^2+c^2) + n(ab+bc+ca),$$
in which m and n are independent, being either zero, or a positive or negative number. To determine them put $c=0$, then the given expression gives
$$\{a^3+b^3+3ab(a+b)\} \div (a+b) = a^2+b^2+2ab,$$
but also $= m(a^2+b^2)+nab$. $\therefore m=1$ and $n=2$.
$\therefore a^3+b^3+c^3+3(a+b)(b+c)(c+a)\} \div (a+b+c) = a^2+b^2+c^2+2(ab+bc+ca) = (a+b+c)^2$.

XXI.

1. Find the value of $x^3 - \left(\dfrac{1}{a} + \dfrac{1}{b}\right)x^2 + \left(\dfrac{1}{b} - \dfrac{1}{a}\right)x + \dfrac{2}{b^2}$ when $a = \frac{1}{4}$, $b = \frac{1}{4}$, $x = 2$. Simplify
$$a^3 - [x^3 - \{(8a^2x - 3ax^2) + a^3\} + 2a^3] + x^3.$$

2. Find, by symmetry, the sum of $(a+b+c)^3 - (a+b-c)^3 - (a-b+c)^3 - (b-a+c)^3$, and of $(a^4 - 4a^3x + 3a^2x^2 - 2ax^3 + 3x^4)^2$ and $(a^4 + 4a^3x + 3a^2x^2 + 2ax^3 + 3x^4)^2$.

3. Explain and illustrate the signs $>$, $<$
Prove: $x^2 + y^2 > 2xy$, $(x+y+z)^2 > 3(xy+yz+zx)$, and $x^3 + y^3 + z^3 > 3xyz$.

4. Determine the value of $x + y - z + 3x^{\frac{1}{3}}y^{\frac{1}{3}}z^{\frac{1}{3}}$, when $x^{\frac{1}{3}} + y^{\frac{1}{3}} - z^{\frac{1}{3}} = 0$, &c.; of $a^7 + 7ax^3 + 8x^2 - 3a^2 - (x^4 + 7ax^3 - 8x^2 - 3a^2)$, when $x = -1$.

5. Show that $(a^m)^{\frac{p}{q}} = a^{\frac{mp}{q}}$.

Simplify $\left\{\left(\dfrac{1}{16}\right)^{\frac{1}{3}}\right\}^{-\frac{1}{2}} \times \left(\dfrac{1}{8}\right)^{-\frac{1}{3}} \times \sqrt[4]{(256)}$, and divide $x^{\frac{2}{3}} - 6ax^{\frac{2}{3}} + 5a^{\frac{2}{3}}x + 2a^2x^{\frac{1}{3}} - 2a^{\frac{2}{3}}$ by $x^{\frac{2}{3}} - 2a^{\frac{1}{3}}x + a^{\frac{2}{3}}$.

6. If $u = \frac{1}{2}\left(x + \dfrac{1}{x}\right)$ and $v = \frac{1}{2}\left(y + \dfrac{1}{y}\right)$ prove that
$$uv - \sqrt{(1-u^2)}\sqrt{(1-v^2)} = \tfrac{1}{2}\left(xy + \dfrac{1}{xy}\right).$$

7. Gold is $19\frac{1}{4}$ times as heavy as water, and silver $10\frac{1}{4}$ times. A mixed mass weighs 4,160 ounces, and displaces 250 ounces of water. What proportion of gold and silver does the mass contain?

8. Shew that $1 + px + qx^2 + rx^3$ is a perfect cube if $p^2 = 3q$, and $q^2 = 3pr$.

9. Solve the equations:

(1) $\sqrt{\dfrac{x-2}{x+2}} + \sqrt{\dfrac{x+2}{x-2}} = 4$.

(2) $(x^4 + y^4)^2 + x^2y^2(x^2 - y^2)^2 + x^2 - y^2 = 328$, $x^2 - y^2 = 3$.

(3) $\dfrac{x^2}{y^2} + \dfrac{2x+y}{\sqrt{y}} = 20 - \dfrac{y^2 + x}{y}$, $x + 8 = 4y$.

10. A person buys two bales of cloth, each containing 80 yards, for $240. By selling the first at a gain of as maeh per cent. as the second cost him, and the second at a loss of as much per cent., he makes a profit of $16 on the whole. Find the cost price per yard of each bale.

SECOND CLASS TEACHERS, 1880.

XXII.

1. Find the value of $x^5 + x^4 - 166x^3 - 166x^2 + 81x + 81$ when $x = -\cdot 7$; and the value of $x^3 - 3px^2 + (3p^2 + q)x - pq$ when $x = a + p$. (Arrange the latter result according to powers of a).

2. What is the condition that $x + b$ shall be a factor of $ax^2 + bx + c$?
Find the factors of
(a). $(a^2 - ab) + 2(b^2 - ab) + 3(a^2 - b^2) + 4(a - b)^2$; and
(b). $(ax + b)(bx + c)(cx + a) - (ax + c)(bx + a)(cx + b)$.

3. What must be the relation among a, b, c, that $ax^2 + bx + c$ may be a perfect square?
(a). Extract the square root of
$$(a-b)^4 - 4(a^2 + b^2)(a-b)^2 + 4(a^4 + b^4) + 8a^2b^2.$$
(b). If 5 be subtracted from the sum of the squares of any four consecutive numbers, the remainder will be a perfect square. (Prove this.)

4. If $\dfrac{a}{b} = \dfrac{c}{d} = \dfrac{e}{f}$ and $\dfrac{h}{k} = \dfrac{l}{m} = \dfrac{n}{p}$

prove that $\dfrac{(a+c+e)(h+l+n)}{(b+d+f)(k+m+p)} = \dfrac{ah+cl+en}{bk+dm+fp}.$

5. (a). Reduce $\dfrac{ab(x^2-y^2) + xy(a^2-b^2)}{ab(x^2+y^2) + xy(a^2+b^2)}$ to its lowest terms.

(b). If $xy + yz + zx = 1$ prove that
$$\dfrac{x}{1-x^2} + \dfrac{y}{1-y^2} + \dfrac{z}{1-z^2} = \dfrac{4xyz}{(1-x^2)(1-y^2)(1-z^2)}.$$

6. Prove that

(a) $\dfrac{2\{x+2+\sqrt{(x^2-4)}\}}{x+2-\sqrt{(x^2-4)}} = x+\sqrt{(x^2-4)}$.

(b) $(b+c-a)a^{\frac{1}{2}}+(c+a-b)b^{\frac{1}{2}}+(a+b-c)c^{\frac{1}{2}} =$
$(a+b+c)(a^{\frac{1}{2}}+b^{\frac{1}{2}}+c^{\frac{1}{2}})-2(a^{\frac{3}{2}}+b^{\frac{3}{2}}+c^{\frac{3}{2}})$.

7. Solve the equations—

(a). $(b-c)(x-a)^3+(c-a)(x-b)^3+(a-b)(x-c)^3 = 0$.

(b). $x+y=4xy$; $y+z=2yz$; $z+x=3zx$.

(c). $x+y+z = 0$.
$ax+by+cz = 0$.
$bcx+cay+abz+(a-b)(b-c)(c-a) = 0$.

(d) $\dfrac{x-1}{x+3} + \dfrac{x-3}{x+1} + 2 = 0$.

FIRST-CLASS TEACHERS, 1876.

XXIII.

1. Investigate Horner's method of division.

Divide $x^9 - 3x^8 - 31x^7 + 25x^6 + 3x^5 - 8x^3 + 19x^2 + 8x + 10$ by $3x^4 - 21x^3 + 9x - 6$, showing the "final remainder."

Find the value of $2x^5 + 803x^4 - 398x^3 + 1605x^2 - 1204x + 422$, when $x = -402$.

2. If $f(x)$, a rational and integral function of x is divided by x^2+px+q, the remainder is $\dfrac{\{f(a)-f(\beta)\}x+a f(\beta)-\beta f(a)}{a-\beta}$, where a, β are the roots of $x^2+px+q = 0$.

Examine the case where $p^2 = 4q$.

3. Show without actual expansion that

$$\dfrac{a^4(b^2-c^2)+b^4(c^2-a^2)+c^4(a^2-b^2)}{a^2(b-c)+b^2(c-a)+c^2(a-b)} = \dfrac{(a^2-b^2)^3+(b^2-c^2)^3+(c^2-a^2)^3}{(a-b)^3+(b-c)^3+(c-a)^3}.$$

4. Find the value of x and y that will render the fraction
$$\frac{2z^2+(x-a)z+2b(x-2c)}{3z^2+(y-b)z+3a(y-3c)}$$ the same for all values of z.

5. Show how to find the sum of n terms of a series in Geometric progression.

(1) Show that the sum of n terms of the series
$$1+r+(1+2r)(1+r)+(1+3r)(1+r)^2+\ldots, \text{ is } n(1+r)^n.$$

(2) Sum to infinity the series $\dfrac{1}{2\cdot 4\cdot 6}+\dfrac{1}{4\cdot 6\cdot 8}+\dfrac{1}{6\cdot 8\cdot 10}+\ldots$

6. Explain the notation of functions :· prove that if
$$f(m)=1+mx+\frac{m(m-1)}{1\cdot 2}x^2+\&c., \text{ then } f(m)\times f(n)=f(m+n).$$

Show that in the expansion of $(1+x)^n$ the sum of the squares of the co-efficients $= \dfrac{1\cdot 2\cdot 3\ldots\ldots 2n}{(1\cdot 2\cdot 3\cdots\cdots n)^2}.$

7. Solve the equations—

(1) $\dfrac{x-a}{b+c}+\dfrac{x-b}{a+c}+\dfrac{x-c}{a+b}=3.$

(2) $x^4-10x^3+35x^2-50x+24=0.$

(3) $\dfrac{1}{21x^2-13x+2}+\dfrac{1}{28x^2-15x+2}=12x^2-7x+1.$

8. Give a brief account of mathematical induction, and show that a square of a multinomial is equal to the square of each term together with twice the product of each term into the sum of all that follow it.

Find the sum of the products of the first n natural numbers taken two and two together?

9. If $\dfrac{x}{a}=y+z, \dfrac{y}{b}=z+x, \dfrac{z}{c}=x+y,$ prove

(1) $\dfrac{1}{a}\cdot\dfrac{1}{b}\cdot\dfrac{1}{c}=\dfrac{1+a}{1-ab}\cdot\dfrac{1+b}{1-bc}\cdot\dfrac{1+c}{1-ca}.$

(2) $\dfrac{x^2}{a(1-bc)}=\dfrac{y^2}{b(1-ca)}=\dfrac{z^2}{c(1-ab)}.$

(3) $\dfrac{\sqrt{1-bc}}{a}+\dfrac{\sqrt{1-ca}}{b}+\dfrac{\sqrt{1-ab}}{c}=\dfrac{\sqrt{1-bc}}{a}\cdot\dfrac{\sqrt{1-ca}}{b}\cdot\dfrac{\sqrt{1-ab}}{c}$

10. AB is divided in C, so that $AB, BC = AC^2$; from CA is cut off a part CD equal to CB; from DC is cut off a part DE equal to DA; from ED is cut off a part equal to EC, and so on *ad inf.* Show that the points of section continually approach a point C' such that $AC' = BC$.

14. Eliminate x, y, z and u from the equations
$$a_1x+b_1y+c_1z+d_1u=0.$$
$$a_2x+b_2y+c_2z+d_2u=0.$$
$$a_3x+b_3y+c_3z+d_3u=0.$$
$$a_4x+b_4y+c_4z+d_4u=0.$$

12. A railway train travels from Toronto to Collingwood. At Newmarket it stops 7 minutes for water, and two minutes after leaving the latter place it meets a special express that left Collingwood when the former was 28 miles on the other side of Newmarket; the express travels at double the rate of the other, and runs the distance from Collingwood to Newmarket in $1\frac{1}{4}$ hour; and if on reaching Toronto it returned at once to Collingwood, it would arrive there three minutes after the first train; find the distance between Toronto, Newmarket and Collingwood.

FIRST CLASS TEACHERS, 1877.

XXIV.

1. Simplify $\left\{\left(\dfrac{x+y}{x-y}\right)^2 + 1\right\} \left\{\left(\dfrac{x+z}{x-z}\right)^2 + 1\right\} \left\{\left(\dfrac{y+z}{y-z}\right)^2 + 1\right\} \times$
$$\dfrac{x^2(y-z)+y^2(z-x)+z^2(x-y)}{x^4y^2+x^2y^4+x^4z^2+x^2z^4+y^4z^2+y^2z^4+2x^2y^2z^2}.$$

2. Solve (1.) $\dfrac{ax+m+1}{ax+m-1} + \dfrac{ax+n}{ax+n-2} = \dfrac{ax+m}{ax+m-2} + \dfrac{ax+n+1}{ax+n-1}.$

(2.) $\sqrt[3]{1 + \sqrt{x}} + \sqrt[3]{1 - \sqrt{x}} = 2.$

3. A, B, and C start from the same place; B, after a quarter of an hour, doubles his rate, and C, after walking 10 minutes, diminishes his rate one-sixth; at the end of half an hour, A is a quarter of a mile before B, and half a mile before C, and it is

observed that the total distance walked by the three, had they continued to walk uniformly from the first, is $6\frac{1}{4}$ miles. Find the original rate of each.

4. (1) Investigate the relations that must exist between the constants in order that $Ax^2 + By^2 + Cz^2 + ayz + bxz + cxy$ shall be a perfect square.

(2) Find the conditions that the values of x and y derived from the equations $ax + by = \dfrac{a^3}{x} + \dfrac{b^2}{y} = c^2$ may be rational.

5. If $x^2 + px + q$ and $x^2 + mx + n$ have a common factor, then will $(n-q)^2 + n(m-p)^2 = m(m-p)(n-q)$.

6. Prove $(a^m)^n = a^{mn}$, whether m and n be positive or negative, integral or fractional.

Show that $(x^{2m} + x^{2n})^{\frac{1}{mn}} = x^{\frac{1}{m} + \frac{1}{n}} \times (x^{m-n} + x^{n-m})^{\frac{1}{mn}}$

7. (1.) If $\dfrac{a}{b} = \dfrac{c}{d}$ then $\sqrt{\dfrac{a^{2n} + b^{2n}}{c^{2n} + d^{2n}}} = \left(\dfrac{a-b}{c-d}\right)^n$,

(2.) If $\dfrac{a^n d^n - b^n c^n}{\frac{1}{4}n(a^n - b^n - c^n + d^n)} = \dfrac{a^n c^n - b^n d^n}{\frac{1}{4}n(a^n - b^n - d^n + c^n)}$, then each of these fractions $= \dfrac{1}{n}(a^n + b^n + c^n + d^n)$.

8. If x be very small, show that—

$$\dfrac{(1+2x)^{\frac{1}{2}} + (1+3x)^{\frac{1}{3}}}{2 + 5x - (1+4x)^{\frac{1}{2}}} = 2 - 4x, \text{ very nearly.}$$

9. Prove that $1 - n^2 + \dfrac{n^2(n^2-1^2)}{1^2 \cdot 2^2} - \dfrac{n^2(n^2-1^2)(n^2-2^2)}{1^2 \cdot 2^2 \cdot 3^3} + \ldots = 0$

10. If a debt \$$a$ at compound interest be discharged in n years by annual payment of \$$\dfrac{a}{m}$, show that $(1+r)^n(1-mr) = 1$, where r is the interest on \$1 for a year.

11. Solve—(1.) $3x^2 - 2xy = 55$.

$x^2 - 5xy + 8y^2 = 7$.

(2) $\dfrac{5}{x^2 - 7x + 10} + \dfrac{5}{x^2 - 13x + 40} = x^2 - 10x + 19$.

(3) $a^2 b^2 x^{\frac{1}{q}} - 4a^{\frac{3}{2}} b^{\frac{3}{2}} x^{\frac{p+q}{2pq}} = (a-b)^2 x^{\frac{1}{p}}$

FIRST CLASS TEACHERS, 1878.

XXV.

1. Simplify $\left(\sqrt{\dfrac{a+x}{x}} - \sqrt{\dfrac{x}{a+x}} \right)^2 - \left(\sqrt{\dfrac{x}{a}} - \sqrt{\dfrac{a}{x}} \right)^2 - \dfrac{x^2}{a(a-x)}$

and $\dfrac{x^2 - (y-z)^2}{(x+z)^2 - y^2} + \dfrac{y^2 - (z-x)^2}{(x+y)^2 - z^2} + \dfrac{z^2 - (x-y)^2}{(y+z)^2 - x^2}$

2. Divide $\dfrac{x}{a} - 1 - \dfrac{b}{a} - \dfrac{b^2}{a^2} + \dfrac{b}{x} + \dfrac{b^2}{x^2}$ by $x - a$;

shew that $(-9a^2)^{\frac{1}{2}} = \frac{1}{2}\{ \sqrt{(6a)} + \sqrt{(-6a)}\}$.

3. If $\dfrac{m}{x} = \dfrac{n}{y} = \dfrac{r}{z}$ and $\dfrac{x^2}{a^2} + \dfrac{y^2}{b^2} + \dfrac{z^2}{c^2} = 1$, prove that

$\dfrac{m^2}{a^2} + \dfrac{n^2}{b^2} + \dfrac{r^2}{c^2} = \dfrac{m^2 + n^2 + r^2}{x^2 + y^2 + z^2}$.

4. Find the relations between the roots and co.efficients of the equation $ax^2 + bx + c = 0$.

If m and n are the roots of the equation $ax^2 + bx + c = 0$, show that the roots of the equation $acx^2 + (2ac - b^2)x + ac = 0$ are $\dfrac{m}{n}$ and $\dfrac{n}{m}$.

5. Solve the equations:

(1) $x^2 + 21\sqrt{x^2 - 2x} = 2x + 8$.

(2) $\dfrac{x^3}{y} - \dfrac{y^3}{x} = 10\frac{5}{6}$, $\dfrac{x}{y} - \dfrac{y}{x} = \frac{5}{6}$.

(3) $xz = y^2$, $x + y + z = 12$, $x^2 + y^2 + z^2 = 91$.

6. Two men start at the same time to meet each other from towns which are 28 miles apart; one takes five minutes longer than the other to walk a mile, and they meet in four hours. Find each man's rate per hour.

7. If P, Q, R be respectively the pth, qth, rth terms of a G.P., shew that
$$P^{q-r} \times Q^{r-p} \times R^{p-q} = 1.$$

Sum to infinity the series $\dfrac{1}{x}+\dfrac{2}{x^2}+\dfrac{3}{x^2}+$ &c.

8. Find the amount of \$$P$ at compound interest for n years, r being the interest on \$1 for one year.

Supposing \$$p$ to be withdrawn at the end of each year, what will be the amount at the end of n years?

9. Determine the number of combinations of n things taken r together.

The number of combinations of n things taken two together exceeds by 6 the number of combinations of $n-1$ things taken two together: find n.

10. (1) Find the limit of $(1+\frac{1}{x})^x$ when x increases without limit.

(2) Find the $(r+1)$th term in the expansion of $(3-5x)^{-\frac{1}{2}}$.

11. Determine the limits between which lies $\dfrac{x^2-3x-3}{2x^2+2x+1}$ for all possible values of x.

FIRST CLASS TEACHERS, 1879.

XXVI.

7. Prove that $2\{(a-b)^7+(b-c)^7+(c-a)^7\} = 7(a-b)(b-c)(c-a)\{(a-b)^4+(b-c)^4+(c-a)^4\}$.

2. Extract the square root of $ab-2a\sqrt{(ab-a^2)}$, and find the simplest real forms of the expression
$$\sqrt{(3+4\sqrt{-1})}+\sqrt{(3-4\sqrt{-1})}.$$

3. Solve the equations:

(1). $2x^4 + x^3 - 11x^2 + x + 2 = 0$.

(2). $x^2 + y^2 + z^2 = a^2$
$yz + zx + xy = b^2$
$x + y - z = c$.

(3). $\sqrt{(x^2 + 5x + 4)} + \sqrt{(x^2 + 3x - 4)} = x + 1$.

4. Prove that the number of positive integral solutions of the equation $ax + by = c$ cannot exceed $\dfrac{c}{ab} + 1$.

In how many ways may £11 15s. be paid in half-guineas and half-crowns?

5. If $xy = ab(a+b)$, and $x^2 - xy + y^2 = a^3 + b^3$, shew that
$$\left(\frac{x}{a} - \frac{y}{b}\right)\left(\frac{x}{b} - \frac{y}{a}\right) = 0.$$

6. Given the sum of an arithmetical series, the first term, and the common difference, shew how to find the number of terms. Explain the negative result. *Ex.* How many terms of the series 6, 10, 14, &c., amount to 96?

7. Find the relation between p and q, when $x^3 + px + x = 0$ has two equal roots, and determine the values of m which will make $a^2 + max + a^2$ a factor of $x^4 - ax^3 + a^2x^2 - a^3x + a^4$.

8. In the scale of relation in which the radix is r, shew that the sum of the digits divided by $r - 1$ gives the same remainder as the number itself divided by $r - 1$.

9. Assuming the Binomial Theorem for a positive integral index, prove it in the case of the index being a positive fraction.

Shew that the sum of the squares of the co-efficients in the expansion of $(1+x)^n$ is $\lfloor 2n \div (\lfloor n\rfloor)^2$, n being a positive integer.

10. Sum the following series:—

(1.) $1 + 3x + 5x^2 + 7x^3 + $ &c, to n terms.

(2.) $\dfrac{1}{3 \times 8} + \dfrac{1}{8 \times 13} + $ &c. to n terms, and to infinity.

11. Shew that $\begin{vmatrix} bc, & -ac, & -ab \\ b^2-c^2, & a^2+2ac, & -a^2-2ab \\ c^2, & c^2, & (a+b)^2 \end{vmatrix}$ is divisible by $abc(a+b+c)$.

FIRST CLASS TEACHERS, 1880—Grade C.

XXVII.

1. If in $ax^2+2bxy+cy^2$, $ku+lv$ be substituted for x and $mu+nv$ for y, the result takes the form $Au^2+2Buv+Cv^2$. Find the value of $(B^2-AC) \div (b^2-ac)$ in terms of k, l, m, n.

2. Resolve $a(b-c)^3+b(c-a)^3+c(a-b)^3$ into factors.

Prove that $\dfrac{Au^3+Bv^3+Cw^3}{uvw} = \dfrac{Ax^3+By^3+Cz^3}{xyz}$

if $u=x(By^3-Cz^3)$, $v=y(Cz^3-Ax^3)$, $w=z(Ax^3-By^3)$.

3. Extract the square root of
$(a-b)^2(b-c)^2+(b-c)^2(c-a)^2+(c-a)^2(a-b)^2$,
and the cube root of
$4(a-b)^6+(b-c)^6+(c-a)^6-3(a-b)^2(b-c)^2(c-a)^2\}.$

4. Eliminate x, y, z from
$ax+by+cz=1 \quad \dfrac{a}{x}=\dfrac{b}{y}=\dfrac{c}{z}$
$k(x^2+y^2+z^2)+2(lx+my+nz)+h=0.$

5. Simplify $\dfrac{a\sqrt{b}+b\sqrt{a}}{\sqrt{a}+\sqrt{b}}$, $\{\sqrt{(4+3j)}+\sqrt{(4-3j)}\}^2$,

and $\left(\dfrac{-1+j\sqrt{3}}{2}\right)^2 + \dfrac{-1+j\sqrt{3}}{2}+1,$

in which $j=\sqrt{(-1)}$.

6. Given the first term, the common difference and the number of terms of an arithmetical progression, find (i.) the sum of the terms, (ii.) the sum of the squares of the terms,

7. Solve the equations

(i.) $(a-x)^3 = (x-b)^3$.

(ii.) $ax+by = \dfrac{a}{x} + \dfrac{b}{y} = 1$.

(iii.) $x(y+z^{-1}) = a,\ y(z+x^{-1}) = b,\ z(x+y^{-1}) = c$.

8. What value (other than 1) must be given to q that one of the roots of $x^2 - 2x + q = 0$, may be the square of the other.

If a, b, c are the roots of $x^3 - px^2 + qx - r$, express

$$\frac{2a^2b^2 + 2b^2c^2 + 2c^2a^2 - a^4 - b^4 - c^4}{2ab + 2bc + 2ca - a^2 - b^2 - c^2}$$

in terms of p, q and r.

9. A vessel makes two runs on a measured mile, one with the tide in m minutes and one against the tide in n minutes. Find the speed of the vessel through the water, and the rate the tide was running at, assuming both to be uniform.

10. Five points, A, B, C, O and P lie on a straight line. The distances of $A, B,$ and C, measured from the point O, are $a, b,$ and c; their distances measured from the point P are x, y, z. Prove that whatever be the positions of the points O and P,

$$x^2(b-c) + y^2(c-a) + z^2(a-b) + (b-c)(c-a)(a-b) = 0.$$

APPENDIX.

SECTION I.—ELEMENTARY THEOREMS ON POLYNOMES.
(See page 39, *et seq.*)

Theorem I. If the polynome $f(x)^n$ be divided by $x-a$, the remainder will be $f(a)^n$.

D'Alembert's Proof. $f(x)^n$ is the dividend, $x-a$ is the divisor; let $f_1(x)^{n-1}$ be the quotient, which is necessarily a polynome of degree $n-1$, and let R be the remainder. Then, since the product of the quotient and the divisor added to the remainder reproduces the dividend,
$$f(x)^n = (x-a)f_1(x)^{n-1} + R.$$
But R does not contain x, hence it will remain the same, not merely in form but *in actual value*, whatever value be given to x. Take the case $x=a$, then $(x-a)f_1(x)^{n-1}$ vanishes for its factor $x-a$ does so, hence $R=f(a)^n$. Thus the remainder is the value of the dividend when x has the value which makes the divisor vanish.

It has been objected to the above proof " Division can be performed only when there is an actual divisor, therefore in assuming R to be the remainder of $f(x)^n \div (x-a)$ it is assumed that x is *not* equal to a, and although R will remain unchanged for all values of x that fulfil this assumption, it cannot thence be inferred that it will do so if the contradictory assumption be made. In such case the only legitimate conclusion is that there being no divisor there is neither quotient nor remainder. Therefore, although $f(a)^n$ may be the remainder in the case in which x is not equal to a, yet the above argument does not prove it." This objection confuses arithmetical or numerical division with algebraic or formal division; division by a definite quantity with division by an undetermined or variable quantity. The following proof does not involve the assumption $x=a$, and consequently is not open to the foregoing objection.

Lagrange's Proof. *Lemma.* $x^n - a^n$ is divisible by $x-a$, if n be a positive integer.

By actual division $\dfrac{x^n - a^n}{x-a} = x^{n-1} - a \cdot \dfrac{x^{n-1} - a^{n-1}}{x-a}$;

∴ $x^n - a^n$ is divisible by $x-a$ if $x^{n-1} - a^{n-1}$ is so divisible, hence $x^{n-1} - a^{n-1}$ " " $x-a$ " $x^{n-2} - a^{n-2}$ ....
Thus we can reduce the exponent unit by unit until at last we arrive at, $x^2 - a^2$ is divisible by $x-a$ if $x-a$ is so divisible. But $x-a$ is certainly divisible by itself, ∴ $x^2 - a^2$ is divisible by $x-a$. ∴ $x^3 - a^3$ is also divisible by $x-a$, ∴ so also is $x^4 - a^4$ and thus we may go on to any positive integral exponent whatsoever.

Theorem. Writing $f(x)^n$ in polynomial form arranged in ascending powers of x,

$f(x)^n = A_0 + A_1 x + A_2 x^2 + A_3 x^3 + \quad\quad + A_n x^n$,
∴ $f(a)^n = A_0 + A_1 a + A_2 a^2 + A_3 a^3 + \quad\quad + A_n a^n$,
∴ $f(x)^n - f(a)^n = A_1(x-a) + A_2(x^2 - a^2) + A_3(x^3 - a^3) + \ldots$
$\quad\quad + A_n(x^n - a^n)$.

But every term of this polynomial is divisible by $x-a$, and the highest power of x in the quotient is x^{n-1} got from the term $A_n(x^n - a^n)$, so the quotient may be represented by $f_1(x)^{n-1}$,

∴ $\{f(x)^n - f(a)^n\} \div (x-a) = f_1(x)^{n-1}$

or $\dfrac{f(x)^n}{x-a} = f_1(x)^{n-1} + \dfrac{f(a)}{x-a}$

Theorem II. If the polynome $f(x)^n$ vanish on substituting for x each of the n different values $a_1, a_2, a_3, \ldots a_n$, then $f(x)^n = A(x-a_1)(x-a_2) \ldots (x-a_n)$, in which A is independent of x and consequently is the coefficient of x^n in $f(x)^n$.

Since $f(a_1) = 0$, ∴ $f(x)^n = (x-a_1)f_1(x)^{n-1}$. In this substitute a_2 for x, ∴ since $f(a_2)^n = 0$, it becomes $0 = (a_2 - a_1)f_1(a_2)^{n-1}$. Of this product the factor $a_2 - a_1$ does not vanish since by hypothesis a_2 is not equal to a_1, therefore the other factor $f_1(a_2)^{n-1}$ must vanish that the product may vanish, and consequently $f_1(x)^{n-1}$ is divisible by $x-a_2$. Let the quotient be denoted by $f_2(x)^{n-2}$, $f(x)^n = (x-a_1)(x-a_2)f_2(x)^{n-2}$. Substitute a_3 for x and proceed as before, and it will be proved that $x-a_3$ is a factor of $f(x)^n$. Continuing to n factors we get a quotient independent of x, since each division reduces the exponent of x by unity, finally

$$f(x)^n = A(x-a_1)(x-a_2) \ldots \ldots (x-a^n).$$

Cor. If $f(x)^n$ and $\varphi(x)^m$ both vanish for the same m different values of x, $f(x)^n$ is algebraically divisible by $\varphi(x)^m$.

Let $a_1, a_2, a_3, \ldots \ldots a_m$ be the m different values of x for which the polynomes vanish,

$$\therefore f(x)^n = (x-a_1)(x-a_2) \ldots \ldots (x-a_m) F(x)^{n-m}$$
and $\varphi(x)^m = A(x-a_1(x-a_2) \ldots \ldots (x-a_m)$
$$\therefore f(x)^n \div \varphi(x)^m = F(x)^{n-m} \div A,$$

which is an integral function of x since A does not contain x.

Theorem III. If the polynome $f(x)^n$ vanish for more than n different values of x it will vanish identically, the coefficient of every term being zero.

Let $a_1, a_2, a_3 \ldots \ldots a_n, a_{n+1}$ be $n+1$ different values of x for which $f(x)^n$ vanishes,

$$\therefore f(x)^n = A(x-a_1)(x-a_2)(x-a_3) \ldots \ldots (x-a_n)$$

Substitute a_{n+1} for x, and since $f(a_{n+1})^n = 0$,

$$\therefore 0 = A(a_{n+1} - a_1)(a_{n+1} - a_2)(a_{n+1} - a_3) \ldots \ldots (a_{n+1} - a_n),$$

But none of the factors $a_{n+1} - a_1$, $a_{n+1} - a_2$, &c. vanishes,
$\therefore A$ must be zero, or

$$f(x)^n = 0(x-a_1)(x-a_2)(x-a_3) \ldots \ldots (x-a_n)$$

and the factor, zero, will be a factor in the coefficients of every term.

Theorem IV. If the polynomes $f(x)^n$, $\varphi(x)^m$ (n not less than m) are equal for more than n different values of x, they are equal for *all* values, and the coefficients of equal powers of x in each are equal to one another.

$$f(x)^n = A_0 + A_1 x + A_2 x^2 + A_3 x_3 + \ldots + A_n x^n$$
$$\varphi(x)^n = B_0 + B_1 x + B_2 x^2 + B_3 x^3 + \ldots + B_m x^m,$$
$$\therefore f(x)^n - \varphi(x)^m = A_0 - B_0 + (A_1 - B_1)x + (A_2 - B_2)x^2 +$$
$$(A_3 - B_3)x^3 + \ldots + (A_m - B_m)x^m$$
$$+ A_{m+1} x^{m+1} + A_{m+2} x^{m+2} \ldots + A_n x^n,$$

and this is a polynome of degree n at most. But $f(x)^n = \varphi(x)^m$ for more than n different values of x, that is $f(x)^n - \varphi(x)^m$ vanishes for these values, $\therefore$ by Theorem III. $f(x)^n - \varphi(x)^m$ vanishes identically, and the coefficients $A_0 - B_0$, $A_1 - B_1$, $A_2 - B_2$, $\ldots$ $A_m - B_m$, A_{m+2}, A_{m+2}, $\ldots$ A_n are all equal to zero,

$\therefore A_0 = B_0$, $A_1 = B_1$, $A_2 = B_2, \ldots A_m = B_m$, $A_{m+1} = 0$, $A_{m+2} = 0 \ldots$

Note to Art. XVII. To find, where such exist, the factors of
$$ax^2 + bxy + cxz + ey^2 + gyz + hz^2.$$
Multiply by $4a$
$$4a^2 x^2 + 4abxy + 4acxz + 4aey^2 + 4agyz + 4ahz^2.$$
Select the terms containing x and complete the square, thus
$$4a^2 x^2 + 4abxy + 4acxz + b^2 y^2 + 2bcxz + c^2 z^2$$
$$- (b^2 - 4ae)y^2 - 2(bc - 2ag)yz - (c^2 - 4ah)z^2 =$$
$$(2ax + by + cz)^2 - \{(b^2 - 4ae)y^2 + 2(bc - 2ag)yz + (c^2 - 4ah)z^2\}$$

If the part within the double bracket is a square say $(my + nz)^2$ the given expression can be written
$$(2ax + by + cz)^2 - (my + nz)^2$$
which can be factored by [4]. Factor and divide the result by $4a$. If the part within the double bracket is not a square, the given expression cannot be factored. If b and c are *both* even, multiply by a instead of by $4a$ and the square can be completed without introducing fractions. If e is less than a it will be easier to multiply by $4e$ instead of by $4a$ and select the terms containing y. A similar remark applies to h.

This method can evidently be extended to quadratic multinomials of any number of terms.

APPENDIX. 245

EXAMPLES.

1. Resolve $x^2 + xy + 2xz - 2y^2 + 7yz - 3z^2$ into factors.

Multiply by 4
$$4x^2 + 4xy + 8xz - 8y^2 + 28yz - 12z^2$$
Complete the square selecting terms in x,
$$4x^2 + 4xy + 8xz + y^2 + 4yz + 4z^2 - 9y^2 + 24yz - 16z^2 =$$
$$(2x + y + 2z)^2 - (3y - 4z)^2 =$$
$$\{(2x+y+2z)+(3y-4z)\}\{(2x+y+2z)-(3y-4z)\} =$$
$$(2x + 4y - 2z)(2x - 2y + 6z) = 4(x+2y-z)(x-y+3z)$$
∴ the factors are $(x + 2y - z)(x - y + 3z)$.

2. $6a^2 - 7ab + 2ac - 20b^2 + 64bc - 48c^2$.

Multiply by $4 \times 6 = 24$
$$144a^2 - 168ab + 48ac - 480b^2 + 1536bc - 1152c^2 =$$
$$(12a - 7b + 2c)^2 - 529b^2 + 1564bc - 1156c^2 =$$
$$(12a - 7b + 2c)^2 - (23b - 34c)^2 =$$
$$(12a + 16b - 32c)(12a - 30b + 36c) =$$
$$24(3a + 4b - 8c)(2a - 5b + 6c),$$
∴ the factors are $3a + 4b - 8c$ and $2a - 5b + 6c$.

3. $x^2 + 12xy + 2xz + 26y^2 - 8yz - 9z^2 =$
$$(x^2 + 12xy + 2xz + 36y^2 + 12yz + z^2) - 10y^2 - 20yz - 10z^2 =$$
$$(x + 6y + z)^2 - \{(y+z)\sqrt{10}\}^2 =$$
$$\{x + (6 + \sqrt{10})y + (\sqrt{10}+1)z\} \times$$
$$\{x + (6 - \sqrt{10})y - (\sqrt{10} - 1)z\}$$

4. $3a^2 + 10ab - 14ac + 12ad - 8b^2 - 8bd + 8c^2 - 8cd$.

Multiply by 3, not 4×3, since the coefficients of the other terms in a, are all even,
$$9a^2 + 30ab - 42ac + 36ad - 24b^2 - 24bd + 24c^2 - 24cd.$$

Select the terms containing a and complete the square
$$(3a+5b-7c+6d)^2-49b^2+$$
$$70bc-84bd-25c^2+60cd-36d^2=$$
$$(3a+5b-7c+6d)^2-(7b-5c+6d)^2=$$
$$(3a+12b-12c+12d)(3a-2b-2c)=$$
$$3(a+4b-4c+4d)(3a-2b-2c),$$
∴ the factors are $a+4b-4c+4d$ and $3a-2b-2c$.
Work Exercise XXIX by this method.

Section II.—Indices and Surds.

The general Index-laws are

$$a^{\frac{m}{n}} \cdot a^{\frac{p}{q}} = a^{\frac{m}{n}+\frac{p}{q}} \qquad (1)$$

$$a^{\frac{m}{n}} \div a^{\frac{p}{q}} = a^{\frac{m}{n}-\frac{p}{q}} \qquad (2)$$

$$(ab)^{\frac{m}{n}} = a^{\frac{m}{n}} \cdot b^{\frac{m}{n}} \qquad (3)$$

$$(a \div b)^{\frac{m}{n}} = a^{\frac{m}{n}} \div b^{\frac{m}{n}} \qquad (4)$$

$$(a^{\frac{m}{n}})^{\frac{p}{q}} = a^{\frac{mp}{nq}} \qquad (5)$$

The law connecting the Index and the Surd symbols is

$$a^{\frac{m}{n}} = \sqrt[n]{(a^m)} \qquad (6)$$

[The indices $\frac{1}{2}$, $\frac{1}{3}$, $\frac{1}{4}$, &c., are generally used to denote ' either square-root,' ' any of the cube-roots,' ' any one of the fourth-roots,' &c.

The surd symbols $\sqrt{\ }$, $\sqrt[3]{\ }$, $\sqrt[4]{\ }$, &c., are by some writers restricted to indicate the arithmetical or absolute roots, sometimes called the positive roots. Thus

$$\sqrt{4}=2, \text{ but } 4^{\frac{1}{2}}=\pm 2, \therefore 4^{\frac{1}{2}}=\pm\sqrt{4}$$

Also, $\sqrt{\{(-2)^2\}} = \sqrt{4} = 2.$

$$\sqrt[3]{27}=3, \text{ but } 27^{\frac{1}{3}}= 3 \text{ or } 3\left(\frac{-1\pm j\sqrt{3}}{2}\right) \therefore 8^{\frac{1}{3}}=(1^{\frac{1}{3}})\sqrt[3]{27}$$

$$\sqrt[4]{16}=2, \text{ but } 16^{\frac{1}{4}}=\pm 2 \text{ or } \pm 2j, \therefore 16^{\frac{1}{4}}=(1^{\frac{1}{4}})\sqrt[4]{16}.$$

With this restriction the general connecting formula would be
$$a^{\frac{m}{n}} = (1^{\frac{m}{n}}) \sqrt[n]{(a^m)}$$
In the following exercises this restriction need not be observed.]

EXERCISE.

1. What is the arithmetical value of each of the following:

$36^{\frac{1}{2}}, 27^{\frac{1}{3}}, 16^{\frac{1}{4}}, 32^{\frac{1}{5}}, 4^{\frac{3}{2}}, 8^{\frac{2}{3}}, 27^{\frac{4}{3}}, 64^{\frac{3}{2}}, 32^{\frac{2}{5}}, 64^{\frac{4}{3}}, 81^{\frac{3}{4}}, (3\frac{3}{8})^{\frac{1}{3}}$,

$(5\frac{1}{16})^{\frac{1}{4}}, (1\frac{9}{16})^{\frac{3}{2}}, (\cdot 25)^{\frac{1}{2}}, (\cdot 027)^{\frac{2}{3}}, 49^{\cdot 5}, 82^{\cdot 2}, 81^{\cdot 75}$

2. Interpret $a^{-2}, a^0, a^{2^2}, (a^2)^{-2}, a^{2^{-2}}, a^{-\frac{1}{2}}, (a^{-\frac{1}{2}})^{-\frac{1}{2}}, a^{\frac{1}{2}}, a^{-\frac{1}{2}}$.

3. What is the arithmetical value of

$36^{-\frac{1}{2}}, 27^{-\frac{1}{3}}, (\cdot 16)^{-\frac{1}{4}}, (\cdot 0016)^{-\frac{3}{4}}, (\frac{1}{4})^{-\frac{1}{2}}, (\frac{4}{25})^{-\frac{1}{2}}, (\frac{9}{16})^{-\frac{1}{2}}, (5\frac{1}{16})^{-\frac{1}{4}}$

4. Prove $(a^m)^n = (a^n)^m$; $(a^m)^{\frac{1}{n}} = (a^{\frac{1}{n}})^m$; $a^{-m} = (a^{-1})^m$;
and express these theorems in words.

5. Simplify $a^{\frac{1}{2}} \cdot a^{\frac{1}{3}}, c^{\frac{1}{3}} \cdot c^{\frac{1}{4}}, m^{\frac{1}{2}} \cdot m^{-1}, n^{\frac{2}{3}} \cdot n^{-\frac{1}{6}}, (7\frac{1}{2})^{\frac{1}{3}} \cdot (2\frac{1}{2})^{\frac{1}{3}} \cdot (3\frac{1}{5})^{\frac{1}{3}}$

$\dfrac{a^{\frac{1}{2}}}{a^{\frac{1}{3}}}, \dfrac{c^{\frac{2}{3}}}{c^{\frac{1}{4}}}, \dfrac{d^{\frac{1}{4}}}{d^{\frac{1}{12}}}, \dfrac{e^{-\frac{1}{6}}}{e^{\frac{1}{3}}}, \dfrac{e^{\frac{1}{4}}}{x^{-\frac{1}{3}}}, (2\frac{2}{5})^{\frac{2}{3}} \cdot (6\frac{2}{3})^{\frac{2}{3}} \div (\frac{1}{4})^{-1\frac{1}{3}}$

6. Remove the brackets from

$(a^6)^{\frac{1}{2}}, (b)^{-\frac{1}{3}}, (c\frac{3}{4})^{-\frac{1}{2}}, (d^3)^{1\frac{1}{3}}, (e^{-\frac{1}{3}})^{\frac{3}{4}}, (f^{-\frac{2}{3}})^{-\frac{2}{3}}$

$(a^2 b^2)^{\frac{1}{2}}, (a^{\frac{2}{3}} b^{\frac{1}{3}})^{\frac{1a}{2}}, (a^2 c^{-1})^{-\frac{1}{2}}, (a^{-5} c^{\frac{1}{3}})^{-\frac{1}{4}}, (x^{\frac{3}{2}} y^{-\frac{1}{3}})^{-6}$.

7. Remove the brackets and simplify

$(x^{\frac{1}{2}-\frac{1}{3}})^{\frac{2}{3}} (x^{\frac{1}{4}-\frac{1}{5}})^{\frac{3}{4}} (x^{\frac{1}{2}-\frac{1}{3}})^{\frac{2}{3}}; x^{(\frac{1}{2}-\frac{1}{3})\frac{2}{3}} x^{(\frac{1}{2}-\frac{1}{3})\frac{1}{2}} x^{(\frac{1}{2}-\frac{1}{3})\frac{2}{3}};$

$x^{(\frac{1}{2}+\frac{1}{16})^{\frac{1}{2}}} x^{(\frac{1}{2}-\frac{1}{16})^{\frac{1}{2}}}; x^{(\frac{1}{2}+\frac{1}{16})^{-\frac{1}{2}}} x^{(\frac{1}{2}-\frac{1}{16})^{\frac{1}{2}}}:$

$\{x^{2^{-2}} \div x^{-2^{-2}}\} \{x^{(-2)^2} \div x^{-2^2}\}$.

248 APPENDIX.

8. Simplify $-x\{x^{-\frac{1}{2}}(-x)^{-1}\}^{\frac{1}{3}}$, $x\{(-x)^{-\frac{3}{2}}(-x)^{-2}\}^{-\frac{2}{3}}$,
$(-x)^{-1}\{x^{-3}x^{-\frac{3}{4}}\}^{\frac{3}{4}}$

9. Determine the commensurable and the surd factors of

$12^{\frac{1}{2}}$, $24^{\frac{1}{3}}$, $18^{-\frac{1}{2}}$, $(-81)^{\frac{1}{3}}$, $12^{\frac{2}{3}}$, $64^{\frac{2}{3}}$, $(\frac{1}{18})^{\frac{2}{3}}$, $(6\frac{1}{4})^{-\frac{3}{2}}$.

(The surd factor must be the incommensurable root of an integer.)

10. Simplify $8^{\frac{1}{2}}+18^{\frac{1}{2}}-50^{\frac{1}{2}}$; $72^{\frac{1}{3}}+(\frac{24}{125})^{\frac{1}{3}}-(\frac{3}{125})^{-\frac{1}{3}}$;

$\{(6+2^{\frac{1}{2}})(6-2^{\frac{1}{2}})\}^{\frac{1}{2}}$; $(2^{\frac{1}{2}}+3^{\frac{1}{2}})^{2}+(2^{\frac{1}{2}}-3^{\frac{1}{2}})^{2}$;

$(2^{\frac{1}{3}}+3^{\frac{1}{3}})(4^{\frac{1}{3}}+9^{\frac{1}{3}}-6^{\frac{1}{3}})$; $(7^{\frac{1}{2}}-3^{\frac{1}{2}})^{\frac{1}{2}}(7^{\frac{1}{2}}+3^{\frac{1}{2}})^{\frac{1}{2}}$.

$[\{(a+x)(x+b)\}^{\frac{1}{2}} - \{(a-x)(x-b)\}^{\frac{1}{2}}]^{2}$;

$\{a^{\frac{3}{2}}+(a^{3}-x^{3})^{\frac{1}{2}}\}^{\frac{1}{3}} \cdot \{a^{\frac{3}{2}}-(a^{3}-x^{3})^{\frac{1}{2}}\}^{\frac{1}{3}}$

Express as surds,

11. $a^{\frac{2}{3}}$, $x^{\frac{3}{5}}$, $p^{3\frac{1}{2}}$, $c^{-\frac{1}{2}}$, $h^{-3\frac{2}{3}}$.

12. $x^{n+\frac{1}{2}}$, $y^{-n+\frac{2}{3}}$, $a^{.25}$, $b^{-n+\frac{1}{m}}$

13. $(ax-b)^{\frac{a}{5}}$, $(x^{2}-4x+1)^{\frac{m-3}{4}}$, $(p-qx)^{\frac{n-3}{3}}$

Express with indices,

14. $\sqrt[3]{a^{2}}$, $\sqrt[4]{c^{3}}$, $\sqrt[n]{x^{m}}$, $\sqrt[3]{y^{m-n}}$, $\sqrt[a]{(ax)}$, $\sqrt{a^{-3}}$.

15. $\sqrt[3]{(a^{3}+b^{3})}$, $\sqrt[3]{(a^{3}+b^{3})^{2}}$, $\{\sqrt[3]{(a^{3}+b^{3})}\}^{2}$, $\sqrt[x]{\{(a-b)x\}}$,
$\sqrt[n]{(a-bx)^{n-1}}$, $\sqrt[n]{(a^{n}-b^{n})^{m-3n}}$

16. $(a^{\frac{3}{3}})^{\frac{2}{4}}$, $(b^{-\frac{1}{3}})^{\frac{2}{3}}$, $(c^{-\frac{2}{3}})^{-\frac{3}{4}}$, $(x^{\frac{1}{3}})^{-\frac{2}{3}}$, $(a^{2}x)^{-\frac{1}{2}}$, $(a^{-3}x^{-\frac{1}{2}})^{-\frac{1}{2}}$,
$(x^{\frac{1}{7}}y^{-\frac{1}{5}})^{14}$.

Simplify the following, expressing the results by both notations.

APPENDIX. 249

17. $a.a^{-\frac{1}{3}}$, $a^0.a^{-\frac{1}{2}}$, $a^{\frac{1}{3}}.a^{-\frac{1}{4}}$, $a.a^{-\frac{1}{3}}$, $a^{-\frac{1}{2}}\cdot\sqrt{a}$, $a^{\frac{2}{3}}\sqrt[3]{a^2}$, $a^{\frac{1}{4}}\sqrt[4]{a}$

$a^{\frac{3}{5}}\sqrt[4]{a^3}$, $a^{\frac{3}{5}}\sqrt[5]{a^{-3}}$, $a^{\frac{5}{8}}\sqrt{a^{-1}}$, $a^{\frac{3}{2}}b^{\frac{1}{3}}c^{-\frac{1}{4}}\cdot a^{\frac{3}{5}}b^{-\frac{1}{2}}c^{\frac{1}{2}}d$,

$a^{\frac{1}{2}}b^{\frac{3}{3}}c^{\frac{1}{4}}\cdot a^{-\frac{2}{3}}b^{-\frac{1}{3}}c^{-\frac{1}{8}}$

18. $\dfrac{a^{\frac{1}{3}}}{\sqrt{a}}$, $\dfrac{a^{\frac{2}{3}}}{\sqrt[3]{a}}$, $\dfrac{\sqrt[a]{c^5}}{\sqrt[3]{c}}$, $\dfrac{\sqrt[4]{x^3}}{x^{\frac{1}{10}}}$, $\dfrac{\sqrt[3]{y^{-2}}}{y^{-\frac{1}{2}}}$, $\dfrac{\sqrt[3]{(24a^{-2})}}{\sqrt[6]{a}}$, $\dfrac{c(ab)^{\frac{1}{2}}-ac}{bc-c(ab)^{\frac{1}{2}}}$

19. $\dfrac{a^{\frac{1}{2}}+a^{-\frac{1}{2}}}{a^{\frac{1}{2}}-a^{-\frac{1}{2}}}$, $\dfrac{a^{\frac{3}{3}}-a^{-\frac{3}{2}}}{a^{\frac{1}{3}}-a^{-\frac{1}{2}}}$, $\dfrac{a^{-\frac{rn}{3}}-a^{\frac{3n}{3}}}{a^{-\frac{3n}{3}}+a^{\frac{3n}{2}}}$, $\dfrac{a^2+1+a^{-2}}{a+1+a^{-1}}$

20. Divide $x-y$ by $x^{\frac{1}{n}}-y^{\frac{1}{n}}$; $x^{\frac{4}{3}}+a^{\frac{3}{3}}x^{\frac{3}{3}}+a^{\frac{4}{3}}$ by $x^{\frac{3}{3}}+a^{\frac{1}{3}}x^{\frac{1}{3}}+a^{\frac{3}{3}}$;

$x+y+z-3x^{\frac{1}{3}}y^{\frac{1}{3}}z^{\frac{1}{3}}$ by $x^{\frac{1}{3}}+y^{\frac{1}{3}}+z^{\frac{1}{3}}$

$2ab+2bc+2ca-a^2-b^2-c^2$ by $a^{\frac{1}{2}}+b^{\frac{1}{2}}+c^{\frac{1}{2}}$

EXERCISE.*

1. Express the following quantities i. as quadratic surds, ii. as cubic surds, iii. as quartic surds.

a, $3a$, $2a^2$, a^2x, x^n, $y^{\frac{1}{2}}$, a^{-m}, $\dfrac{x}{y}$, $mx^{-\frac{n}{p}}$, $\cdot 1$, $\cdot 01$, $1\cdot 1x^0$.

2. Reduce to entire surds,

$x\sqrt{x}$, $a\sqrt[3]{a}$, $b^2\sqrt[3]{b^2}$, $3\sqrt[3]{3}$, $4\sqrt[3]{2}$, $\frac{1}{2}\sqrt{2}$, $\frac{1}{2}\sqrt[3]{4}$, $\frac{1}{3}\sqrt[3]{9}$, $3\sqrt[3]{\frac{1}{3}}$,

$a\sqrt{\left(\dfrac{b}{a}\right)}$, $\dfrac{a}{b}\sqrt{b}$, $\dfrac{a}{b}\sqrt{\left(\dfrac{b}{a}\right)}$, $\dfrac{a}{b}\sqrt{\left(\dfrac{a}{b}\right)}$,

$\dfrac{x}{y}\sqrt[3]{\left(\dfrac{y}{x}\right)}$, $\dfrac{x}{y}\sqrt[3]{\left(\dfrac{y}{x}\right)^2}$, $\dfrac{x^2}{y}\sqrt[3]{(x^2y^{-1})}$,

$a\sqrt[n]{b}$, $a\sqrt[n]{(a^m)}$, $(a+x)\sqrt[n]{(a+x)^m}$, $(a+x)\sqrt[n]{(a-x)^{n+1}}$,

$\dfrac{x+y}{m}\sqrt{\left(\dfrac{x+y}{m}\right)}$, $(x+y)\sqrt{\left(\dfrac{x-y}{x+y}\right)}$, $\dfrac{a-b}{a+b}\sqrt[3]{\left(\dfrac{a+b}{a-b}\right)^2}$,

$(x-y)^{-2}\sqrt[5]{(x^2+2xy+y^2)^{-4}}$, $(x-x^{-1})\sqrt[3]{(x^2+1)^2}$.

250   APPENDIX.

3. Reduce to their simplest form

$\sqrt{12}$, $\sqrt{8}$, $\sqrt{50}$, $\sqrt[3]{16}$, $4\sqrt[3]{250}$, $\sqrt{\tfrac{1}{2}}$, $\sqrt[3]{\tfrac{1}{4}}$, $\sqrt{\tfrac{8}{27}}$, $5\sqrt[3]{(-320)}$,

$\sqrt[4]{(1 - \tfrac{1}{81})}$, $\sqrt{a^3}$, $\sqrt{(a^3 b^7)}$, $\sqrt[3]{a^5}$, $3\tfrac{1}{3}\sqrt[3]{(54x^9)}$, $\sqrt[4]{(x^5 y^7 z^9)}$,

$\sqrt{\{a^3(1-x^2)\}}$, $\sqrt[3]{\{a^2(a^2-1)^4\}}$, $\sqrt{(ab)}$, $\sqrt[n]{a^{n+1}}$, $\sqrt[n]{a^{m+n}}$,

$\sqrt[n]{a^{2n+3}}$, $\sqrt[n]{a^{3m-2}}$, $\sqrt{(a^2 x + a^3)}$, $\sqrt[3]{(a^3 + 2a^4 x + a^5 x^2)}$,

$\sqrt{\{(x-1)(x^2-1)\}}$, $\sqrt[3]{\{(a^2 + 2ax + x^2)(a^3 + x^3)\}}$,

$\sqrt[3]{\{(x^2-a^2)^2(x-a)\}}$, $\sqrt{(4x^3 - 8x^2 + 4x)}$,

$\sqrt{(8x^2 - 16x + 8)}$, $\sqrt[3]{\{(x^2 - 2 + x^{-2})(x^4 - 2x^2 + 1)\}}$,

$\sqrt{\left(\dfrac{2x - 2 + 2x^{-1}}{x + 2 + x^{-1}}\right)}$ $\sqrt{\left(\dfrac{3x^3 - 6x^2 + 3x}{27x^2 + 18x + 3}\right)}$ $\sqrt{\left(\dfrac{(a^2 - ab)^2 + 4a^3 b}{a - b}\right)}$

4. Compare the following quantities by reducing them to the same surd index:

$2 : \sqrt{3}$; $2 : \sqrt[3]{9}$; $\sqrt{2} : \sqrt[3]{3}$; $\sqrt{10} : \sqrt[3]{30}$; $2\sqrt{2} : \sqrt[3]{22}$;

$a^2 : \sqrt{a^3}$; $\sqrt[6]{x} : \sqrt[8]{y}$; $\sqrt[m]{x} : \sqrt[n]{y}$; $\sqrt[m]{x^n} : \sqrt[n]{x^m}$; $\sqrt[4]{a} : \sqrt[6]{b} : \sqrt[8]{c}$;

$\sqrt[m]{a} : \sqrt[n]{b} : \sqrt[p]{c}$; $\sqrt[m]{a^n} : \sqrt[n]{b^p} : \sqrt[p]{c^m}$.

5. Reduce to simple surds with lowest integral surd index

$\sqrt{(\sqrt[3]{a})}$, $\sqrt[3]{(\sqrt[4]{b})}$, $\sqrt[3]{(\sqrt{c})}$, $\sqrt[3]{(\sqrt{x^3})}$, $\sqrt[4]{(\sqrt[3]{x^2})}$, $\sqrt[5]{(\sqrt[3]{x^{10}})}$,

$\sqrt[3]{(\sqrt[4]{x^{13}})}$, $\sqrt[3]{(\sqrt{27})}$, $\sqrt{(\sqrt[3]{81})}$, $\sqrt[4]{(\sqrt[3]{81})}$, $\sqrt{(a\sqrt{a})}$,

$\sqrt[3]{(a\sqrt{a})}$, $\sqrt{(x\sqrt[3]{x})}$, $\sqrt[3]{(x^2 \sqrt[3]{x})}$, $\sqrt[3]{(5\sqrt{5})}$, $\sqrt{(3\sqrt[3]{3})}$,

$\sqrt[4]{(3\sqrt[3]{3})}$, $\sqrt[6]{(x \sqrt[3]{x})}$, $\sqrt[n]{\{a \sqrt[n]{(b \sqrt[n]{c})}\}}$, $x\sqrt{(x^{-1} \sqrt{x^{-1}})}$,

$y\sqrt[3]{(y^{-2} \sqrt[3]{y^{-2}})}$, $z\sqrt[4]{(z^{-3} \sqrt[3]{z^{-2}})}$, $\sqrt[m]{\{x\sqrt[n]{(y\sqrt[p]{z})}\}}$, $x^{-1}\sqrt[3]{(x^2 \sqrt{x^3})}$

6. In the following quantities, combine the terms involving the same radical;

$3\sqrt{2} + 5\sqrt{2} - 7\sqrt{2}$; $\sqrt{8} - \sqrt{2}$; $\sqrt[3]{16} + 3\sqrt[3]{2}$;

$\sqrt[3]{16} + \sqrt{2}$; $a\sqrt{x} - \sqrt{x}$; $a\sqrt[3]{x} - b\sqrt[3]{x}$;

$8\sqrt{a} + 5\sqrt{x} - 7\sqrt{a} + \sqrt{(4a)} - 3\sqrt{(4x)} + 4\sqrt{(9x)}$;

$\sqrt{x} + 3\sqrt{(2x)} - 2\sqrt{(3x)} + \sqrt{(4x)} - \sqrt{(8x)} + \sqrt{(12x)}$;

APPENDIX.   251

$7x - 3\sqrt{x} + 5\sqrt[3]{x} - 2\sqrt[4]{x^2} + \sqrt[6]{x^2}$;

$4\sqrt{(a^2x)} + 2\sqrt{(b^2x)} - 3\sqrt{\{(a+b)^2x\}}$;

$\sqrt{\{(a-b)^2x\}} + \sqrt{\{(a+b)^2x\}} - \sqrt{(a^2x)} + \sqrt{\{(1-a)^2x\}} - \sqrt{x}$;

$\sqrt{(a-b)} + \sqrt{(16a-16b)} + \sqrt{(ax^2-bx^2)} - \sqrt{\{9(a-b)\}}$;

$\sqrt{(a^3+a^2b)} - \sqrt{(b^3+ab^2)}$;

$\sqrt{(a^3+2a^2b+ab^2)} - \sqrt{(a^3-2a^2b+ab^2)} - \sqrt{(4ab^2)}$.

7. In the following quantities, perform, as far as possible, the indicated multiplications and divisions, expressing the results in their simplest forms:

$\sqrt{2}.\sqrt{6}$; $\sqrt{3}.\sqrt{12}$; $\sqrt{14}.\sqrt{35}.\sqrt{10}$; $\sqrt{a}.\sqrt{(3a)}$;

$\sqrt{c}.\sqrt{(12c)}$; $\sqrt{(6x)}.\sqrt{(8x)}$; $\sqrt{y^3}.\sqrt{y^3}$; $\sqrt[3]{y^5}.\sqrt[3]{y^7}$;

$\sqrt[3]{a}.\sqrt[3]{a^2}.\sqrt{b}$; $\sqrt{a}.\sqrt{\left(\frac{x}{a}\right)}$; $\sqrt{a}.\sqrt{\left(\frac{a}{x}\right)}$; $\sqrt{3a}.\sqrt{\left(\frac{5c}{6a}\right)}$;

$\sqrt{a^{n+1}}.\sqrt{a^{n+1}}$; $\sqrt[3]{b^{n+1}}.\sqrt[3]{b^{2n+1}}$; $\sqrt{12} \div \sqrt{3}$; $\sqrt{(6x)} \div \sqrt{(2x)}$;

$a \div \sqrt[3]{a}$; $a^2 \div \sqrt[4]{a^3}$; $a \div \sqrt[n]{a^{n-1}}$; $a^p \div \sqrt[n]{a^{n-m}}$;

$(a+x) \div \sqrt{(a+x)}$; $(a^2-x^2) \div \sqrt{(a-x)}$; $(x^2-1) \div \sqrt[3]{(x+1)^2}$;

$(3\sqrt{8} - 5\sqrt{2} + \sqrt{18} + \sqrt{32} + \sqrt{72} - 2\sqrt{50}).\sqrt{2}$;

$(7\sqrt{2} - 5\sqrt{6} - 3\sqrt{8} + 4\sqrt{20})\sqrt{18}$; $(\sqrt{5}+\sqrt{3})(\sqrt{5}-\sqrt{3})$ ·

$(\sqrt{2}+1)(\sqrt{6}-\sqrt{3})$; $(3-\sqrt{2})(2+3\sqrt{2})$;

$(5\sqrt{3}+\sqrt{6})(5\sqrt{2}-2)$; $(\sqrt{a}-\sqrt{b})(\sqrt{a}+\sqrt{b})$;

$(a\sqrt{b}+b\sqrt{a})(b\sqrt{a}-a\sqrt{b})$;

$\{\sqrt{(x+1)}+\sqrt{(x-1)}\}\{\sqrt{(x+1)}-\sqrt{(x-1)}\}$;

$\{\sqrt{(3a-b)}+\sqrt{(3b-a)}\}\{\sqrt{(3a-b)}-\sqrt{(3b-a)}\}$;

$\sqrt{(a+\sqrt{b})}.\sqrt{(a-\sqrt{b})}$; $\sqrt{(\sqrt{x}+\sqrt{y})}.\sqrt{(\sqrt{x}-\sqrt{y})}$;

$\sqrt{\{a+\sqrt{(a^2-x^2)}\}}.\sqrt{\{a-\sqrt{(a^2-x^2)}\}}$;

$\sqrt[3]{\{x-\sqrt{(x^2-1)}\}}.\sqrt[3]{\{x+\sqrt{(x^2-1)}\}}$;

$\sqrt[3]{\{a\sqrt{a}-\sqrt{(a^3-x^3)}\}}.\sqrt[3]{\{\sqrt{(a^3-x^3)}+a\sqrt{a}\}}$;

252   APPENDIX.

$\sqrt[3]{(8+3\sqrt{7})} \cdot \sqrt[3]{(8-3\sqrt{7})}$; $(\sqrt{a+\sqrt{b}})^2$; $(\sqrt{a}+\sqrt{b})^3$;

$(a-c\sqrt{x})^2$; $(\sqrt{x}+\sqrt{x^{-1}})^2$;

$\{a+\sqrt{(1-a^2)}\}^2$; $\{\sqrt{(a+b-x)}-\sqrt{(a-b+x)}\}^2$;

$[\{\sqrt{(a+x)(x-b)}\}+\sqrt{\{(a-x)(x+b)\}}]^2$; $\left\{\sqrt{\frac{2a}{\ }} - \sqrt{\frac{3b}{2a}}\right\}^2$

$[\sqrt{\{(a+x)(x+b)\}}+\sqrt{\{(a-x)(x-b)\}}]^2$;

$\left\{\sqrt{\left(\frac{a-x}{x-b}\right)} - \sqrt{\left(\frac{x-b}{a-x}\right)}\right\}^2$; $\sqrt{(x^2-1)} \cdot \sqrt{\left(\frac{x+1}{x-1}\right)}$;

$(\sqrt[3]{a}+\sqrt[3]{b})^3$; $\{\sqrt{(\sqrt{a}+\sqrt{b})}+\sqrt{(\sqrt{a}-\sqrt{b})}\}^2$;

$\{\sqrt{(\sqrt{10}+1)}-\sqrt{(\sqrt{10}-1)}\}^2$;

$[\sqrt{\{a+\sqrt{(a^2-x^2)}\}}+\sqrt{\{a-\sqrt{(a^2-x^2)}\}}]^2$;

$(\sqrt{x}+\sqrt{y})^4+(\sqrt{x}-\sqrt{y})^4$: $(a^2+ab\sqrt{2}+b^2)(a^2-ab\sqrt{2}+b^2)$;

$(\sqrt[3]{a}+\sqrt[3]{c})(\sqrt[3]{a^2}-\sqrt[3]{(ac)}+\sqrt[3]{c^2})$;

$(\sqrt{a}+\sqrt{b}+\sqrt{c})(\sqrt{b}+\sqrt{c}-\sqrt{a})(\sqrt{c}+\sqrt{a}-\sqrt{b})(\sqrt{a}+\sqrt{b}-\sqrt{c})$;

$\left(\sqrt{\frac{x}{a}}+\sqrt{\frac{y}{b}}+1\right)\left(\sqrt{\frac{x}{a}}+\sqrt{\frac{y}{b}}-1\right) \times$

$\left(\sqrt{\frac{x}{a}}-\sqrt{\frac{y}{b}}+1\right)\left(-\sqrt{\frac{x}{a}}+\sqrt{\frac{y}{b}}+1\right)$

$(\sqrt[3]{a}+\sqrt[3]{b}+\sqrt[3]{c})\{\sqrt[3]{a^2}+\sqrt[3]{b^2}+\sqrt[3]{c^2}-2\sqrt[3]{(bc)}-2\sqrt[3]{(ca)}-2\sqrt[3]{(ab)}\}$.

8. Find rationalizing multipliers for the following expressions, and also the products of multiplication by these:

$a+\sqrt{b}$, $\sqrt{a}+b\sqrt{c}$, $a\sqrt{b}-b\sqrt{a}$, $a+\sqrt{(a^2-x^2)}$,

$\sqrt{(a-x)}-\sqrt{(a+x)}$, $\sqrt{(a^2+\sqrt{c})}+\sqrt{(a^2-\sqrt{c})}$,

$\sqrt{\{8+\sqrt{(24+\sqrt{5})}\}} - \sqrt{\{8+\sqrt{(24-\sqrt{5})}\}}$, $\sqrt{a}+\sqrt{b}+\sqrt{c}$,

$3+\sqrt{2}+\sqrt{7}$, $\sqrt{6}+\sqrt{5}-\sqrt{3}-\sqrt{2}$, $\sqrt{a}+\sqrt{b}+\sqrt{c}+\sqrt{d}$,

$\sqrt{(1+a)}-\sqrt{(1-a)}+\sqrt{(1+b)}-\sqrt{(1-c)}$, $\sqrt[3]{a}+\sqrt[3]{c}$,

$\sqrt[3]{a^2}-\sqrt[3]{c^2}$, $\sqrt[4]{a}+\sqrt[4]{c}$, $\sqrt{a}-\sqrt[3]{b}$, $\sqrt[4]{a}+\sqrt{a}$, $\sqrt{x}+\sqrt[4]{y^3}$,

$\sqrt{x+1}+\sqrt{x^{-1}}$, $\sqrt{(ab^{-1})}-\sqrt{(a^{-1}b)}$, $\sqrt[3]{2}+\sqrt[3]{3}-\sqrt[3]{5}$,

$\sqrt[3]{a}+\sqrt[3]{b}+\sqrt[3]{c}$, $a+\sqrt{b}+\sqrt[3]{c}$.

APPENDIX. 253

9. Rationalize the divisors and the denominators in the following, and reduce the results to their simplest form:

$1 \div (2 - \sqrt{3})$, $3 \div (3 + \sqrt{6})$, $5 \div (\sqrt{2} + \sqrt{7})$,

$(\sqrt{3} + \sqrt{2}) \div (\sqrt{3} - \sqrt{2})$, $(7\sqrt{5} + 5\sqrt{7}) \div (\sqrt{5} + \sqrt{7})$,

$a \div (\sqrt{a} + a)$, $(x - a) \div (\sqrt{x} - \sqrt{a})$,

$(a^2 + ab + b^2) \div \{a + \sqrt{(ab)} + b\}$, $(x + a) \div (\sqrt[3]{x} + \sqrt[3]{a})$,

$$\frac{a\sqrt{x} + b\sqrt{y}}{c\sqrt{x} - e\sqrt{y}}, \quad \frac{2\sqrt{6}}{\sqrt{2} + \sqrt{3} - \sqrt{5}}, \quad \frac{1 + 3\sqrt{2} - 2\sqrt{3}}{\sqrt{2} + \sqrt{3} + \sqrt{6}},$$

$$\frac{\sqrt{6} - \sqrt{5} - \sqrt{3} + \sqrt{2}}{\sqrt{6} + \sqrt{5} - \sqrt{3} - \sqrt{2}}, \quad \frac{2}{\sqrt{(a+1)} - \sqrt{(a-1)}},$$

$$\frac{2c}{\sqrt{(a+c)} + \sqrt{(a-c)}}, \quad \frac{a + x + \sqrt{(a^2 + x^2)}}{a + x - \sqrt{(a^2 + x^2)}},$$

$$\frac{\sqrt{(a+x)} + \sqrt{(a-x)}}{\sqrt{(a+x)} - \sqrt{(a-x)}}, \quad \frac{1}{a\sqrt{(1+b^2)} + b\sqrt{(1+a^2)}},$$

$$\frac{a\sqrt{(1-b^2)} - b\sqrt{(1-a^2)}}{\sqrt{(1-b^2)} + \sqrt{(1-a^2)}}, \quad \frac{a\sqrt{(1-a^2)} + c\sqrt{(1-c^2)}}{a\sqrt{(1-c^2)} + c\sqrt{(1-a^2)}},$$

$$\frac{\sqrt{\{(1+a)(1+b)\}} - \sqrt{\{(1-a)(1-b)\}}}{\sqrt{\{(1+a)(1+b)\}} + \sqrt{\{(1-a)(1-b)\}}},$$

$$\frac{(a-x)\sqrt{(b^2+y^2)} - (b-y)\sqrt{(a^2+x^2)}}{(a+x)\sqrt{(b^2+y^2)} + (b+y)\sqrt{(a^2+x^2)}},$$

$$\frac{\sqrt{(1+a)} - \sqrt{(1-a)} + \sqrt{(1+b)} - \sqrt{(1-b)}}{\sqrt{(1+a)} + \sqrt{(1-a)} + \sqrt{(1+b)} + \sqrt{(1-b)}},$$

$$\frac{\sqrt{(x+a)} - \sqrt{(x-a)} - \sqrt{(x+b)} + \sqrt{(x-b)}}{\sqrt{(x+a)} + \sqrt{(x-a)} + \sqrt{(x+b)} + \sqrt{(x-b)}},$$

$$\frac{\sqrt{a}}{\sqrt{b}} + \frac{\sqrt{b}}{\sqrt{a}}, \quad \sqrt{\left(\frac{a+x}{a-x}\right)} - \sqrt{\left(\frac{a-x}{a+x}\right)}, \quad \sqrt{\frac{a+\sqrt{x}}{a-\sqrt{x}}}, \quad \sqrt{\frac{\sqrt{x}-\sqrt{y}}{\sqrt{x}+\sqrt{y}}},$$

$$\sqrt{\frac{a+\sqrt{(a^2-1)}}{a-\sqrt{(a^2-1)}}}, \quad \frac{\frac{1}{\sqrt{x}} - \frac{1}{\sqrt{y}}}{\frac{1}{\sqrt{x}} + \frac{1}{\sqrt{y}}}, \quad \frac{\frac{\sqrt{a}}{\sqrt{x}} - \frac{\sqrt{x}}{\sqrt{a}}}{\frac{\sqrt{a}}{\sqrt{x}} + \frac{\sqrt{x}}{\sqrt{a}}}.$$

10. Find the values of the following expressions for $n = 1, 2, 3, 4, 5$, respectively.

$$\frac{1}{\sqrt{5}}\left\{\left(\frac{1+\sqrt{5}}{2}\right)^n - \left(\frac{1-\sqrt{5}}{2}\right)^n\right\}.$$

$$\frac{1}{2\sqrt{6}}\left\{\frac{(2+\sqrt{6})^{n+1}-(2+\sqrt{6})}{1+\sqrt{6}} - \frac{(2-\sqrt{6})^{n+1}-(2-\sqrt{6})}{1-\sqrt{6}}\right\}$$

11. Show that

$$\frac{1}{2(x-1)}[\{x+\sqrt{(x^2-1)}\}^{4n}{\pm 1} + \{x-\sqrt{(x^2-1)}\}^{4n\pm 1} \mp 2]$$

is a square for $n = 1, 2$, or 3 respectively.

12. Extract the square roots of

$x+y-2\sqrt{(xy)}$, $a+c+e+2\sqrt{(ac+ce)}$,

$a+2c+e+2\sqrt{\{(a+c)(c+e)\}}$, $2a+2\sqrt{(a^2-c^2)}$,

$2\{a^2+b^2-\sqrt{(a^4+a^2b^2+b^4)}\}$, $x-2+x^{-1}$,

$\sqrt{x}+2+\sqrt{x^{-1}}$, $x+3x^2+x^3+2x\sqrt{x}+2x^2\sqrt{x}$,

$x^2-xy+\tfrac{1}{4}y^2+\sqrt{(4x^3y-8x^2y^2+xy^3)}$, $2x+\sqrt{(3x^2-y^2)}$,

$5-2\sqrt{6}$, $10+2\sqrt{21}$, $9+4\sqrt{5}$, $4-\sqrt{15}$, $7+4\sqrt{3}$,

$12-5\sqrt{6}$, $70+3\sqrt{451}$, $4-\sqrt{15}$,

$9+2\sqrt{6}+4(\sqrt{2}+\sqrt{3})$, $15.25 - 5\sqrt{.6}$.

13. Find the value of

$\dfrac{(a+b)xy}{ay^2+bx^2}$, given $x = \dfrac{a\sqrt{a}}{\sqrt{(a+b)}}$ and $y = \dfrac{b\sqrt{b}}{\sqrt{(a+b)}}$;

$\sqrt{(x^2+y^2)}$, given $x = \sqrt[3]{(a^2c)}$ $y = \sqrt[3]{(a^2e)}$;

$\dfrac{x+\sqrt{(x^2+1)}}{x-\sqrt{(x^2+1)}}$, given $x = \tfrac{1}{2}\left\{\sqrt{\dfrac{a}{c}} - \sqrt{\dfrac{c}{a}}\right\}$

$\dfrac{\sqrt{(1+x)}-\sqrt{(1-x)}}{\sqrt{(1+x)}+\sqrt{(1-x)}}$, given $x = \dfrac{2ab}{a^2+b^2}$;

$\dfrac{2a\sqrt{(1+x^2)}}{x+\sqrt{(1+x^2)}}$, given $x = \tfrac{1}{2}\left\{\sqrt{\dfrac{a}{e}} - \sqrt{\dfrac{e}{a}}\right\}$.

APPENDIX.

14. If $\sqrt{(x+a+b)} + \sqrt{(x+c+d)} = \sqrt{(x+a-c)} + \sqrt{(x-b+d)}$,
$\therefore b+c = 0$.

15. Simplify $\dfrac{\frac{1}{2}(1+\sqrt{5})x-2}{x^2 - \frac{1}{2}(1+\sqrt{5})x+1} + \dfrac{\frac{1}{2}(1-\sqrt{5})x-2}{x^2 - \frac{1}{2}(1-\sqrt{5})x+1}$.

COMPLEX QUANTITIES.

Quantities of the form $a+b\sqrt{-1}$ in which neither a nor b involves $\sqrt{-1}$, are called complex quantities. The letter j (or i) is frequently used as the symbol of the ditensive unit $\sqrt{-1}$, so that $a+b\sqrt{-1}$ would be written $a+bj$. So also $\sqrt{-x} = j\sqrt{x}$, $\sqrt{-x}.\sqrt{-y} = j^2\sqrt{(xy)} = -\sqrt{xy}$, and $j^3 = -j$

EXERCISE.

Simplify the following, writing j for $\sqrt{-1}$ in any result in which the latter occurs:

1. $\sqrt{-4}$, $\sqrt{-36}$, $\sqrt{-81}$, $\sqrt{-8}$, $\sqrt{-12}$, $\sqrt{-72}$, $\sqrt[3]{-8}$, $\sqrt{-5}.\sqrt{-6}$, $\sqrt{-6}.\sqrt{-8}$, $\sqrt{-8}.\sqrt{12}$, $\sqrt{-8}.\sqrt[3]{-8}$, $\sqrt{-5}.\sqrt{-20}$.

2. $\sqrt{-x}$, $\sqrt{-x^2}$, $\sqrt{-a^3}$, $\sqrt{-a^{2n}}$, $\sqrt{(-a)^2}$, $\sqrt{(-a)^3}$, $\sqrt{(-3ax^3)}$, $\sqrt{-a}.\sqrt{a^3}$, $\sqrt{-x^3}.\sqrt{-y^3}$, $\sqrt{-a}.\sqrt{-1}$, $\sqrt{5}.\sqrt{-a}$.

3. $j^2, j^3, j^4, j^5, j^9, j^{15}, j^{16}, j^{17}, j^{18}, j^{4n}, j^{4n+1}, j^{4n+2}, j^{4n+3}$,

4. $aj.bj$, $j\sqrt{x}.j\sqrt{y}$, $5j$, $j^2\sqrt{5}$, $j\sqrt{-a}$, $j\sqrt{-a^2}$, $j\sqrt{a}.\sqrt{-a}$.

5. $\sqrt{-j^2}$, $\sqrt{-j^3}$, $\sqrt{-j^4}$, $\sqrt{-j^5}$, $\sqrt{-j^{2n}}$, $\sqrt{-j^{4n}}$.

6. $\dfrac{\sqrt{-6}}{\sqrt{3}}$, $\dfrac{\sqrt{-6}}{\sqrt{-3}}$, $\dfrac{\sqrt{6}}{\sqrt{-3}}$, $\dfrac{\sqrt{a}}{\sqrt{-b}}$, $\dfrac{\sqrt{-a}}{\sqrt{-b}}$, $\dfrac{1}{\sqrt{-1}}$,

$\dfrac{a}{\sqrt{-a}}$, $\dfrac{a^2}{\sqrt{-a^2}}$, $\dfrac{\sqrt{(-ax)}}{\sqrt{-x}}$, $\dfrac{-\sqrt{-1}}{\sqrt{-a}}$, $\dfrac{a^3}{\sqrt[3]{-a^3}}$,

$\dfrac{-c}{\sqrt{-c^3}}$, $\dfrac{\sqrt{(-a)^{2n+1}}}{\sqrt{(-a)^{2n-1}}}$.

256   APPENDIX.

7. $\dfrac{1}{j}$, $\dfrac{1}{j^2}$, $\dfrac{1}{j^3}$, $\dfrac{-1}{j}$, $\dfrac{1}{j^6}$, $\dfrac{1}{j^{4n+1}}$, $\dfrac{-1}{j^{4n+1}}$, $\dfrac{1}{j^{4n-1}}$,

$\dfrac{a^2 j}{\sqrt{-a^3}}$, $\dfrac{x}{j\sqrt{x}}$, $\dfrac{-y}{j\sqrt{-y^2}}$, $-\dfrac{cj^3}{\sqrt{-c^2}}$.

8. $\sqrt{(a-b)} \cdot \sqrt{(b-a)}$, $\sqrt{(3x-4y)} \cdot \sqrt{(4y-3x)}$, $(8+5j)(7+4j)$,

$(8-9j)(8-7j)$, $(7-j\sqrt{5})(7+j\sqrt{10})$, $(\sqrt{3}-j\sqrt{6})(\sqrt{2}-j\sqrt{6})$,

$(a+bj)(c+ej)$, $\{a+(a-1)j\}\{a+(a+1)j\}$,

$(\sqrt{a}+j\sqrt{b})(\sqrt{a}-j\sqrt{c})$ $(a+bj)(a-bj)$, $(aj+b)(aj-b)$,

$(\sqrt{a}+j\sqrt{b})(\sqrt{a}-j\sqrt{b})$, $(a\sqrt{b}+cj\sqrt{x})(a\sqrt{b}-cj\sqrt{x})$,

$\sqrt{(1+j)} \cdot \sqrt{(1-j)}$, $\sqrt{(3+4j)} \cdot \sqrt{(3-4j)}$,

$\sqrt{(12+5j)} \cdot \sqrt{(12-5j)}$, $(1+j)^2$, $(\sqrt{a}-j\sqrt{b})^2$, $(5-2j\sqrt{6})^2$,

$(a+bj)^2+(a-bj)^2$, $(a+bj)^2-(a-bj)^2$, $(a+bj)^2+(aj-b)^2$,

$\{\sqrt{(4+3j)}+\sqrt{(4-3j)}\}^2$, $\{\sqrt{(3-4j)}-\sqrt{(3+4j)}\}^2$,

$\{\sqrt{(1+j)}+\sqrt{(1-j)}\}^2$, $(1+j)^3$, $(1+j)^4$,

$(a+bj)^4$, $(a+bj)^3+(a-bj)^3$, $(a+bj)^3-(a-bj)^3$,

$\left(\dfrac{1+j\sqrt{3}}{2}\right)^3$, $\left(\dfrac{-1+j\sqrt{3}}{2}\right)^3$, $\left(\dfrac{-1-j\sqrt{3}}{2}\right)^3$, $\left(\dfrac{1+j}{\sqrt{2}}\right)^4$, $\left(\dfrac{1-j}{\sqrt{2}}\right)^4$,

$(x+jy)^4+(x-jy)^4$, $(x+jy)^4-(jx+y)^4$,

$(1+j\sqrt{5})^4+(1-j\sqrt{5})^4$, $(a+bj)^5+(a-bj)^5$,

$(a+bj)^5-(a-bj)^5$, $(1+j)^5+(1-j)^5$, $(1+j\sqrt{2})^5+(1-j\sqrt{2})^5$

$\left(\dfrac{j+\sqrt{3}}{2}\right)^6$, $\left(\dfrac{j-\sqrt{3}}{2}\right)^6$,

$[\tfrac{1}{8}\{\sqrt{(30-6\sqrt{5})}-1-\sqrt{5}\}+\tfrac{1}{8}j\{\sqrt{15}+\sqrt{3}+\sqrt{(10-2\sqrt{5})}\}]^n$

for all positive integral values of n.

9. $\dfrac{4}{1+j\sqrt{3}}$, $\dfrac{64}{1-j\sqrt{7}}$, $\dfrac{21}{4+3j\sqrt{6}}$, $\dfrac{5}{\sqrt{2}+j\sqrt{3}}$, $\dfrac{1-20j\sqrt{5}}{7-2j\sqrt{5}}$,

$\dfrac{1-j\sqrt{3}}{1+j\sqrt{3}}$, $\dfrac{1+j}{1-j}$, $\dfrac{1+j^3}{1+j}$, $\dfrac{1-j^3}{1-j}$, $\dfrac{1-j^3}{(1+j)^3}$, $\dfrac{1+j^3}{1-j}$, $\dfrac{x+iy}{x-yj}$,

$\dfrac{a+j\sqrt{x}}{a-j\sqrt{x}}$, $\dfrac{j\sqrt{a}+\sqrt{-b}}{\sqrt{-a}-j\sqrt{b}}$, $\dfrac{a-bj}{aj+b}$, $\dfrac{a+j\sqrt{(1-x^2)}}{a-j\sqrt{(1-x^2)}}$.

APPENDIX. 257

$$\frac{\sqrt{(x-y)}-\sqrt{(y-x)}}{\sqrt{(y-x)}+\sqrt{(x-y)}}, \quad \frac{1}{1+j}+\frac{1}{1-j}, \quad \frac{1+j}{1-j}+\frac{1-j}{1+j},$$

$$\frac{1}{(1+j)^2}+\frac{1}{(1-j)^2}, \quad \frac{1}{(1+j)^4}-\frac{1}{(1-j)^4}, \quad \frac{x+yj}{a+bj}+\frac{x-yj}{a-bj}.$$

$$\frac{x+yj}{a+bj}-\frac{x-yj}{a-bj}, \quad \frac{\sqrt{x}+j\sqrt{y}}{\sqrt{x}-j\sqrt{y}}-\frac{\sqrt{y}+j\sqrt{x}}{\sqrt{y}-j\sqrt{x}},$$

$$\frac{\sqrt{(1+a)}+j\sqrt{(1-a)}}{\sqrt{(1+a)}-j\sqrt{(1-a)}}-\frac{\sqrt{(1-a)}+j\sqrt{(1+a)}}{\sqrt{(1-a)}-j\sqrt{(1+a)}}.$$

10. $\sqrt{(3+4j)}+\sqrt{(3-4j)}, \quad \sqrt{(3+4j)}-\sqrt{(3-4j)},$

$\sqrt{(4+3j)}\pm\sqrt{(4-3j)}, \quad \sqrt{(1+2j\sqrt{6})}\pm\sqrt{(1-2j\sqrt{6})}$

$\sqrt{(5+2j\sqrt{6})}\pm\sqrt{(5-2j\sqrt{6})},$

$\sqrt{(2\sqrt{15}+30j)}\pm\sqrt{(2\sqrt{15}-30j)},$

$\sqrt{(\sqrt{3}+j\sqrt{105})}\pm\sqrt{(\sqrt{3}-j\sqrt{105})},$

$\sqrt{\{a+j\sqrt{(x^2-a^2)}\}}\pm\sqrt{\{a-j\sqrt{(x^2-a^2)}\}},$

$\sqrt{\{a^2+jx\sqrt{(x^2+2a^2)}\}}\pm\sqrt{\{a^2-jx\sqrt{(x^2+2a^2)}\}}.$

11. Prove that both $\frac{1}{2}(-1+j\sqrt{3})$ and $\frac{1}{2}(-1-j\sqrt{3})$ satisfy the equation $\dfrac{x^3-1}{x-1}=0,$

that $(x+wy+w^2z)^3 = x^3+y^3+z^3+3(x+wy)(y+wz)(z+wx)$
and that $(x+y+z)(x+wy+w^2z)(x+w^2y+wz) =$
$x^3+y^3+z^3-3xyz$, in which w represents *either* of the preceding complex quantities.

Hence, prove that

(i) $\{2a-b-c+(b-c)j\sqrt{3}\}^3 = \{2b-c-a+(c-a)j\sqrt{3})\}^3 =$
$\{2c-a-b+(a-b)j\sqrt{3}\}^3 ;$

(ii) $u^3+v^3+w^3-3uvw = (a^3+b^3+c^3-3abc)x$
$(x^3+y^3+z^3 - 3xyz),$ if $u=ax+by+cz, \quad v=ay+bz+cx,$
$w=az+bx+cy,$ or if $u=ax+cy+bz,$
$v=cx+by+az, \quad w=bx+ay+cz.$

12. Prove that $\frac{1}{4}\{\sqrt{5}+1+j\sqrt{(10-2\sqrt{5})}\}$ satisfies the equation
$$\frac{x^5+1}{x+1}=0.$$
Writing w for the preceding complex quantity, prove that
$$(7+w+w^2+3w^3)(7-w^4-w^3-3w^2)=71,$$
and $(x+y+z)(x+w^2y-w^3z)(x-w^3y-wz)(x-wz+w^4z)$
$(x+w^4y+w^2z) = x^5 + y^5 + z^5 - 5x^3yz + 5xy^2z^2.$

Prove that $\{4a+(b-c)(\sqrt{5}-1)+(b+c)j\sqrt{(10+2\sqrt{5})}\}^5 =$
$\{[(a+b)\{-1+j\sqrt{(\sqrt{5}+2)}\}+(a-b)\{\sqrt{5}+j\sqrt{(\sqrt{5}-2)}\}]$
$\times \sqrt[4]{5}-4c\}^5.$

Section III.—Pure Quadratics.

Examples.

1. $\dfrac{x+3(a-b)}{x-3(a-b)} = \dfrac{a(3x+9a-7b)}{b(3x-7a+9b)}.$

Apply, if $\dfrac{m}{n} = \dfrac{p}{q}$, $\therefore \dfrac{m+n}{m-n} = \dfrac{p+q}{p-q}$;

$\therefore \dfrac{x}{3(a-b)} = \dfrac{3x(a+b)+9a^2-14ab+9b^2}{3x(a-b)+9(a^2-b^2)}.$

Dividing the denominators by $3(a-b)$

$\therefore x\{+3(a+b)\} = 3x(a+b)+9a^2-14ab+9b^2,$

$\therefore x^2 = 9a^2-14ab+9b^2$

2. $\left(\dfrac{x-2a+4b}{x+4a-2b}\right)^2 = \dfrac{5x-9a+3b}{5x+3a-9b}.$

Apply, if $\dfrac{m}{n} = \dfrac{p}{q}$, $\therefore \dfrac{n-m}{n} = \dfrac{q-p}{p}$, and factor the numerator

$(x+4a-2b)^2 - (x-2a+4b)^2,$

APPENDIX. 259

$$\therefore \frac{12(x+a+b)(a-b)}{(x+4a-2b)^2} = \frac{12(a-b)}{5x+3a-9b},$$

$$\therefore \frac{x+a+b}{x+4a-2b} = \frac{x+4a-2b}{5x+3a-9b} = \frac{3(a-b)}{4x-a-7b},$$ by taking difference of numerators and difference of denominators. To the first and third of these fractions, apply if

$$\frac{m}{n} = \frac{p}{q}, \therefore \frac{m}{n-m} = \frac{p}{q-p},$$

$$\therefore \frac{x+a+b}{3(a-b)} = \frac{3(a-b)}{4x-4a-4b},$$

$$\therefore 4\{x^2 - (a+b)^2\} = 9(a-b)^2,$$

$$\therefore x^2 = \tfrac{1}{4}\{4(a+b)^2\} + 9(a-b)^2\}.$$

3. $\dfrac{\sqrt{(3x^2-1)}+\sqrt{(3-x^2)}}{\sqrt{(3x^2-1)}-\sqrt{(3-x^2)}} = \dfrac{a}{b},$

$$\therefore \sqrt{\frac{3x^2-1}{3-x^2}} = \frac{a+b}{a-b},$$

$$\therefore \frac{3x^2-1}{3-x^2} = \frac{(a+b)^2}{(a-b)^2},$$

$$\therefore x^2 = \frac{3(a+b)^2 + (a-b)^2}{(a+b)^2 + 3(a-b)^2} = \frac{a^2+ab+b^2}{a^2-ab+b^2}.$$

4. $m\sqrt{(1+x)} - n\sqrt{(1-x)} = \sqrt{(m^2+n^2)}$ \hfill (1)

Square both members and reduce

$$\therefore (m^2-n^2)x - 2mn\sqrt{(1-x^2)} = 0. \hfill (2)$$

Transfer the radical term and square both members,

$$\therefore (m^2-n^2)^2 x^2 = 4m^2n^2(1-x^2) \hfill (3)$$

$$\therefore (m^2+n^2)^2 x^2 = 4m^2n^2 \hfill (4)$$

$$\therefore \quad x = \frac{\pm 2mn}{m^2+n^2}. \hfill (4)$$

The above follows the usual mode of solving equations involving radicals, viz., make a radical term the right-hand member gathering all the other terms into the left-hand member, square each member, repeat, if necessary, until all radicals are rationalized. This method is convenient but it does not explain the difficulty that only one of the values of x in (4) satisfies (1) viz. $\dfrac{+2mn}{m^2+n^2}$

The other value, $\dfrac{-2mn}{m^2+n^2}$ satisfies the equation

$$m\sqrt{(1+x)}+n\sqrt{(1-x)} = \sqrt{(m^2+n^2)}.$$

The explanation is simple. Squaring both members of (1) is really equivalent to substituting for (1) the conjoint equation

$$\{m\sqrt{(1+x)} - n\sqrt{(1-x)} - \sqrt{(m^2+n^2)}\}$$
$$\{m\sqrt{(1+x)} + n\sqrt{(1-x)} - \sqrt{(m^2+n^2)}\} = 0 \qquad (5)$$

which reduces to (2) above.

Treating (5) or (2) by transferring and squaring is equivalent to substituting for it, the equation

$$\{m\sqrt{(1+x)} - n\sqrt{(1-x)} - \sqrt{(m^2+n^2)}\} \times$$
$$\{m\sqrt{(1+x)} - n\sqrt{(1-x)} + \sqrt{(m^2+n^2)}\} \times$$
$$\{m\sqrt{(1+x)} + n\sqrt{(1-x)} - \sqrt{(m^2+n^2)}\} \times$$
$$\{m\sqrt{(1+x)} + n\sqrt{(1-x)} + \sqrt{(m^2+n^2)}\} = 0 \qquad (6)$$

which reduces to

$$\{(m^2-n^2)x - 2mn\sqrt{(1-x^2)}\}\{m^2-n^2)x + 2mn\sqrt{(1-x^2)}\} = 0 \qquad (7)$$

which further reduces to (3)

Thus the whole process of solving (1) is equivalent to reducing it to an equation of the type $A=0$ and then multiplying the member A by rationalizing factors. Thus instead of solving (1) we !really solve (6), *i.e.*, a conjoint equation equivalent to *four* disjunctive equations. (See page 140, Art xl) Now the values given in (4) will satisfy (6), the positive value making the first factor vanish, the negative value making the third factor vanish, while no values can be found that will make either the second or the fourth factor vanish,

Hence, if one of such a set of disjunctive equations is proposed for solution, the conjoint equation must be solved, and if there be a value of x which satisfies the particular equation proposed, that value must be retained and the others rejected.

(This process is the opposite to that given in Arts. XL. and XLV.: there a conjoint equation is solved by resolving it into its equivalent disjunctive equations. The two processes are related somewhat as involution and evolution are).

Further, it should be noticed that just as there are four factors in (6) while there are only two values in (4), it will in general be possible to form more disjunctive equations than there are values of x that satisfy the conjoint equation, and consequently it will be possible to select disjunctive equations that are not satisfied by any value of x, or, in other words, whose solution is impossible.

This will perhaps be better understood by considering the following problem.

Find a number such that if it be increased by 4 and also diminished by 4 the difference of the square-roots of the results shall be 4.

Reduced to an equation this is
$$\sqrt{(x+4)} - \sqrt{(x-4)} = 4 \qquad (8)$$
Rationalizing this becomes
$$\{4-\sqrt{(x+4)}+\sqrt{(x-4)}\}\{4-\sqrt{(x+4)}-\sqrt{(x-4)}\} \times$$
$$\{4+\sqrt{(x+4)}+\sqrt{(x-4)}\}\{4+\sqrt{(x+4)}-\sqrt{(x-4)}\} = 0 \qquad (9)$$
which reduces to
$$\{24-8\sqrt{(x+4)}\}\{24+8\sqrt{(x+4)}\} = 0$$
i.e. $9-(x+4)=0$, or $x=5$.

Now $x = 5$ satisfies (9) because it makes the factor
$$4-\sqrt{(x+4)}-\sqrt{(x-4)}$$
vanish and *it is the only finite value of x that does satisfy* (9), or, in other words, there are no values of x which will make any of the factors

$4 - \sqrt{(x+4)} + \sqrt{(x-4)}$, $4 + \sqrt{(x+4)} + \sqrt{(x-4)}$,
or $4 + \sqrt{(x+4)} - \sqrt{(x-4)}$

vanish. There is, therefore, no number that will satisfy the conditions of the problem.

[It will be found that as x increases, $\sqrt{(x+4)} - \sqrt{(x-4)}$ decreases, hence as 4 is the least value that can be given to x without involving the square-root of a negative, the greatest real value of $\sqrt{(x+4)} - \sqrt{(x-4)}$ is $\sqrt{8}$ which is less than 4. We see by this that our method of solution fails for (8) simply because (8) is *impossible*].

5. $\sqrt{\{(a+x)(b+x)\}} - \sqrt{\{(a-x)(b-x)\}} =$
$\sqrt{\{(a-x)(b+x)\}} - \sqrt{\{(a+x)(b-x)\}}$ (1)

Collecting the terms involving $\sqrt{(a+x)}$ and $\sqrt{(a-x)}$ respectively the equation becomes

$\{\sqrt{(a+x)} - \sqrt{(a-x)}\}\{\sqrt{(b+x)} + \sqrt{(b-x)}\} = 0$ (2)

This is satisfied if either

$\sqrt{(a+x)} - \sqrt{(a-x)} = 0$ (3)
or $\sqrt{(b+x)} + \sqrt{(b-x)} = 0$ (4)

The rational form of (3) is $(a+x) - (a-x) = 0$ which is satisfied by $x = 0$ and this also satisfies (3).

The rational form of (4) is $(b+x) - (b-x) = 0$ which requires $x = 0$, but this does not satisfy (4). Hence the second factor of the left-hand member of (2) cannot vanish.

Therefore the only solution of (2) and $\therefore$ of (1) is $x = 0$, derived from (3).

6. $\sqrt[3]{(a+x)} + \sqrt[3]{(a-x)} = \sqrt[3]{(2a)}$

Cube by the formula $(u+v)^3 = u^3 + v^3 + 3uv(u+v)$

$\therefore (a+x) + (a-x) + 3\sqrt[3]{\{2a(a^2 - x^2)\}} = 2a$
$\therefore 2a(a^2 - x^2) = 0$,
$\therefore x = \pm a$.

Both these values belong to the proposed equation.

APPENDIX.   263

The rationalizing factors of
$$\sqrt[3]{(a+x)} + \sqrt[3]{(a-x)} - \sqrt[3]{(2a)} = 0$$
are $\sqrt[3]{(a+x)} + \omega\sqrt[3]{(a-x)} - \omega^2\sqrt[3]{(2a)}$,
and $\sqrt[3]{(a+x)} + \omega^2\sqrt[3]{(a-x)} - \omega\sqrt[3]{(2a)}$. See page 257.

The remarks on Ex. 4, will apply *mutatis mutandis* to equations of this type.

7. $\dfrac{\sqrt[3]{(a+x)^2} + \sqrt[3]{(a^2-x^2)} + \sqrt[3]{(a-x)^2}}{\sqrt[3]{(a+x)^2} - \sqrt[3]{(a^2-x^2)} + \sqrt[3]{(a-x)^2}} = c.$ (1)

Assume $\sqrt[3]{(a+x)} = u$ and $\sqrt[3]{(a-x)} = v$

$\therefore u^3 + v^3 = 2a$ and $u^3 - v^3 = 2x$,

and $\therefore \dfrac{u^3 - v^3}{u^3 + v^3} = \dfrac{x}{a}$ (2)

Also (1) becomes
$$\dfrac{u^2 + uv + v^2}{u^2 - uv + v^2} = c$$ (3)

Multiply both members by $\dfrac{u-v}{u+v}$

$\therefore \dfrac{u^3 - v^3}{u^3 + v^3} = c\dfrac{u-v}{u+v}$, $\therefore$ by (2) $\dfrac{x}{a} = c\dfrac{u-v}{u+v}$ (4)

Again adding and subtracting denominators and numerators in (3)
$$\dfrac{u^2 + v^2}{uv} = \dfrac{c+1}{c-1}.$$

Adding and subtracting 2 (denominators) and numerators in this
$$\dfrac{u^2 - 2uv + v^2}{u^2 + 2uv + v^2} = \dfrac{3-c}{3c-1}, \text{ or } \left(\dfrac{u-v}{u+v}\right)^2 = \dfrac{3-c}{3c-1}.$$

$\therefore$ substituting by (4), $\dfrac{x^2}{a^2} = c^2\dfrac{3-c}{3c-1}$,

$\therefore x = ac\sqrt{\dfrac{3-c}{3c-1}}.$

8. $\{\sqrt[4]{(x+a)}+\sqrt[4]{(x-a)}\}^3\{\sqrt[4]{(x+a)}-\sqrt[4]{(x-a)}\} = 2c$. (1)

Assume $u = \sqrt[4]{(x+a)}$ and $v = \sqrt[4]{(x-a)}$, and (1) becomes

$(u+v)^3(u-v) = 2c$ or $(u+v)^2(u^2-v^2) = 2c$ (2)

Also $u^4 - v^4 = 2a$ or $(u^2+v^2)(u^2-v^2) = 2a$ (3)

and $u^4 + v^4 = 2x$. (4)

From (2) and (3), $(u-v)^2(u^2-v^2) = 4a - 2c$ (5)

$\therefore$ (2)×(5), $(u^2-v^2)^2(u^2-v^2)^2$ or $(u^2-v^2)^4 = 4c(2a-c)$ (6)

Also (3)²+(6),

$\{(u^2+v^2)^2+(u^2-v^2)^2\}(u^2-v^2)^2 = 4(a^2+2ac-c^2)$

or $(u^4+v^4)(u^2-v^2)^2 = 2(a^2+2ac-c^2)$

Substituting by (4) and (6)

$2x\sqrt{(2ac-c^2)} = a^2+2ac-c^2$.

Exercise.

1. $(x+a+b)(x-a+b)+(x+a-b)(x-a-b) = 0$.
2. $(a+bx)(b-ax)+(b+cx)(c-bx)+(c+ax)(a-cx) = 0$.
3. $(a+bx)(ax-b)+(b+cx)(bx-c)+(c+ax)(cx-a)$
   $= \frac{1}{2}(a^2+b^2+c^2)$.
4. $(a+x)(b-x)+(1+ax)(1-bx) = (a+b)(1+x^2)$.
5. $(a+x)(b+x)(c-x)+(a+x)(b-x)(c+x)+(a-x)(b+x)(c+x)$
   $+(a-x)(b-x)(c+x)+(a-x)(b+x)(c-x)+$
   $(a+x)(b-x)(c-x) = 5abc$.
6. $(a+x)(b+x)(c+x)+(a+x)(b+x)(c-x)+(a+x)(b-x)(c+x)$
   $+(a-x)(b+x)(c+x)+(a+x)(b-x)(c-x)+(a-x)(b+x)(c-x)$
   $+(a-x)(b-x)(c+x)+(a-x)(b-x)(c-x) = 8x^3$
7. $(a+5b+x)(5a+b+x) = 3(a+b+x)^2$.
8. $(a+17b+x)(17a+b+x) = 9(a+b+x)^2$.
9. $(9a-7b+8x)(9b-7a+8x) = (3a+3b+x)^2$.

10. $\dfrac{ab}{a^2-b^2x^2}+\dfrac{cd}{c^2-d^2x^2}=0.$ 11. $\dfrac{x-a}{x+1}+\dfrac{x+a}{x-1}=2c.$

12. $\dfrac{a+x}{a-x}=\dfrac{x+b}{x-b}.$ 13. $\dfrac{ax+b}{a+bx}=\dfrac{cx+d}{c+dx}.$

14. $\dfrac{a-x}{1-ax}=\dfrac{1-bx}{b-x}.$ 15. $\dfrac{a-x}{1-ax}=\dfrac{4-x}{1-bx}.$

16. $\dfrac{x+a+2b}{x+a-2b}=\dfrac{b-2a+2x}{b+2a-2x}.$ 17. $\dfrac{a+4b+x}{a-4b+x}=\dfrac{3b-a+x}{3b+a-x}.$

18. $\dfrac{x+5a+b}{x-3a+b}=\dfrac{x-a+b}{a-x+3b}.$ 19. $\dfrac{a-7b+x}{7a-b-x}=\dfrac{a+5b+x}{5a+b+x}.$

20. $\dfrac{3a-b-x}{a-3b+x}=\dfrac{5b-3a+x}{5a-3b+x}.$ 21. $\dfrac{3a-2b+3x}{a-2b+x}=\dfrac{x-a+2b}{3x-3a+2b}.$

22. $\dfrac{3a-2b+3x}{a-2b+x}=\dfrac{x-7a+8b}{3x-5a+4b}.$

23. $\dfrac{5a-6b+x}{a+x}=\dfrac{3a-5b+3x}{a+b+x}.$ 24. $\dfrac{a+b-x}{3a-b-3x}=\dfrac{3(a-b+x)}{a-5b+x}.$

25. $\dfrac{7a+b-x}{5a+3b-3x}=\dfrac{3(a-b+x)}{a-17b+x}.$

26. $\dfrac{5a-b+x}{2(a+2b-x)}=\dfrac{2(2a-b+x)}{a+11b-x}.$

27. $\dfrac{7a-b+x}{7b-a+x}=\dfrac{a(a+5b+x)}{b(5a+b+x)}.$ 28. $\dfrac{x+a-b}{x-a+b}=\dfrac{a(x+a+5b)}{b(x+5a+b)}.$

29. $\left(\dfrac{5a-3b+x}{5b-3a+x}\right)^2=\dfrac{7a-9b+3x}{7b-9a+3x}.$

30. $\left(\dfrac{a+5b+x}{5a+b+x}\right)^2=\dfrac{a+17b+x}{17a+b+x}.$

31. $\left(\dfrac{7a-b+x}{7b-a+x}\right)^2=\dfrac{17a+b-x}{17b+a-x}.$

32. $\dfrac{17a+b-x}{a+17b-x}=\dfrac{a^2(a+17b+x)}{b^2(17a+b+x)}.$

33. $\dfrac{(x+7a+b)(x-a+b)}{(5x+3a-11b)(x-a+17b)} = \dfrac{x-5a+b}{5x+7a-59b}$.

34. $\dfrac{(1+3x+5x^2)(x^2+3x+5)}{(1+2x+3x^2)(x^2+2x+3)} = \dfrac{9}{4}$.

35. $\dfrac{\sqrt{(1+x^2)}+\sqrt{(1-x^2)}}{\sqrt{(1+x^2)}-\sqrt{(1-x^2)}} = \dfrac{a}{b}$.

36. $\dfrac{\sqrt[3]{(1+x^2)}+\sqrt[3]{(1-x^2)}}{\sqrt[3]{(1+x^2)}-\sqrt[3]{(1-x^2)}} = \dfrac{a}{b}$.

37. $\dfrac{\sqrt[4]{(1+x^2)}+\sqrt[4]{(1-x^2)}}{\sqrt[4]{(1+x^2)}-\sqrt[4]{(1-x^2)}} = \dfrac{a}{b}$.

38. $\dfrac{\sqrt[5]{(1+x^2)}+\sqrt[5]{(1-x^2)}}{\sqrt[5]{(1+x^2)}-\sqrt[5]{(1-x^2)}} = \dfrac{a}{b}$.

39. $\dfrac{\sqrt[6]{(1+x^2)}+\sqrt[6]{(x^2-1)}}{\sqrt[6]{(1+x^2)}-\sqrt[6]{(x^2-1)}} = \dfrac{a}{b}$.

40. $\dfrac{\sqrt[n]{(x^2+1)}+\sqrt[n]{(x^2-1)}}{\sqrt[n]{(x^2+1)}-\sqrt[n]{(x^2-1)}} = \dfrac{a}{b}$.

41. $\sqrt{(4a+b-4x)} - 2\sqrt{(a+b-2x)} = \sqrt{b}$.

42. $\sqrt{(3a-2b+2x)} - \sqrt{(3a-2b-2x)} = 2\sqrt{a}$.

43. $\sqrt{(2a-b+2x)} - \sqrt{(10a-9b-6x)} = 4\sqrt{(a-b)}$.

44. $\sqrt{(3a-4b+5x)} + \sqrt{(x-a)} = 2\sqrt{(x+a)}$.

45. $\sqrt{(3a-4b+5x)} + \sqrt{(x-a)} = 2\sqrt{(2x-2b)}$.

46. $\sqrt{(5x-3a+4b)} + \sqrt{(5x-3a-4b)} = 2\sqrt{(x+a)}$.

47. $\sqrt{(2a+b+2x)} + \sqrt{(10a+9b-6x)} = 2\sqrt{(2a+b-2x)}$.

48. $2\sqrt{(2a+b+2x)} + \sqrt{(10a+b-6x)} = \sqrt{(10a+9b-6x)}$.

49. $\sqrt{(2a-13b+14x)} + \sqrt{\{3(b-2a+2x)\}} = 2\sqrt{(2a-b+2x)}$.

50. $\sqrt{\{3(7a+b+x)\}} - \sqrt{(a+7b-x)} = 2\sqrt{(7a+b-x)}$.

51. $\sqrt{\{(a+x)(x+b)\}} + \sqrt{\{(a-x)(x-b)\}} = 2\sqrt{(ax)}$.

52. $\sqrt{\{(a+x)(x+b)\}} - \sqrt{\{(a-x)(x-b)\}} = 2\sqrt{(bx)}$.

53. $\sqrt{(ax+x^2)} - \sqrt{(ax-x^2)} = \sqrt{(2ax-a^2)}$.

APPENDIX.

54. $\sqrt{(ax-x^2)} + \sqrt{(ax+x^2)} = \sqrt{(2ax+a^2)}$.

55. $\dfrac{1}{1+\sqrt{(1-x)}} + \dfrac{1}{1-\sqrt{(1-x)}} = \dfrac{2}{9}x$.

56. $\dfrac{x+\sqrt{(ax)}}{a-\sqrt{(ax)}} + \dfrac{a+\sqrt{(ax)}}{x-\sqrt{(ax)}} = \dfrac{x-a}{a}$.

57. $\dfrac{\sqrt{\{(a+x)(x+b)\}} + \sqrt{\{(a-x)(x-b)\}}}{\sqrt{\{(a+x)(x+b)\}} - \sqrt{\{(a-x)(x-b)\}}} = \sqrt{\dfrac{a}{b}}$.

58. $\sqrt{\dfrac{3a-2b+2x}{3a-2b-2x}} = \dfrac{\{\sqrt{a} + \sqrt{(2a-2b)}\}^2}{2b-a}$.

59. $\sqrt[3]{(a+x)} + \sqrt[3]{(a-x)} = 2\sqrt[3]{a}$.

60. $\sqrt[3]{(a+x)^2} - \sqrt[3]{(a^2-x^2)} + \sqrt[3]{(a-x)^2} = \sqrt[3]{a^2}$.

61. $\dfrac{\sqrt[3]{(1+x)^2} + \sqrt[3]{(1-x^2)} + \sqrt[3]{(1-x)^2}}{\sqrt[3]{(1+x)^2} - \sqrt[3]{(1-x^2)} + \sqrt[3]{(1-x)^2}} = 2\tfrac{1}{3}$.

62. $\sqrt[3]{(1+x)^2} + \sqrt[3]{(1-x)^2} = 2\tfrac{1}{2}\sqrt[3]{(1-x^2)}$.

63. $\sqrt[3]{(3+x)} + \sqrt[3]{(3-x)} = \sqrt[3]{6}$.

64. $\sqrt[3]{(1+x)^2} + \sqrt[3]{(1-x)^2} = 5\{\sqrt[3]{(1+x)} + \sqrt[3]{(1-x)}\}^2$.

65. $\sqrt[3]{(14+x)^2} - \sqrt[3]{(196-x^2)} + \sqrt[3]{(14-x)^2} = 7$.

66. $\{\sqrt[3]{(9+x)} + \sqrt[3]{(9-x)}\}\sqrt[3]{(81-x^2)} = 12$.

67. $\{\sqrt[3]{(14+x)^2} - \sqrt[3]{(14-x)^2}\}\{\sqrt[3]{(14+x)} - \sqrt[3]{(14-x)}\} = 16$.

68. $\{\sqrt[3]{(57+x)^2} + \sqrt[3]{(57-x)^2}\}\{\sqrt[3]{(57-x)} + \sqrt[3]{(57+x)}\} = 100$.

69. $5\{\sqrt[4]{(41+x)} + \sqrt[4]{(41-x)}\}^2 = 8\{\sqrt{(41+x)} + \sqrt{(41-x)}\}$.

70. $\{\sqrt[4]{(x+5)} + \sqrt[4]{(x-5)}\}^3\{\sqrt[4]{(x+5)} - \sqrt[4]{(x-5)}\} = 2$.

71. $\{\sqrt[4]{(x+1)} + \sqrt[4]{(x-1)}\}\{\sqrt{(x+1)} + \sqrt{(x-1)}\} =$
    $26\{\sqrt[4]{(x+1)} - \sqrt[4]{(x-1)}\}$.

72. $\sqrt[3]{\dfrac{1+x}{1-x}} + \sqrt[3]{\dfrac{1-x}{1+x}} = a$. $[y+y^{-1}=a]$.

73. $2\{\sqrt[3]{(1+x)^2} + \sqrt[3]{(1-x^2)}\} = (c^2+1)\{\sqrt[3]{(1+x)} + \sqrt[3]{(1-x)}\}^2$.

268   APPENDIX.

74. $\sqrt[3]{(a+x)} + \sqrt[3]{(a-x)} = \sqrt[3]{c}$.

75. $\{\sqrt[3]{(a+x)} + \sqrt[3]{(a-x)}\}\sqrt[3]{(a^2-x^2)} = c$.

76. $\sqrt[3]{(a+x)^2} = \sqrt[3]{(a^2-x^2)} + \sqrt[3]{(a-x)^2} = \sqrt[3]{c^2}$.

77. $\{\sqrt[3]{(a+x)^2} - \sqrt[3]{(a-x)^2}\}\{\sqrt[3]{(a+x)} - \sqrt[3]{(a-x)}\} = c$.

78. $\{\sqrt[3]{(a+x)^2} + \sqrt[3]{(a-x)^2}\}\{\sqrt[3]{(a+x)} + \sqrt[3]{(a-x)}\} = c$.

79. $(a+x)\sqrt[3]{(a-x)} - (a-x)\sqrt[3]{(a+x)} = c\{\sqrt[3]{(a+x)} - \sqrt[3]{(a-x)}\}$.

80. $(a+x)\sqrt[3]{(a+x)} - (a-x)\sqrt[3]{(a-x)} = c\{\sqrt[3]{(a+x)} - \sqrt[3]{(a-x)}\}$.

81. $\{\sqrt[3]{(a+x)^2} - \sqrt[3]{(a^2-x^2)} + \sqrt[3]{(a-x)^2}\}^2 =$
    $c\{\sqrt[3]{(a+x)} + \sqrt[3]{(a-x)}\}$.

82. $\{\sqrt[4]{(a+x)} + \sqrt[4]{(a-x)}\}^2 = (c+1)\{\sqrt{(a+x)} + \sqrt{(a-x)}\}$.

83. $\{\sqrt[4]{(x+a)} - \sqrt[4]{(x-a)}\}\{\sqrt{(x+a)} + \sqrt{(x-a)}\}^2 =$
    $c\{\sqrt[4]{(x+a)} + \sqrt[4]{(x-a)}\}$.

SECTION IV.—QUADRATIC EQUATIONS AND EQUATIONS THAT CAN BE RESOLVED AS QUADRATICS.

EXAMPLES.

1. $x^4 + (ab+1)^2 = (a^2+b^2)(x^2+1) + 2(a^2-b^2)x + 1$,

  $\therefore x^4 + a^2b^2 = (a^2+b^2)x^2 + 2(a^2-b^2)x + (a-b)^2$

  $\therefore x^4 + 2abx^2 + a^2b^2 = (a+b)^2x^2 + 2(a^2-b^2)x + (a-b)^2$

  $\therefore x^2 + ab = \pm\{(a+b)x + (a-b)\}$,

  or $x^2 \mp (a+b)x + ab = \pm(a-b)$,

  $\therefore x^2 \mp (a+b)x + \tfrac{1}{4}(a+b)^2 = \tfrac{1}{4}(a-b)^2 \pm (a-b)$,

  $\therefore x \mp \tfrac{1}{2}(a+b) = \tfrac{1}{2}\sqrt{\{(a-b)^2 \pm 4(a-b)\}}$.

APPENDIX. 269

2. $\dfrac{(a-x)^2\sqrt{(a-x)}+(x-b)^2\sqrt{(x-b)}}{(a-x)\sqrt{(a-x)}+(x-b)\sqrt{(x-b)}}=a-b.$

Write $a-b$ in the form $(a-x)+(x-b)$ and multiply by the denominator of the left-hand member,

$\therefore (a-x)^2\sqrt{(a-x)}+(x-b)^2\sqrt{(x-b)}=$
$(a-x)^2\sqrt{(a-x)}+(a-x)(x-b)\{\sqrt{(a-x)}+\sqrt{(x-b)}\}+$
$(x-b)^2\sqrt{(x-b)},$

$\therefore (a-x)(x-b)\{\sqrt{(a-x)}+\sqrt{(x-b)}\}=0,$
$\therefore (a-x)=0,$ or $x-b=0,$
or $\sqrt{(a-x)}+\sqrt{(x-b)}=0.$

$x_1 = a,\ x_2 = b,$

The equation $\sqrt{(a-x)}+\sqrt{(x-b)}=0$ has no solution for the sum of two *positive* square-roots, cannot vanish.

The solution $x=\tfrac{1}{2}(a+b)$) belongs to the equation

$\sqrt{(a-x)}-\sqrt{(x-b)}=0.$

3. $\dfrac{ax+b}{bx+a}=\dfrac{mx-n}{nx-m}.$

Add and subtract Numerators and Denominators

$\dfrac{(a+b)(x+1)}{(a-b)(x-1)}=\dfrac{(m+n)(x-1)}{(m-n)(x+1)},$

$\therefore \left(\dfrac{x+1}{x-1}\right)^2=\dfrac{(a-b)(m+n)}{(a+b)(m-n)}=s^2$ say,

$\therefore x_1=\dfrac{s+1}{s-1},\ x_2=\dfrac{s-1}{s+1}.$

4. $\sqrt[3]{\dfrac{a-x}{b+x}}+\sqrt[3]{\dfrac{b+x}{a-x}}=c.$

Square both members, subtract 4 and extract the square-root.

$\therefore \sqrt[3]{\dfrac{a-x}{b+x}}-\sqrt[3]{\dfrac{b+x}{a-x}}=\pm\sqrt{(c^2-4)}$

APPENDIX.

$$\therefore \sqrt[3]{\frac{a-x}{b+x}} = \tfrac{1}{2}\{c \pm \surd(c^2-4)\} = e \text{ say},$$

$$\therefore \frac{a-x}{b+x} = e^3 \quad \therefore \quad \frac{2x-(a-b)}{a+b} = \frac{1-e^3}{1+e^3},$$

$$\therefore x = \tfrac{1}{2}\left\{(a-b)+(a+b)\frac{1-e^3}{1+e^3}\right\}.$$

Or thus, cube both members,

$$\therefore \frac{a-x}{b+x} + 3c + \frac{b+x}{a-x} = c^3$$

$$\therefore \frac{(a-x)^2+(b+x)^2}{2(a-x)(b+x)} = c^3 - 3c$$

$$\therefore \left\{\frac{(b+x)-(a-x)}{(b+x)+(a-x)}\right\}^2 = \frac{c^3-3c-2}{c^3-3c+2} = \frac{(c+1)^2(c-2)}{(c-1)^2(c+2)}$$

$$\therefore \frac{2x-(a-b)}{a+b} = \frac{c+1}{c-1}\sqrt{\frac{c-2}{c+2}}.$$

(Prove that $\dfrac{1-e^3}{1+e^3} = \dfrac{c+1}{c-1}\sqrt{\dfrac{c-2}{c+2}}$, if $2e = c \pm \surd(c^2-4)$.

5. $\dfrac{\surd(a-x) - \surd(b-x)}{\surd(a-x)+\surd(b-x)} = \dfrac{\surd\{(a-x)(b-x)\}}{c}$. Rationalize Denom.

$$\therefore \frac{\{\surd(a-x)-\surd(b-x)\}^2}{(a-b)} = \frac{\surd\{(a-x)(b-x)\}}{c},$$

or $\dfrac{\{\surd(a-x)-\surd(b-x)\}^2}{\surd\{(a-x)(b-x)\}} = \dfrac{a-b}{c},$ \hfill (A)

$$\therefore \left\{\frac{\surd(a-x)-\surd(b-x)}{\surd(a-x)+\surd(b-x)}\right\}^2 = \frac{a-b}{a-b+4c};$$

$$\therefore \frac{\surd\{(a-x)(b-x)\}}{c} = \sqrt{\frac{a-b}{a-b+4c}}. \hfill (B)$$

Also from (A),

$$\frac{a+b-2x}{\surd\{(a-x)(b-x)\}} = \frac{a-b+2c}{c},$$

Multiply (B) and (C) member by member

$$a+b-2c = (a-b+2c)\sqrt{\frac{a-b}{a-b+4c}}.$$

$$x = \tfrac{1}{2}\left\{a+b-(a-b+2c)\sqrt{\frac{a-b}{a-b+4c}}\right\}.$$

6. $x^4 - 4 = \dfrac{x^2+20}{x^2-2}$; $x^6 - 2x^4 - 5x^2 - 12 = 0$.

Find the rational linear factors of the left-hand member by the method of Art. XXVII., page 90.

$\therefore (x-2)(x+2)(x^4+2x^2+3) = 0$,

$\therefore x-2=0$, or $x+2=0$, or $x^4+2x^2+3=0$.

The last of these equations may be solved as a quadratic giving

$x^2 = -1 \pm 2\sqrt{-2}$, $\therefore x = \pm 1 \pm \sqrt{-2}$,

$\therefore x_1 = 2$, $x_2 = -2$, $x_3 = 1+\sqrt{-2}$, $x_4 = 1-\sqrt{-2}$,

$x_5 = -1+\sqrt{-2}$, $x_6 = -1-\sqrt{-2}$.

N.B.—In solving numerical equations of the higher orders, *the rational linear factors should always be found and separated as disjunctive equations*, before other methods of reduction are applied. Such separation may always be effected by the methods of Arts. XXVII. to XXX., and unless it is done the application of the higher methods may actually fail. Thus, if it be attempted to solve as a cubic the equation,

$$x^3 - 9x - 10 = 0$$

the result is $x = \{5+\sqrt{-2}\}^{\frac{1}{3}} + \{5-\sqrt{-2}\}^{\frac{1}{3}}$, which can be reduced only by trial. The left-hand member can however be easily factored by the method of Art. XXVII., and the equation reduces to

$$(x+2)(x^2-2x-5) = 0,$$

which gives $x = 2$ or $1 \pm \sqrt{6}$.

7. $(x-2)^7 - x^7 + 2^7 = 0$.

Factor, (See No. 20, p. 89), rejecting constant factors,

∴ $x(x-2)(x^2 - 2x + 4)^2 = 0$

∴ $x = 0$, or $x - 2 = 0$, or $x^2 - 2x + 4 = 0$.

The last equation gives $x = 1 \pm \sqrt{-3}$.

Exercise.

Solve the following equations:

1. $(x+a+b)^3 = x^3 + a^3 + b^3$. 2. $(x+a+b)^5 = x^5 + a^5 + b^5$.
3. $(a-b)x^3 + (b-x)a^3 + (x-a)b^3 = 0$.
4. $(a-b)x^2 + (x-b)a^2 + (x+a)b^2 = 2abx$.
5. $(x-a)^5 + (a-b)^5 + (b-x)^5 = 0$.
6. $(x-a)^7 + (a-b)^7 + (b-x)^7 = 0$.
7. $(a^3-b)x^4 + (x^3-a)b^4 + (b^3-x)a^4 = abx(a^2b^2x^2 - 1)$.
8. $(x-a)(x-b)(a-b) + (x-b)(x-c)(b-c) + (x-c)(x-a)(c-a) = 0$.
9. $\dfrac{x^5 - 1}{x - 1} = 0$. 10. $\dfrac{x^{12} - 1}{x^4 - 1} = 0$.
11. $\dfrac{x^{16} - 1}{x^4 - 1} = 0$. 12. $\dfrac{x^{20} - 1}{x^4 - 1} = 0$.
13. $x^4 + 5x^3 - 16x^2 + 20x - 16 = 0$. (See Art. XXII.)
14. $x^4 - 3x^3 + 5x^2 + 6x + 4 = 0$.
15. $(x-a)^4 + x^4 + a^4 = 0$. 16. $2x^3 = (x-6)^2$.
17. $x(x-2)^2(x+2) = 2$. 18. $(4x^2 - 17)x + 12 = 0$.
19. $x^4 + (ab+1)^2 = (a^2+b^2)(x^2+1) + 2(a^2-b^2)x + 1$.
20. $x^2(x-169)^2 + 17x = x^2 - 3540$.

21. $6x(x^2+1)^2+(2x^2+5)^3=150x+1$.
22. $2x(x-1)^2+2=(x+1)^2$. 23. $x^4=12x+5$.
24. $5x^4=12x^3+1$. 25. $(x+4)^3=3(2x-1)^2$.
26. $\sqrt{(x^2+m^2)}+\sqrt{\{(n-x)^2+m^2\}}=\sqrt{\{(x-\tfrac{1}{2}n)^2+(\tfrac{1}{2}n\sqrt{3}-m)\}}$.

27. $\dfrac{(x+1)^4}{(x^2+1)(x-1)^2}=\dfrac{m}{n}$. 28. $\dfrac{(x+1)^5}{x(x^3+1)}=\dfrac{m}{n}$.

29. $\dfrac{(x^2+1)(x^3+1)}{(x+1)(x^4+1)}=\dfrac{m}{n}$. 30. $\dfrac{(x^2+1)(x^3+1)}{(x^2-1)(x^3-1)}=\dfrac{m}{n}$.

31. $\dfrac{(x^2+1)(x^3+1)}{x^2(x+1)}=\dfrac{m}{n}$. 32. $\dfrac{(x^3-1)^2}{x(x^2+1)(x-1)^2}=\dfrac{m}{n}$.

33. $\dfrac{x(x+1)^2}{(x^2+1)(x-1)^2}=\dfrac{n(n-m)}{2m(2m-n)}$.

34. $\dfrac{(x^3+1)^2}{x(x^2-1)^2}=\dfrac{4m^2}{m^2-n^2}$.

35. $\dfrac{(x-1)(x^2+1)^2}{(x^3-1)(x+1)^2}=\dfrac{2(m-n)^2}{mn}$.

36. $\dfrac{x^6-1}{(x+1)(x^5-1)}=\dfrac{2m}{2m-n}$.

37. $\dfrac{(x^3-1)(x+1)^3}{(x^3+1)(x-1)^3}=\dfrac{m+n}{m-n}$. 38. $\dfrac{(x+1)(x^4+1)}{(x-1)(x^4-1)}=\dfrac{m+n}{m-n}$.

39. $x^3=\dfrac{ax-b}{bx-a}$. 40. $x^4=\dfrac{ax-b}{bx-a}$.

41. $x^5=\dfrac{ax-b}{bx-a}$. 42. $x^4=\dfrac{ax^2+bx+c}{a+bx+cx^2}$.

43. $x^2=(x-1)^2(x^2+1)$. 44. $a^2x^2=(a-x)^2(a^2-x^2)$.

45. $x^2=(x-a)^2(x^2-1)$.

46. $a\sqrt{(x^2+1)}-x\sqrt{(x^2+a^2)}=cx$.

274   APPENDIX.

47. $\sqrt[3]{(a^3+x^3)}+\sqrt[3]{(a^3-x^3)}=\sqrt[3]{(a^6-x^6)^2}.$

48. $m(x+m-n)(x-m+7n)^2 = n(x-m+n)(x+7m-n)^2.$

49. $m^2(x+m+17n)(x-m-5n)^2 = n^2(x+17m+n)(x-5m+n)^2.$

50. $m^2(x+m+17n)(x-m+7n)^2 = n^2(x+17m+n)(x+7m-n)^2.$

51. $\dfrac{\sqrt{(x-a)}+\sqrt{(x-b)}}{\sqrt{(x-a)}-\sqrt{(x-b)}} = \sqrt{\dfrac{x-a}{x-b}}.$

52. $\dfrac{\sqrt{(x-a)}+\sqrt{(x-b)}}{\sqrt{(x-a)}-\sqrt{(x-b)}} = \sqrt{\dfrac{a-x}{x-b}}.$

53. $\sqrt{\dfrac{a-x}{b+x}}-\sqrt{\dfrac{b+x}{a-x}}=c.$ 54. $\sqrt{\dfrac{a-x}{b-x}}+\sqrt{\dfrac{b-x}{a-x}}=c.$

55. $\sqrt[3]{\dfrac{a-x}{b+x}}-\sqrt[3]{\dfrac{b+x}{a-x}}=c.$ 56. $\sqrt[3]{\left(\dfrac{a-x}{b-x}\right)^2}-\sqrt[3]{\left(\dfrac{b-x}{a-x}\right)^2}=c.$

57. $\sqrt[4]{\dfrac{a-x}{b+x}}+\sqrt[4]{\dfrac{b+x}{a-x}}=c.$ 58. $\sqrt[4]{\dfrac{a-x}{b-x}}-\sqrt[4]{\dfrac{b-x}{a-x}}=c.$

59. $\sqrt[5]{\dfrac{a-x}{b+x}}+\sqrt[5]{\dfrac{b+x}{a-x}}=c.$ 60. $\sqrt[5]{\dfrac{a-x}{b-x}}-\sqrt[5]{\dfrac{b-x}{a-x}}=c.$

61. $\sqrt[6]{\dfrac{a-x}{b+x}}+\sqrt[6]{\dfrac{b+x}{a-x}}=c.$ 62. $\sqrt[6]{\dfrac{a-x}{b-x}}-\sqrt[6]{\dfrac{b-x}{a-x}}=c.$

63. $\dfrac{\sqrt{(a-x)^3}+\sqrt{(b-x)^3}}{\sqrt{(a-x)}+\sqrt{(b-x)}}=c.$

64. $\dfrac{\sqrt{(a-x)^3}+\sqrt{(b-x)^3}}{\{\sqrt{(a-x)}+\sqrt{(b-x)}\}^3}=c.$

65. $\dfrac{\sqrt{(a-x)^3}+\sqrt{(b-x)^3}}{\sqrt{(a-x)}-\sqrt{(b-x)}}=c.$

66. $\dfrac{\{\sqrt{(a-x)}+\sqrt{(b-x)}\}^3}{\sqrt{(a-x)}-\sqrt{(b-x)}}=c.$

67. $\dfrac{\sqrt{(a-x)^5}+\sqrt{(x-b)^5}}{\sqrt{(a-x)}+\sqrt{(x-b)}}=c.$

68. $\dfrac{\sqrt{(a-x)^5} - \sqrt{(x-b)^5}}{\{\sqrt{(a-x)} - \sqrt{(x-b)}\}^5} = c.$

69. $\dfrac{\sqrt{(a-x)^5} + \sqrt{(x+b)^5}}{\sqrt{(a-x)^3} + \sqrt{(x+b)^3}} = \sqrt{\{(a-x)(x+b)\}}.$

70. $\dfrac{\sqrt{(a-x)^3} + \sqrt{(x+b)^3}}{\sqrt{(a-x)} + \sqrt{(x+b)}} = \dfrac{(a+b)^2}{4\sqrt{\{(a-x)(x+b)\}}}.$

71. $\dfrac{x^3 + (a-x^2)\sqrt{(a-x^2)}}{x + \sqrt{(a-x^2)}} = c.$

72. $\dfrac{x^3 + (a^2 - x^2)\sqrt{(a^2 - x^2)}}{x + \sqrt{(a^2 - x^2)}} = cx\sqrt{(a^2 - x^2)}.$

73. $\sqrt[3]{(a-x)^2} - \sqrt[3]{\{(a-x)(x-b)\}} + \sqrt[3]{(x+b)^2} = \sqrt[3]{(a^2 - ab + b^2)}$

74. $\dfrac{b\sqrt{(a-x)} + a\sqrt{(x-b)}}{\sqrt{(a-x)} + \sqrt{(x-b)}} = x.$

75. $\dfrac{a\sqrt{(a-x)} + b\sqrt{(x-b)}}{\sqrt{(a-x)} + \sqrt{(x-b)}} = x.$

76. $\dfrac{\sqrt{(x-a)} + \sqrt{(x+a)} - \sqrt{(2a)}}{\sqrt{(x-a)} - \sqrt{(x+a)} + \sqrt{(2a)}} = \sqrt{\dfrac{x+c}{x-c}}.$

77. $\dfrac{\sqrt{(a-x)} + \sqrt{c}}{\sqrt{(x-b)} + \sqrt{c}} = \sqrt{\dfrac{a-x}{x-b}}.$

78. $\sqrt[3]{(a-x)^2} - \sqrt[3]{\{(a-x)(x+b)\}} + \sqrt[3]{(x+b)^2} = \sqrt[3]{(a^2 - ab + b^2)}.$

79. $\{\sqrt[3]{(a-x)^2} - \sqrt[3]{[(a-x)(x-b)]} + \sqrt[3]{(x-b^2)^2}\}^2 = (a-b)\{\sqrt[3]{(a-x)} + \sqrt[3]{(x-b)}\}.$

80. $\{\sqrt[3]{(a-x)^2} + \sqrt[3]{(b+x)^2}\}^2 = (a+b)\{\sqrt[3]{(a-x)} + \sqrt[3]{(b+x)}\}.$

81. $\sqrt[3]{(a-x)} + \sqrt[3]{(x-b)} = \sqrt[3]{c}.$

82. $\sqrt[3]{(a+x)^2} - \sqrt[3]{(a-x)^2} = \sqrt[3]{(2cx)}.$

83. $\sqrt[3]{(a-x)^2} + \sqrt[3]{\{(a-x)(b-x)\}} + \sqrt[3]{(b-x)^2} = \sqrt[3]{c^2}.$

84. $\sqrt[3]{(a-x)^2} - \sqrt[3]{\{(a-x)(x+b)\}} + \sqrt[3]{(x+b)^2} = c\{\sqrt[3]{(a-x)} + \sqrt[3]{(x+b)}\}.$

APPENDIX.

85. $\{\sqrt[3]{(a-x)} + \sqrt[3]{(x+b)}\}\sqrt[3]{\{(a-x)(x+b)\}} = c.$

86. $\sqrt[3]{(a-x)^2} + \sqrt[3]{(x-b)^2} = c\{\sqrt[3]{(a-x)} + \sqrt[3]{(x-b)}\}^2.$

87. $x + \sqrt[3]{(a^3 - x^3)} = \dfrac{c^3}{x\sqrt[3]{(a^3 - x^3)}}.$

88. $\dfrac{a^3}{x^3 - b^3} = \dfrac{x + \sqrt[3]{(2b^3 - x^3)}}{x - \sqrt[3]{(2b^3 - x^3)}}.$

89. $(a+x)\sqrt[4]{(a+x)} + (a-x)\sqrt[4]{(a-x)} = a\{\sqrt[4]{(a+x)} + \sqrt[4]{(a-x)}\}.$

90. $(a+x)\sqrt[4]{(a-x)} + (a-x)\sqrt[4]{(a+x)} = a\{\sqrt[4]{(a+x)} + \sqrt[4]{(a-x)}\}.$

91. $\sqrt[4]{(26-x)} + \sqrt[4]{(x-10)} = 2.$

92. $\{\sqrt[4]{(a-x)} + \sqrt[4]{(x-b)}\}^2 = c\{\sqrt{(a-x)} + \sqrt{(x-b)}\}.$

93. $(a-x)\sqrt[4]{(a-x)} + (x-b)\sqrt[4]{(x-b)} =$
$\qquad (a-b)\{\sqrt[4]{(a-x)} + \sqrt[4]{(x-b)}\}.$

94. $\{\sqrt[4]{(a-x)} + \sqrt[4]{(x-b)}\}^2 \{\sqrt{(a-x)} + \sqrt{(x-b)}\} = c(a+b-2x).$

95. $\{\sqrt[4]{(a-x)} + \sqrt[4]{(b-x)}\}\{\sqrt{(a-x)} + \sqrt{(b-x)}\}^2 =$
$\qquad c\{\sqrt[4]{(a-x)} - \sqrt[4]{(b-x)}\}.$

96. $a\sqrt{(1+x^2)} - x\sqrt{(x^2+a^2)} = e.$

97. $(a-x)\sqrt[3]{(x-b)} + (x-b)\sqrt[3]{(a-x)} = c\{\sqrt[3]{(a-x)} + \sqrt[3]{(x-b)}\}^4.$

98. $\{\sqrt[3]{(a-x)} + \sqrt[3]{(b+x)}\}^5 = c\{\sqrt[3]{(a-x)^2} + \sqrt[3]{(b+x)^2}\}.$

99. $\{\sqrt[3]{(a-x)} + \sqrt[3]{(b+x)}\}^5 = c\sqrt[3]{\{(a-x)(b+x)\}}.$

100. $\sqrt[3]{(a-x)^2} - \sqrt[3]{(b-x)^2} = c\sqrt[3]{(a+b-2x)}.$

101. $\sqrt[4]{(a-x)} + \sqrt[4]{(x-b)} = \sqrt[4]{c}.$

102. $\sqrt[5]{(a-x)} + \sqrt[5]{(x-b)} = \sqrt[5]{c}.$

103. $\dfrac{(a-x)\sqrt[4]{(a-x)} + (x-b)\sqrt[4]{(x-b)}}{(a-x)\sqrt[4]{(x-b)} + (x-b)\sqrt[4]{(a-x)}} = c.$

104. $\dfrac{(a-x)\sqrt[4]{(b-x)} + (b-x)\sqrt[4]{(a-x)}}{\sqrt[4]{(a-x)} - \sqrt[4]{(b-x)}} = c.$

105. $\dfrac{\sqrt[4]{(a-x)} + \sqrt[4]{(x-b)}}{\sqrt[4]{(a-x)} - \sqrt[4]{(x-b)}} = \dfrac{c}{a+b-2x}.$

106. $\dfrac{\{\sqrt[4]{(a-x)} + \sqrt[4]{(b-x)}\}^5}{\sqrt[4]{(a-x)} \cdot \sqrt[4]{(b-x)}} = c.$

107. $(a-x)\sqrt[5]{(a-x)} - (x-b)\sqrt[5]{(x-b)} = c\{\sqrt[5]{(a-x)} - \sqrt[5]{(x-b)}\}.$

108. $(a-x)\sqrt[5]{(x+b)} - (x+b)\sqrt[5]{(x-a)} = c\{\sqrt[5]{(a-x)} - \sqrt[5]{(x+b)}\}.$

109. $\{\sqrt[5]{(a-x)^3} + \sqrt[5]{(x-b)^3}\}\sqrt[5]{\{(a-x)(x-b)\}} = c.$

110. $\{\sqrt[5]{(a-x)} - \sqrt[5]{(x-b)}\}^3\{\sqrt[5]{(a-x)^2} - \sqrt[5]{(x-b)^2}\} = c.$

111. $\{\sqrt[5]{(a-x)^2} - \sqrt[5]{(x-b)^2}\}^2\{\sqrt{(a-x)} + \sqrt[5]{(x-b)}\} = c.$

112. $\{\sqrt[5]{(a-x)^3} + \sqrt[5]{(x+b)^3}\}^2 = c\{\sqrt[5]{(a-x)} + \sqrt[5]{(x+b)}\}.$

SECTION V.—QUADRATIC EQUATIONS INVOLVING TWO OR MORE VARIABLES.

1. $(x+y)(x^2+y^2) = a,$ I.
$x^2y + xy^2 = c.$ II.

I+2II. $\therefore (x+y)^3 = a+2c$

$\therefore x+y = \sqrt[3]{(a+2c)}.$ (Any one of the three cube-roots). III.

I ÷ II $\dfrac{x^2+y^2}{xy} = \dfrac{a}{c}; \therefore \left(\dfrac{x-y}{x+y}\right)^2 = \dfrac{a-2c}{a+2c}.$

By III. $x - y = \dfrac{\sqrt{(a-2c)}}{\sqrt[6]{(a+2c)}}.$

Also $x + y = \dfrac{\sqrt{(a+2c)}}{\sqrt[6]{(a+2c)}},$

$\therefore x = \dfrac{\sqrt{(a+2c)} + \sqrt{(a-2c)}}{2\sqrt[6]{(a+2c)}}$

$y = \dfrac{\sqrt{(a+2c)} - \sqrt{(a-2c)}}{2\sqrt[6]{(a+2c)}}.$

(Not any one of the six sixth-roots of $a+2c$ may be used indifferently in the denominator, but only any cube-root of whichever square-root of $a+2c$ is used in the numerator. Thus if the radical sign be restricted to denote merely the arithmetical root, if k be defined by the equation $k^2 - k + 1 = 0$, and if m and n indicate any integers whatever, equal or unequal, the value of x may be written

$$\{k^{2m}\sqrt{(a+2c)} + k^{3n-m}\sqrt{(a-2c)}\} \div 2\sqrt[6]{(a+2c)}.$$

2. $8x^2 - 5xy + 3y^2 = 9(x+y)$ I.

    $11x^2 - 8xy + 5y^2 = 13(x+y)$ II.

1st Method. Eliminate $(x+y)$.

$\therefore 104x^2 - 65xy + 39y^2 = 99x^2 - 72xy + 45y^2$,

$\therefore 5x^2 + 7xy - 6y^2 = 0$,

$\therefore (5x - 3y)(x + 2y) = 0$,

$\therefore x = \tfrac{3}{5}y$ or $-2y$.

Substitute these values for x in I,

$\therefore 72y^2 = 360y$ or $45y^2 = -9y$

$\therefore y = 0$, or 5, or $-\tfrac{1}{5}$,

and $x = 0$, or 3, or $\tfrac{2}{5}$.

2nd Method. Take the sum of the products of I. and II. by arbitrary multipliers k and l,

$k(8x^2 - 5xy + 3y^2) + l(11x^2 - 8xy + 5y^2) = (9k + 13l)(x+y)$.   III.

Determine k and l so that the left-hand member of III. may, like the right-hand member, be a multiple of $x+y$. This may be done by putting $x = -y$ in III. from which

$$16k + 24l = 0, \therefore 2k = -3l$$

$\therefore$ if $k = 3$, $l = -2$.

Substituting these values in III., it becomes

$$2x^2 + xy - y^2 = x + y$$

$\therefore (x+y)(2x-y) = x+y$, or $(x+y)(2x-y-1) = 0$.

$\therefore$ either $x+y = 0$, or $2x-y-1 = 0$,

$\therefore y = -x$, or $2x - 1$.

APPENDIX. 279

Substituting these values for y in I., it becomes
$$16x^2 = 0, \text{ or } 10x^2 - 7x + 3 = 27x - 9,$$
$$\therefore x = 0, \text{ or } 3, \text{ or } \tfrac{2}{5};$$
and $y = 0$, or 5, or $-\tfrac{1}{5}$.

3. $\dfrac{x^5 + y^5}{x^3 + y^3} = \dfrac{a^5 + b^5}{a^3 + b^3}$ I.

$x^2 + xy + y^2 = a^2 + ab + b^2$ II.

I.$\div$II., $\therefore \dfrac{x^4 - x^3 y + x^2 y^2 - xy^3 + y^4}{(x^2 + y^2)^2 + x^2 y^2} =$

$\dfrac{a^4 - a^3 b + a^2 b^2 - ab^3 + b^4}{(a^2 + b^2)^2 - a^2 b^2}$,

$\therefore \dfrac{x^3 y + xy^3}{(x^2 + y^2)^2 - x^2 y^2} = \dfrac{a^3 b + ab^3}{(a^2 + b^2)^2 - a^2 b^2}$ III.

Write z for $\dfrac{xy}{x^2 + y^2}$ and k for $\dfrac{ab}{a^2 + b^2}$

III., $\therefore \dfrac{z}{1 - z^2} = \dfrac{k}{1 - k^2}$, $\therefore z = k$ or $-\dfrac{1}{k}$,

$\therefore \dfrac{xy}{x^2 + y^2} = \dfrac{ab}{a^2 + b^2}$ or $\dfrac{a^2 + b^2}{-ab}$,

$\therefore \dfrac{xy}{x^2 + xy + y^2} = \dfrac{ab}{a^2 + ab + b^2}$, or $\dfrac{a^2 + b^2}{a^2 - ab + b^2}$.

II., $\therefore xy = ab$, or $(a^2 + b^2) \dfrac{a^2 + ab + b^2}{a^2 - ab + b^2}$ IV.

$\sqrt{}$(II.+IV.), $\therefore x + y = \pm(a + b)$
or $\sqrt{}(2a^2 - ab + 2b^2) \sqrt{\dfrac{a^2 + ab + b^2}{a^2 - ab + b^2}}$;

$\sqrt{}$(II.$-$3IV.) and $x - y = \pm(a - b)$,
$j\sqrt{}(2a^2 + ab + 2b^2)\sqrt{\dfrac{a^2 + ab + b^2}{a^2 - ab + b^2}}$.

$\therefore x = \pm a, \pm b$ or

$\frac{1}{2}\{\sqrt{(2a^2-ab+b^2)}+j\sqrt{(2a^2+ab+2b^2)}\}\sqrt{\dfrac{a^2+ab+b^2}{a^2-ab+b^2}}.$

$y = \pm b\,;\ \pm a$ or

$\frac{1}{2}\{\sqrt{(2a^2-ab+b^2)}-j\sqrt{(2a^2+ab+2b^2)}\}\sqrt{\dfrac{a^2+ab+b^2}{a^2-ab+b^2}}.$

4. $(x^2+y^2)(x^3+y^3) = a,$ I.

$(x+y)(x^4+y^4) = b.$ II.

Put $z = \dfrac{xy}{x^2+y^2},\ \therefore\ \dfrac{1-z}{1-2z^2} = \dfrac{a}{b}$

$\therefore\ 2az^2 - bz - (a-b) = 0$

$\therefore\ 4az^2 = b \pm \sqrt{(8a^2-8ab+b^2)} = b+r$ say.

$\therefore\ \dfrac{xy}{x^2+y^2} = \dfrac{b+r}{4a}$ III.

$\therefore\ \dfrac{x+y}{x-y} = \pm\sqrt{\dfrac{2a+b+r}{2a-b-r}}$

$\therefore\ \dfrac{x}{y} = \dfrac{\sqrt{(2a+b+r)}+\sqrt{(2a-b-r)}}{\sqrt{(2a+b+r)}-\sqrt{(2a-b-r)}}$

$= \dfrac{\{\sqrt{(2a+b+r)}+\sqrt{(2a-b-r)}\}^2}{2(b+r)}$ IV.

I². $(x^2+y^2+2xy)(x^2+y^2)^2\{(x^2+y^2)-xy\}^2 = a^3.$

III. $\therefore (xy)^5 \left\{\dfrac{4a+2b+2r}{b+r}\right\}\left(\dfrac{4a}{b+r}\right)^2\left\{\dfrac{4a-b-r}{b+r}\right\}^2 = a^3,$

$\therefore\ x^{10}\left(\dfrac{y}{x}\right)^5\left\{\dfrac{32a^2(2a+b+r)(4a-b-r)^2}{(b+r)^5}\right\} = a$

$\therefore\ x^{10} = \left(\dfrac{x}{y}\right)^5\left\{\dfrac{(b+r)^5}{32(2a+b+r)(4a-b-r)^2}\right\}$

IV $= \dfrac{\{\sqrt{(2a+b+r)}+\sqrt{(2a-b-r)}\}^{10}}{1024(2a+b+r)(4a-b-r)^2}$

APPENDIX. 281

$$\therefore x = \frac{\sqrt{(2a+b+r)} + \sqrt{(2a-b-r)}}{2\sqrt[10]{\{(2a+b+r)(4a-b-r)^2\}}}$$

in which $r = \pm\sqrt{(8a^2 - 8ab + b^2)}$.

The value of y may be derived from that of x by the first form in IV.

5. $x^4 = ax - by$, I.
   $y^4 = ay - bx$. II.

$x.\text{I.} - y.\text{II.}$ $x^5 - y^5 = a(x^2 - y^2)$

$y.\text{I.} - x.\text{II.}$ $xy(x^3 - y^3) = b(x^2 - y^2)$,

$\therefore$ either $x - y = 0$ from which $x = y = 0$, or $\sqrt[3]{(a-b)}$ III.

or $x^4 + x^3y + x^2y^2 + xy^3 + y^4 = a(x+y)$ IV.

and $xy(x^2 + xy + y^2) = b(x+y)$ V.

(IV. + V.) $(x+y)^2(x^2+y^2) = a+b$ VI.

V. $(x+y)^4 - (x^2+y^2)^2 = 4b(x+y)$ VII.

$\sqrt{(\text{VII}^2 + 4.\text{VI})}$. $(x+y)^4 + (x^2+y^2)^2 = 2t(x+y)$ VIII.

in which $t = \sqrt{\{(a+b)^2 + 4b^2\}}$. IX.

$\frac{1}{2}(\text{VII.} + \text{VIII.})$, $\therefore (x+y)^4 = (2b+t)(x+y)$

$\therefore (x+y)^3 = 2b+t$

$\therefore (x+y) = \sqrt[3]{(2b+t)}$ X.

VI.$\div$X^2 $\therefore x^2 + y^2 = \frac{a+b}{\sqrt[3]{(2b+t)}}$ XI.

2.XI. $-$ X^2 $\therefore (x-y)^2 = \frac{2(a+b)}{\sqrt[3]{(2b+t)}} - \sqrt[3]{(2b+t)^2} = \frac{2a-t}{\sqrt[3]{(2b+t)}}$.

$\therefore x - y = \frac{\sqrt{(2a-t)}}{\sqrt[6]{(2b+t)}}$

X. and $x+y = \frac{\sqrt[3]{(2b+t)}}{\sqrt[6]{(2b+t)}}$

$$\therefore x = \frac{\sqrt{(2b+t)} + \sqrt{(2a-t)}}{\sqrt[6]{(2b+t)}}$$

$$\text{and } y = \frac{\sqrt{(2b+t)} - \sqrt{(2a-t)}}{\sqrt[6]{(2b+t)}};$$

in which $t = \sqrt{(a^2 + 2ab + 5b^2)}$.

6. $x^4 - c^4 = m(x+y)^4$;

$y^4 + c^4 = n(x-y)^4$.   II.

Let $z = \dfrac{x+y}{x-y}$, $\therefore z+1 = \dfrac{2x}{x-y}$ and $z-1 = \dfrac{2y}{x-y}$ III.

I. + II. $x^4 + y^4 = m(x+y)^4 + n(x-y)^4$

$\therefore (z+1)^4 + (z-1)^4 = 16(mz^4 + n)$

$\therefore (8m-1)z^4 - 6z^2 + (8n-1) = 0$,

$$\therefore z = \sqrt{\frac{3 + \sqrt{\{9 - (8m-1)(8n-1)\}}}{8m-1}} \quad \text{IV.}$$

II. & III. $(z-1)^4(x-y)^4 + 16c^4 = 16n(x-y)^4$

$$\therefore x - y = \frac{2c}{\sqrt[4]{\{16n - (z-1)^4\}}} \quad \text{V.}$$

$$\text{and } x + y = \frac{2cz}{\sqrt[4]{\{16n - (z-1)^4\}}}.$$

$$\therefore x = \frac{c(z+1)}{\sqrt[4]{\{16n - (z-1)^4\}}} = \frac{c(z+1)}{\sqrt[4]{\{z+1)^4 - 16mz^4\}}},$$

and $y = \dfrac{c(z-1)}{\sqrt[4]{\{16n - (z-1)^4\}}}$, and the value of z is given by IV.

7. $x^2 + y^2 = \tfrac{1}{3}(2m + n^2)$,

$x^3 + y^3 = mn$.

$\therefore (x+y)^2 - 2xy = \tfrac{1}{3}(2m + n^2)$

and $(x+y)^3 - 3xy(x+y) = mn$.

Let $u = x+y$ and $v = xy$, and the equations become
$$u^2 - 2v = \tfrac{1}{2}(2m+n^2);$$
$$u^3 - 3uv = mn.$$

Eliminate v, $\therefore u^3 - (2m+n^2)u + 2mn = 0$,

$\therefore u^4 - (2m+n^2)u^2 + 2mn\,u = 0$,

$\therefore u^4 - 2mu^2 + m^2 = n^2u^2 - 2mnu + m^2$.

$\therefore u^2 - m = \pm(nu - m)$,

$\therefore u = n$, (the value $u = 0$ was introduced by the multiplication by u),

or $u^2 + nu - 2m = 0$,

$\therefore u = \tfrac{1}{2}\{-n \pm \sqrt{(n^2 + 8m)}\}$

$\therefore v = \tfrac{1}{2}(n^2 - m)$ or $\tfrac{1}{8}\{n^2 + 8m \mp 8n\sqrt{(n^2+8m)}\}$

$\therefore u$ and v are completely determined.

Also $x+y = u$, $x-y = \sqrt{(u^2 - 4v)}$

$\therefore x = \tfrac{1}{2}\{u + \sqrt{(u^2 - 4v)}\}$;

$\therefore y = \tfrac{1}{2}\{u - \sqrt{(u^2 - 4v)}\}$.

If $m = 7$ and $n = 5$, the above equations become
$$x^2 + y^2 = 13, \text{ and } x^3 + y^3 = 35.$$

Solving, as above, gives

$u = 5$, or 2, or -7,

$2v = 12$, or -9, or 86,

$\therefore x+y = 5$, or 2, or -7,

$x-y = \pm 1$, or $\pm\sqrt{22}$, or $\pm j\sqrt{23}$.

$\therefore x = 3$, 2, $\tfrac{1}{2}(2\pm\sqrt{22})$ or $\tfrac{1}{2}(-7\pm j\sqrt{23})$;

$y = 2$, 3, $\tfrac{1}{2}(2\mp\sqrt{22})$ or $\tfrac{1}{2}(-7\mp j\sqrt{23})$.

8. $x^2+y=\frac{17}{16}$;

   $x+y^2=\frac{5}{4}$.

   $\therefore \frac{17}{16}-y=(\frac{5}{4}-y^2)^2$,

   $\therefore y^4-\frac{5}{2}y^2+y+\frac{1}{2}=0$.

Testing this for rational linear factors it is easily reduced to

$$(y-1)^2(y^2+2y+\frac{1}{2})=0,$$

$\therefore y=1$ or $\frac{1}{2}(-2\pm\sqrt{2})$;

$x=\frac{1}{4}$ or $\frac{1}{4}(-1\pm 4\sqrt{2})$.

9. $(2x-y+z)(x+y+z)=9$; I.
   $(x+2y-z)(x+y+z)=1$; II.
   $(x+y-2z)(x+y+z)=4$. III.

Let $s=x+y+z$ and the equations may be written

$(s+x-2y)s=9$ IV.
$(s+y-2z)s=1$ V.
$(s-3z)s=4$. VI.

IV.+8.V. $(4s+x+y-6z)s=12$, or $(5s-7z)s=12$ VII.

8.VII−7.VI. $\{(15s-21z)-(7s-21z)\}s=8$,

$\therefore 8s^2=8$, $\therefore s=\pm 1$.

Substituting in I, II. and III. they become

$2x-y+z=\pm 9$, $x+2y-z=\pm 1$, $x+y-2z=\pm 4$,

$\therefore x=\pm 4$, $y=\mp 2$, $z=\mp 1$.

10 $x^2+y^2=a$;

   $u^2+v=b$;

   $xy+uv=c$;

   $xu+yv=e$.

Let $t=xy-uv$.

$\therefore (x+y)^2=a+c+t$, $\therefore x=\frac{1}{2}\{\sqrt{(a+c+t)}+\sqrt{(a+c-t)}\}$

$(x-y)^2=a-c-t$, $y=\frac{1}{2}\{\sqrt{(a+c+t)}-\sqrt{(a-c-t)}\}$

APPENDIX. 285

$$(u+v)^2 = b+c-t, \quad u = \tfrac{1}{2}\{\sqrt{(b+c-t)}+\sqrt{(b-c+t)}\}$$
$$(u-v)^2 = b-c+t, \quad v = \tfrac{1}{2}\{\sqrt{(b+c-t)}-\sqrt{(b-c+t)}\}$$

Also $2(xu+yv) = (x+y)(u+v)+(x-y)(u-v) = 2e$,

$\therefore \sqrt{\{(a+c+t)(b+c-t)\}} + \sqrt{\{(a-c-t)(b-c+t)\}} = 2e$,

$\therefore \{4e^2+(a-c-t)(b-c+t)-(a+c+t)(b+c-t)\}^2 =$
$$16e^2(a-c-t)(b-c+t).$$

$\therefore \{(a-b)^2+4e^2\}t^2 - 2(a^2-b^2)ct +$
$$(a+b)^2c^2 - 4e^2(ab+c^2)+4e^4 = 0,$$

$\therefore t = \dfrac{(a^2-b^2)c \pm 2c\sqrt{[(ab-e^2)\{(a-b)^2-4(c^2-e^2)\}]}}{(a-b)^2+4e^2}.$

11. $\qquad xy = uv \qquad\qquad$ I.

$\qquad x+y+u+v = a \qquad\qquad$ II.

$\qquad x^3+y^3+u^3+v^3 = b^3 \qquad\qquad$ III

$\qquad x^5+y^5+u^5+v^5 = c^5 \qquad\qquad$ IV.

Let $x+y = \tfrac{1}{2}(a+z)$. $\therefore u+v = \tfrac{1}{2}(a-z)$. $\qquad$ V.

Also let $r = xy = uv \qquad\qquad$ VI.

$\qquad (x+y)^3 = x^3+y^3+3xy(x+y)$
$\qquad (u+v)^3 = u^3+v^3+3uv(u+v)$

$\therefore a(3z^2+a^2) = 4(b^3+3ar) \qquad\qquad$ VII.

Also $(x+y)^5 = x^5+y^5+5xy(x^3+y^3)+10x^2y^2(x+y)$
$\qquad (u+v)^5 = u^5+v^5+5uv(u^3+v^3)+10u^2v^2(u+v)$

$\therefore a(5z^4+10a^2z^2+a^4) = 16\{c^5+5b^3r+10ar^2\} \qquad$ VIII.

Eliminating r between VII. and VIII,

$45a^2z^4 - 30a(a^3+2b^3)z^2 + a^6 - 20a^3b^3 - 80b^6 + 144ac^5 = 0$

$\therefore 15az^2 - 5(a^3+2b^3) = \pm 2\sqrt{\{5(a^3+5b^3)^2 - 180ac^5\}} \qquad$ IX.

$\therefore z = \sqrt{\dfrac{a^3+2b^3 \pm 2\sqrt{[\tfrac{1}{5}\{(a^3+5b^3)^2-36ac^5\}]}}{3a}} \qquad$ X.

VII. & IX. $12ar = a^3 - 4b^3 + 3az^2$

$$= 2a^3 - 2b^3 \pm 2\sqrt{[\tfrac{1}{5}\{(a^3+5b^3)^2 - 36ac^5\}]}$$

$\therefore r = \dfrac{5(a^3-b^3) \pm \sqrt{\{5(a^3+5b^3)^2 - 180ac^5\}}}{30a}$.   XI.

X. and XI. give the values of z and r which may now be treated as known in ✓ and ✓.

$$x+y = \tfrac{1}{2}(a+z), \text{ and } xy = r$$

$\therefore x - y = \tfrac{1}{2}\sqrt{\{(a+z)^3 - 16r\}}$

$\therefore \quad x = \tfrac{1}{4}(a+z \pm \sqrt{\{(a+z)^2 - 16r\}})$;

$\quad\quad y = \tfrac{1}{4}(a+z \mp \sqrt{\{(a+z)^2 - 16r\}})$.

The values of u and v may be obtained from those of x and y respectively by changing z into $-z$.

EXERCISE.

1. $6\{(7-x)^2 + y^2\} = 13(7-x)y$, $x^2 + 4y = y^2 + 4$.
2. $10x^2 - 9y^2 = 2x^3$, $8x^2 - 6y^2 = 13x$.
3. $xy = (3-x)^2 = (2-y)^2$.   4. $x^2 + y^2 = 8x + 9y = 144$.
5. $x^2 + y^2 = x + y + 12$, $xy + 8 = 2(x+y)$.
6. $x + xy + y = 5$, $x^2 + xy + y^2 = 7$.
7. $x^3 + y^3 = 7xy = 28(x+y)$.   8. $x^2 + xy + y^2 = \dfrac{35}{x^2+y^2} = \dfrac{28}{xy}$.
9. $x^4 + x^2y^2 + y^4 = 133$, $x^3y + x^2y^2 + xy^3 = 114$,
10. $(x+y)(x^2+y^2) = 17xy$, $(x-y)(x^2-y^2) = 9xy$.
11. $25(x^3+y^3) = 7(x+y)^3 = 175xy$.
12. $2x^2 - y^2 = 14(x^2 - 2y^2) = 14(x-y)$.
13. $2x^2 - 3xy = 9(x-3y)$, $3(x^2-3y^2) = 2(2x^2 - 3xy)$.
14. $2x^2 - xy + 5y^2 = 10(x+y)$, $x^2 + 4xy + 3y^2 = 14(x+y)$.
15. $(2x-3y)(3x+4y) = 39(x-2y)$, $(3x+2y)(4x-3y) = (99(x-2y)$

16. $(x+2y)(x+3y) = 3(x+y)$, $(2x+y)(3x+y) = 28(x+y)$.
17. $x+y=8$, $x^4+y^4=706$. 18. $x+y=5$, $x^5+y^5=275$.
19. $x+y=2$, $13(x^5+y^5) = 121(x^3+y^3)$.
20. $x+y=4$, $41(x^5+y^5) = 122(x^4+y^4)$.
21. $x^2-5xy+y^2+5=0$, $xy = x+y-1$.
22. $x^2+y = 5(x-y)$, $x+y^2 = 2(x-y)$.
23. $3(x^2+y) = 3(x+y^2) = 13xy$.
24. $10(x^2+y) = 10(x+y^2) = 13(x^2+y^2)$.
25. $x^2+y = \tfrac{16}{9}$, $x+y^2 = \tfrac{22}{9}$. 26. $9(x^2+y) = 3(x+y^2) = 7$.
27. $x+xy+y = 5$, $x^3+xy+y^3 = 17$.
28. $x+y=2$, $(x+1)^5+(y-2)^5 = 211$.
29. $3(x-1)(y+1) = 4(x+1)(y-1)$, $\dfrac{x^2+x+1}{y^2+y+1} = \dfrac{31}{39}\left(\dfrac{x^2-x+1}{y^2-y+1}\right)$
30. $x+y = \dfrac{1}{xy}$, $x-y = xy$.
31. $x+y+1=0$, $x^6+y^6+2=0$.
32. $x+y=1$, $3(x^8+y^8)=7$.
33. $4xy^2 = 5(5-x)$, $2(x^2+y^2) = 5$.
34. $27xy = 17$, $9(x^3+y^3) = -3$.
35. $(x^2+y^2)^2 + 4x^2y^2 = 5-12y$, $y(x^2+y^2)+3=0$.
36. $x+y = xy$, $x^2+y^2 = x^3+y^3$.
37. $x^4 - 6x^2\sqrt{(y^2-x^2)^{\frac{1}{2}}} - 16y^2 = 9x^2$,
    $(x^2+2)^3 = 4\{2+x^2\sqrt{(y^2+x^2)}-y^2\}$.
38. $x(y^2+3y-1) = 2y^2+2y+3$, $y(x^2+3x-1) = 2x^2+2x+3$.

39. $\dfrac{x^3}{a^3} + \dfrac{y^3}{b^3} = 2c^3$, $\dfrac{x}{a} + \dfrac{y}{b} = e\left(\dfrac{x}{a} - \dfrac{y}{b}\right)$.

40. $x^3 + xy^2 = a$, $y^3 + x^2 y = b$.

41. $x + y = a$, $\dfrac{x}{b-y} + \dfrac{b-y}{x} = c$.

42. $x^2 + ay^2 = \dfrac{a+1}{a-1}$, $ax^2 + y^2 = (a^2 - 1)y$.

43. $x + y^2 = ax$, $x^2 + y = by$.   44. $x + y^2 = ay^2$, $x^2 + y = bx^2$.

45. $x^4 - y^4 = a^2(x-y)^2$, $x^3 - x^2 y + xy^2 - y^3 = b^2(x+y)$.

46. $(x+y)(x^2 + 3y^2) = m$, $(x-y)(x^2 + 3y^2) = n$.

47. $x^2 y^2 = y(a-x)^3 = x(b-y)^3$.

48. $x^3(b-y) = y^3(a-x) = (a-x)^2(b-y)^2$.

49. $a^2(x^2 + e^2) = b^2(x+y)^2$, $a^2(y^2 + e^2) = c^2(x+y)^2$.

50. $x^3 - y^3 = a(x^2 - y^2)$, $x^2 + y^2 = b(x+y)$.

51. $x + y = a$, $x^3 + y^3 = bxy$.

52. $\sqrt{\dfrac{x}{y}} - \sqrt{\dfrac{y}{x}} = \dfrac{x-y}{a}$, $\dfrac{x(c^2 + xy)}{y(c^2 - xy)} = b^2$.

53. $x + y = xy = x^2 + y^2$.   54. $x - y = \dfrac{x}{y} = x^2 - y^2$.

55. $x^3(1+y^2)(1+y^4) = a$, $x^3(1-y^2)(1-y^4) = b$.

56. $\dfrac{x^2 + xy + y^2}{x^2 - xy + y^2} = \dfrac{x^2 + y^2}{a} = \dfrac{xy}{b}$.

57. $x^2 y + xy^2 = \dfrac{a}{x^2 + y^2}$, $x^4 y + xy^4 = b$.

58. $x^2 y + xy^2 = a(x^2 + y^2)$, $x^2 y - xy^2 = b(x^2 - y^2)$.

59. $\left(\dfrac{x}{y} + \dfrac{y}{x}\right)(x+y) = a$, $\dfrac{x^2}{y} + \dfrac{y^2}{x} = b$.

APPENDIX.

60. $x^2+y^2=ax^2y^2=xy(x+y)$.

61. $abxy=a(x^3+y^3)=b(x+y)^3$.

62. $xy(x+y)=a, \quad x^3y^3(x^3+y^3)=b$.

63. $\left(\dfrac{1}{x}+\dfrac{1}{y}\right)(x^3-y^3)=a, \quad \left(\dfrac{1}{x}-\dfrac{1}{y}\right)(x^3+y^3)=b$.

64. $x^4+y^4=m(x^2+y^2), \quad x^2+xy+y^2=n$.

65. $ab(x+y)=xy(a+b), \quad x^2+y^2=a^2+b^2$.

66. $x^3+y^3=a(x+y), \quad x^4+y^4=b(x+y)^2$.

67. $x^2+y^2=a, \quad x^5+y^5=b(x^3+y^3)$.

68. $xy=a, \quad x^5+y^5=b(x^3+y^3)$.

69. $(x-y)(x^3+y^3)=(a-b)(a^3+b^3), \quad x^2-y^2=a^2-b^2$.

70. $x^2-y^2=a, \quad x^3+y^3=b(x-y)$.

71. $x+y=a, \quad x^4+y^4=b$.   72. $x+y=a, \quad x^5+y^5$

73. $x+y=a, \quad x^2+y^2=b^2x^2y^2$.

74. $x+y=a+b, \quad (a-b)^2(x^4+y^4)=(x-y)^2(a^4+b^4)$.

75. $x+y=a, \quad c(x^4+y^4)=xy(x^3+y^3)$.

76. $(x+y)^3=a(x^2+y^2), \quad xy=c(x+y)$.

77. $x^2y+xy^2=a^3, \quad c^3(x^3+y^3)=x^3y^3$.

78. $x^3=a(x^2+y^2)-cxy, \quad y^3=c(x^2+y^2)-axy$.

79. $x^2-y^2=a^2, \quad x^3-y^3=c^4\left(\dfrac{1}{x}-\dfrac{1}{y}\right)$.

80. $x^4-y^4=a^2xy, \quad (x^2+y^2)^2=b^2(x^2-y^2)$.

81. $(x+y)x^2y^2=a, \quad x^5+y^5=b$.

82. $(x+y)xy=a, \quad x^5+y^5=bxy$.

83. $x^4 + y^4 = a(x+y)^2, \quad x^5 + y^5 = b(x+y)^3.$

84. $x^4 + x^2y^2 + y^4 = a, \quad x^2 - xy + y^2 = 1.$

85. $(x^2 + y^2)xy = x^2 - y^2, \quad \dfrac{x^4(1+x^2y^2)}{y^4(1+xy)^2} = \dfrac{a}{b} \cdot \dfrac{1+xy}{1-xy}.$

86. $x+y = (x-y)\sqrt{(xy)}, \quad \dfrac{\sqrt{(x^7y^5)} - x}{\sqrt{(x^5y^7)} + y} = a.$

87. $\dfrac{1}{x} + \dfrac{1}{y} = \dfrac{1}{a}, \quad \sqrt{(1-x)} - \sqrt{(1-y)} = b.$

88. $x^2 + y^2 = a(x+y); \quad x^4 + y^4 = b(x^3 + y^3).$

89. $x^3 + y^3 = a, \quad (x+y)(x^4 + y^4) = b(x^2 + y^2).$

90. $(x^2+y^2)(x^3+y^3) = axy, \quad (x+y)(x^4+y^4) = bxy.$

91. $(x+y)^2(x^2+y^2) = a, \quad (x^2+y^2)^2(x^4+y^4) = b.$

92. $(x-y)(x^2-y^2)(x^4-y^4) = 4axy,$
$(x+y)(x^2+y^2)(x^3+y^3) = b(x-y).$

93. $x^4y + xy^4 = a(x^3y + xy^3) = b(x^4 + y^4).$

94. $a(x^5 + y^5) = ab(x+y) = bxy(x^3 + y^3).$

95. $\dfrac{x^3 - y^3}{x^2 - y^2} = \dfrac{a^3 - b^3}{a^2 - b^2}, \quad \dfrac{x^5 - y^5}{x^4 - y^4} = \dfrac{a^5 - b^5}{a^4 - b^4}.$

96. $\dfrac{x^3 + y^3}{x^2 - y^2} = \dfrac{a^3 + b^3}{a^2 - b^2}, \quad \dfrac{x^4 + y^4}{x^3 - y^3} = \dfrac{a^4 + b^4}{a^3 - b^3}.$

97. $x^5 = 2ax - by, \quad y^5 = 2ay - bx.$

98. $(x+y)(x^3+y^3) = a, \quad (x-y)(x^3-y^3) = b.$

99. $\dfrac{(x+y)^4(x^2+xy+y^2)}{(x^2+y^2)(x^2-xy+y^2)} = 8m^2,$

$\dfrac{(x-y)^4(x^2-xy+y^2)}{(x^2+y^2)(x^2+xy+y^2)} = 8n^2.$

APPENDIX.

100. $(x+y)(x^3+y^3) = axy$, $(x-y)(x^3-y^3) = bxy$.

101. $(x+y)(x^3+y^3) = a(x^2+y^2)$, $(x-y)(x^3-y^3) = b(x^2+y^2)$.

102. $\dfrac{(x+y)^3(x^3+y^3)}{(x^2+xy+y^2)(x^2+y^2)} = a^2$,

$\dfrac{(x-y)^3(x^3-y^3)}{(x^2-xy+y^2)(x^2+y^2)} = b^2$.

103. $\dfrac{(x^3+y^3)(x+y)^3}{x^2+xy+y^2} = 2a^2$, $\dfrac{(x^3-y^3)(x-y)^3}{x^2-xy+y^2} = 2b^2$.

104. $\dfrac{(x^3+y^3)(x+y)^5}{(x^2+xy+y^2)^2} = 8a^2$, $\dfrac{(x^3-y^3)(x-y)^5}{(x^2-xy+y^2)^2} = 8b^2$.

105. $xy(x+y)(x^3+y^3) = a$, $xy(x-y)(x^3-y^3) = b$.

106. $x(x+y)(x+2y)(x+3y) = a^2$, $(x+y)^2 + (x+2y)^2 = b$.

107. $\sqrt{(x-xy)} + \sqrt{(y-xy)} = a$, $\sqrt{(x-x^2)} + \sqrt{(y-y^2)} = b$.

108. $(x+1)(y-1) = a(x-1)(y+1)$,

$(x^5+1)(y-1)^5 = b^3(x-1)^5(y^5+1)$.

109. $x+y = a$, $\sqrt[4]{(x-k)} + \sqrt[4]{(y-k)} = c$.

110. $x+y = a(1+xy)$, $(x+y)^4 = b^4(1+x^4y^4)$.

111. $x+y = a(1+xy)$, $x^5+y^5 = b^3(1+x^5y^5)$.

112. $(x+1)(y-1) = a(x-1)(y+1)$,

$(x^5-1)(y-1) = b^2(y^5-1)(x-1)$.

113. $\dfrac{(1+x)(1+y)}{(1-x)(1-y)} = a$, $\dfrac{(1+x)^3(1+y)^3}{(1-x^3)(1-y^3)} = b$.

114. $\dfrac{(c+x)(c+y)}{(c-x)(c-y)} = a$, $\dfrac{(c^4+x^4)(c^4+y^4)}{(c^4-x^4)(c^4-y^4)} = b$.

115. $\dfrac{(x+m)(y+n)}{(x-m)(y-n)} = a$, $\dfrac{(x^4+m^4)(y^4+n^4)}{(x-m)^4(y-n)^4} = b$.

116. $\dfrac{(x+1)(y+1)}{(x-1)(y-1)} = \dfrac{a}{c}$, $\dfrac{(x^5+1)(y^5+1)}{(x^5-1)(y^5-1)} = \dfrac{b}{c}$.

117. $\dfrac{y(1+x^2)}{x(1+y^2)} = a$, $\dfrac{y^2(1+x^4)}{x^2(1+y^4)} = b$.

118. $\dfrac{1+x}{1+y} = a\sqrt{\dfrac{x}{y}}$, $\dfrac{y(1+x+x^2)}{x(1+y+y^2)} = b$.

119. $\dfrac{y(1+x^2)}{x(1+y^2)} = a$, $\dfrac{y^3(1+x^6)}{x^3(1+y^6)} = b$.

120. $\dfrac{y(1+x^2)}{x(1+y^2)} = a$, $\dfrac{y^4(1+x^8)}{x^4(1+y^8)} = b$.

121. $\dfrac{y(1+x^2)}{x(1+y^2)} = a$, $\dfrac{y^5(1+x^{10})}{x^5(1+y^{10})} = b$.

122. $\dfrac{(x+y)(xy+1)}{(x-y)(xy-1)} = \dfrac{a^2+b^2}{2ab}$, $\dfrac{x(y^2+1)}{y(x^2-1)} = \dfrac{a-b}{a+b}$.

123. $\dfrac{(x+y)(1+xy)}{(x-y)(1-xy)} = a^2(b^2-1)$, $\dfrac{x(1-y^2)}{y(1-x^2)} = b$.

124. $\dfrac{(x+y)(1+xy)}{(x-y)(1-xy)} = a$, $\dfrac{(x^2+y^2)(1+x^2y^2)}{(x^2-y^2)(1-x^2y^2)} = b$.

125. $\dfrac{(x+y)(1+xy)}{(x-y)(1-xy)} = a$, $\dfrac{(x^3+y^3)(1+x^3y^3)}{(x^3-y^3)(1-x^3y^3)} = b$.

126. $\dfrac{(x^4+y^4)(1+x^4y^4)}{(x^4-y^4)(1-x^4y^4)} = a$, $\dfrac{(x+y)(1+xy)}{(x-y)(1-xy)} = b$.

127. $\dfrac{(x+y)(1+xy)}{(x-y)(1-xy)} = a$, $\dfrac{(x^5+y^5)(1+x^5y^5)}{(x^5-y^5)(1-x^5y^5)} = b$.

128. $\dfrac{(x^2+xy+y^2)(1+xy+x^2y^2)}{(x+y)^2(1+xy)^2} = a$.

$\dfrac{(x^2-xy+y^2)(1-xy+x^2y^2)}{(x-y)^2(1-xy)^2} = b$.

129. $x^4 - 3x^2y + 5a^3x + y^2 = 0$, $y^3 - x^2y^2 - 2a^5x = 0$.

APPENDIX.

130. $2x(y^2-2x)^2 = a$, $y(y^2-2x)^2\sqrt{(y^2-4x)} = b$.

(Hence deduce the solution of $x^5-5x^2+2=0$).

131. $2xy(x^2+y^2)^2 = a$, $(x^2-y^2)(x^2+y^2)^2 = b$.

132. $\sqrt{(x^2+y^2)} + \sqrt{\{(a-x)^2+y^2\}} = \sqrt{\{(\tfrac{1}{2}a\sqrt{3}-y)^2 + (\tfrac{1}{2}a-x)^2\}}$.
$6(x^2-y^2) = a(6x-2y\sqrt{3}+a)$.

Exercise.

1. $(2x+y-4z)(x+y+z) = 24$,
$(x+2y-2z)(x+y+z) = 6$,
$(-2x+3y+5z)(x+y+z) = 30$.

2. $x^2 - yz = 1$,
$y^2 - xz = 2$,
$z^2 - xy = 3$.

3. $(x+2y-3z)(x+y+z) - 2(xy+yz+zx) = -12$,
$(2x-3y+z)(x+y+z) + (xy+yz+zx) = 61$,
$(3x-y+2z)(x+y+z) - 5(xy+yz+zx) = 5$.

4. $x^2 - yz = 0$,
$x+y+z = 7$,
$x^2+y^2+z^2 = 21$.

5. $(x^5+y^5+z^5)^3 + (x+y)^2 = 31$.
$(x^5+y^5+z^5)^3 + (x+y+z)^3 = 729$.
$(x+y)^2 + (x+y+z)^3 = 81$.

6. $x^2 - yz = 0$,
$x+y+z = 21$,
$(x-y)^2+(y-z)^2+(z-x)^2 = 126$.

7. $x+yz = 14$,
$y+zx = 11$,
$z+xy = 10$.

8. $x+y = 8z$,
$x^3+y^3 = 134z^3$,
$x^2+y^2+z^3 = 134$.

9. $x+y = 5z$,
$x^2+y^2 = 39z$,
$x^3+y^3 = 105z^2$.

10. $x+y = 7z$,
$x^2+y^2 = 25z^2$,
$x^4+y^4 = 674z^3$.

11. $x+y = 7z$,
$x^2+y^2 = 25z^2$,
$x^5+y^5 = 20272z$.

12. $x+y : y+z : z+x :: a:b:c$,
$(a+b+c)xyz = 2$.

13. $x+y : y+z : z+x :: a:b:c$,
$(a+b+c)xyz = 2(x+y+z)$.

14. $ax = by = cz = \dfrac{1}{x} + \dfrac{1}{y} + \dfrac{1}{z}.$ 16. $z\left(\dfrac{x}{y} + \dfrac{y}{x}\right) = a,$

15. $(x+y-z)x = a,$
    $(x-y+z)y = b,$
    $(-x+y+z)z = c.$

$y\left(\dfrac{x}{z} + \dfrac{z}{x}\right) = b,$

$x\left(\dfrac{y}{z} + \dfrac{z}{y}\right) = c.$

17. $(y+z)(2x+y+z) = a,$ 18. $x(y+z) : y(z+x) : z(x+y) =$
    $(z+x)(x+2y+z) = b,$ $b+c : c+a : a+b,$
    $(x+y)(x+y+2z) = c,$ $xy + yz + zx = (a+b+c)(x+y+z).$

19. $(a+b)x + (b+c)y + (c+a)z = (a+b+c)(x+y+z),$
    $a(x+y) = c(y+z),$
    $(x+y)^2 + (y+z)^2 + (z+x)^2 = 4(a^2 + b^2 + c^2).$

20. $c(x+y) + b(x-z) - a(y+z) = 0,$
    $b(x-z) = (a-c)y,$
    $x^2 + y^2 + z^2 = a^2 + b^2 + c^2.$

21. $x + y - az = x - by + z = -cx + y + z = xyz.$

22. $(a+b+c)(x-y) + a(x+z) - b(y+z) = 0,$
    $(a+b+c)(x-z) + a(x+y) - c(y+z) = 0,$
    $\dfrac{ax^2}{(b+c)^2} + \dfrac{by^2}{(c+a)^2} + \dfrac{cz^2}{(a+b)^2} = 1.$

23. $xy + \dfrac{x}{z} = a, \quad yz + \dfrac{y}{x} = b, \quad zx + \dfrac{z}{y} = c.$

24. $y+z : z+x : x+y :: b+c : c+a : a+b,$
    $(x+y+z)(xyz) = (a+b+c)(xy+yz+zx).$

25. $x^2 - yz = a, \quad y^2 - xz = b, \quad z^2 - xy = c.$

26. $x^2 + (y-z)^2 = a^2, \quad y^2 + (z-x)^2 = b^2, \quad z^2 + (x-y)^2 = c^2.$

27. $x^2+xy+y^2=a^2$, $y^2+yz+z^2=b^2$, $z^2+zx+x^2=c^2$.

28. $x^3+y^3-z^3+3xyz=a(x+y-z)$
$x^3-y^3+z^3+3xyz=b(x-y+z)$,
$-x^3+y^3+z^3+3xyz=c(-x+y+z)$.

29. $x+y+2az=0$,
$x^2+y^2-2b^2z^2=0$,
$x^n+y^n+z^n=c^n$.

30. $x+y-az=0$,
$y(x^2+y^2)=b^2$,
$x^3+y^3=c^3$.

31. $x(y-1)(z-1)=2a$,
$x^2(y^2-1)(z^2-1)=4bz$,
$x^3(y^3-1)(z^3-1)=6cz^2$

32. $x(y-1)=a(z-1)$,
$x^2(y^2-1)=b^2(z^2-1)$,
$x^3(y^3-1)=c^3(z^3-1)$.

33. $x(y-1)=a(z-1)$,
$x^2(y^2-1)=b^2(z^2-1)$,
$x^4(y^4-1)=c^4(z^4-1)$.

34. $x(y-1)=a(z-1)$,
$x^2(y^2-1)=b^2(z^2-1)$,
$x^4(y^4+1)=c^4(z^4+1)$.

35. $x(y-1)=a(z-1)$,
$x^3(y^3-1)=b^3(z^3-1)$,
$x^5(y^5-1)=c^5(z^5-1)$.

36. $(x-y)^2=az(x+y)$,
$(x^3-y^3=bz(x+y)^3$,
$(x-y)^3=cz(x^3+y^3)$

37. $x-y=a$,
$u-v=b$,
$xy=uv$,
$x^5-y^5+u^5-v^5=c(a+b)$.

38. $x+y=a$,
$u+v=b$,
$x^2+u^2=c^2$,
$y^2+v^2=e^2$.

39. $xy=uv=a^2$,
$x+y+u+x=b$,
$x^3+y^3+u^3+v^3=c^3$.

40. $xy=uv=a^2$,
$x+y+u+v=b$,
$x^4+y^4+u^4+v^4=c^4$.

41. $xy=uv=a^2$,

42. $xy=uv=a^2$,

$x+y+u+v=b,$ $\qquad$ $x+y+u+v=b,$
$x^5+y^5+u^5+v^5=c^5.$ $\qquad$ $(x+u)^3+(y+v)^3=c^3.$

43. $xy=uv=a^2,$ $\quad x+y+u+v=b,$ $\quad (x+u)^4+(y+v)^4=c^4.$

44. $xy=uv,$ $\qquad\qquad$ 45. $xy=uv,$
$x+y+u+v=a,$ $\qquad$ $x+y+u+v=a,$
$x^2+y^2+u^2+v^2=b^2,$ $\qquad$ $x^2+y^2+u^2+v^2=b^2,$
$x^3+y^3+u^3+v^3=c^3.$ $\qquad$ $x^4+y^4+u^4+v^4=c^4.$

46. $\quad xy=uv,$ $\qquad$ 47. $\quad xy=uv,$
$x+y+u+v=a,$ $\qquad$ $x+y+u+v=a,$
$x^2+y^2+u^2+v^2=b^2,$ $\qquad$ $x^3+y^3+u^3+v^3=b^3,$
$x^5+y^5+u^5+v^5=c^5.$ $\qquad$ $x^4+y^4+u^4+v^4=c^4.$

48. $xy-uv=0,$ $\qquad$ 49. $x^2+y^2=a^2,$
$xu+yv=a^2,$ $\qquad\qquad$ $u^2+v^2=m^2,$
$x+y+u+v=b,$ $\qquad\qquad$ $ux+vy=c^2,$
$x^3+y^3+u^3+v^3=c^3.$ $\qquad$ $vx+uy=n^2.$

50. $x+y+u+v=a,$ $\qquad$ 51. $y(1+x^2)=2x,$
$xy+uv=b^2,$ $\qquad\qquad$ $u(1+y^2)=2y,$
$x^2+y^2=m^2,$ $\qquad\qquad$ $v(1+u^2)=2u,$
$u^2+v^2=n^2.$ $\qquad\qquad$ $x(1+v^2)=2v.$

52. $x+y+u+v=a,$ $\quad (x+y)^2+(u+v)^2=b^2,$
$(x+u)^2+(y+v)^2=c^2,$ $\quad (x+v)^2+(y+u)^2=e^2.$

53. $\dfrac{x}{y+z}=\dfrac{2a-u}{a-2u},$ $\quad \dfrac{y}{z+x}=\dfrac{2b-u}{b-2u},$ $\quad \dfrac{z}{x+y}=\dfrac{2e-u}{c-2u},$
$x^2+y^2+z^2=u^2.$

ANSWERS.

Exercise i.

1. $9, -69, 1, 0, 1206, -29, 1\frac{3}{8}$. 2. $-160, 106, 41, 108$
3. $-\frac{5}{19}, 1\frac{3}{35}, -25, 125, \frac{7}{19}, -81, -4\frac{1}{81}, 0, -1$. 4. $9, 8,$
$7, -\frac{1}{22}$. 5. $176, 82, 254\frac{26}{27}, -87 \div 7\frac{3}{3}$. 6. 18 each.
7. $146, 14, -72, -270, 396$. 8. Each $= 0$.

Exercise ii.

1. -1. 2. -166542. 3. 100. 4. -2967511.
5. 968. 6. -162. 7. 10. 8. -8. 9. 0.
10. -20. 11. 706440254900. 12. 0 each. 13. Each 0.

Exercise iii.

1. $0, 16a^4$. 2. $a, a\sqrt{3}$. 3. $2a, 0$. 4. $26a^6, -26a^6$.
5. 0. 6. $4a^4$. 7. $6a^4$. 8. $\frac{3}{2}$. 9. c. 10. 0.
11. $a \div (a+b)$. 12. $a^2c(b+2c) \div b^2$. 13. $a^2+b^2+c^2$.
14. 0. 15. $(12a^2b - 24ab^2 + 28b^3) \div (3b-a)^3$. 16. 0.
17. 0. 18. $-b^2c$. 19, 20, 21 and 22, each 0.
25. $2(b+l)h, 4x^2$. 32. $d^2 = 3l^2$. 33. $l = \sqrt{(\frac{1}{3}d^2)}$.
35. $\pi r^2, \pi(r+r')(r-r')$.

Exercise iv.

1. $2(bx+cy)$. 2. $3(ax-by)$.
3. $a^2(x-z) - ab(x-y) - b^2(y-z)$.
4. $(x+y+z)(a+b+c)$. 5. $(a+b+c)(x^2+y^2+z^2)$.
6. $2(x+y+z) \times (a^2+b^2+c^2-ab-bc-ca)$. 7. 0.
8. $2(ax+by+cz)$. 9. $a^2+b^2+c^2$.
10. $2x^n(a-2b)$. 11. $a+b-c$.

Exercise v.

1. $2(x^2+9y^4)$, $4a^2b^2$.   4. $4(a^2-b^2)^2$.
5. x^2+4x, $-3\frac{1}{4}x^4-4x^2y^2+3\frac{1}{4}y^4$.   6. a^2.
8. $x^2-6x^3+9x^4+2xy-6xy^2-6x^2y+18x^2y^2+y^2-6y^3+9y^4$.
9. $4xy(x^2-y^2)$, $2(1+12x^2+16x^4)$.
10. $\frac{1}{16}c^2$.   11. a^2-2b^2; $8ab(a+b)^2$.   12. $2(a-c)(b-d)$.
13. $\frac{1}{4}x^2+\frac{1}{4}y^2+\frac{1}{4}z^2+\frac{1}{2}(xy+yz+zx)$.   15. $(1+x^2)^2$.
16. $4(xy+yz+zx)-2(x^2+y^2+z^2)$.   17. x^2.
18. $(a^2+2b^2-2c^2)^2$.   19. $16x^2y^2$.   20. $-4ab$.
21. $4(a+b+c)^2$.   23. $4(1+x^2+x^4+x^6)$.
24. $(a^2x^2+b^2y^2)^2$.

Exercise vi.

1. $1-4x+10x^2-20x^3+25x^4-24x^5+16x^6$,
   $1-2x+3x^2-4x^3+3x^4-2x^5+x^6$.
2. $1-4x+8x^2-14x^3+14x^4-8x^5+5x^6+6x^7+x^8$,
   $1+6x+15x^2+20x^3+15x^4+6x^5+x^6$.
3. $4a^2+b^2+c^4+1-4ab-4ac^2-4a+2bc^2+2b+2c^2$,
   $1+x^2+y^2+z^2-2x+2y+2z-2xy-2xz+2yz$,
   $\frac{1}{4}x^2+\frac{1}{9}y^2+36z^2-\frac{1}{3}xy+6xz-4yz$.
4. $x^6-2x^5y+3x^4y^2-4x^3y^3+3x^2y^4-2xy^5+y^6$,
   $a^2x^2+2abx^3+(2ac+b^2)x^4+2(ad+bc)x^5+(2bd+c^2)x^6+2cdx^7+d^2x^8$.   8. $3(a^2+b^2+c^2)-2(ab+bc+ca)$.
11. $4a^2+\frac{1}{4}b^2x^2+\frac{1}{16}c^2x^2+4d^2x^2-2abx-acx+8adx+\frac{1}{4}bcx^2-2bdx^2-cdx^2$.

Exercise vii.

1. $(a^2-b^2)^2$.   2. $\frac{1}{4}x^4+y^4$.   3. $a^4+3a^2b^2+4b^4$.
4. x^4-y^4.   5. x^2.   6. $16x^2$.   7. 0.
8. $4a^4-9b^4-16c^4+24b^2c^2$.   9. $b^2-9c^2-4a^2+12ac$,
   $9c^2-4a^2-b^2+4ab$.   10. x^8-y^8.   11. $x^8+x^4y^4+y^8$.

ANSWERS.

12. $a^2 - a^2b^2 + b^2 - 1$.   14. $x^4 + y^4 + \frac{7}{10}x^2y^2$.
15. $x^8 + 2x^6 + 3x^4 + 2x^2 + 1$.
16. $4a^4x^2 - 4a^4xy + a^4y^2 - a^2x^4 - 2a^2x^3y + 2ax^5 + 2ax^4y - x^6$.
20. $(x^2 + y^2 - 2xy - z^2)^2$.   21. $x^8 - y^8$.
22. $\frac{1}{4} - 6a^2 + 27a^4$.   23. $(m+p)^2 - (n+q)^2$.
24. $2x^2 + x^4 + 2x^6 - x^8 - 1$.   25. $a^8 - b^{16}$.

Exercise viii.

1. $x^4 + 4x^3 + 3x^2 - 2x - 12$, $x^2 + y^2 - 2xy + 8xz - 8yz + 15z^2$.
2. $x^4 + 12x^3 + 49x^2 + 78x + 40$, $x^6 + bx^3 - a^3 + 3ab - 2b^3$.
3. $a^8 + 8a^6 - 10a^4 - 104a^2 + 105$, $x^8 + 2x^6 - x^2 - 2$.
4. $x^4 + 5x^3y^2 - 12x^2y^2 + 5xy^3 + y^4$.
5. $x^{2n} - 2x^n - a^2 - 16a - 63$, $\dfrac{x^2}{y^2} + \dfrac{y^2}{x^2} + \dfrac{2x}{y} + \dfrac{2y}{x} - 1$.
6. $n^2x^2 + 2nxy + y^2 + 10nx + 10y + 21$.
7. $(x+a)^2 + 2y(x+a) - 3y^2$.   8. $x^{4n} + 2x^{3n} + x^{2n}(1-a-b) - x^n(a+b) + ab$.   9. $\frac{1}{4}x^8 - x^4y^2 + y^4 - x^4 + 2y^2 - 8$.
10. $\left(\dfrac{1}{x} + \dfrac{1}{y}\right)^2 = 2\left(\dfrac{1}{x} + \dfrac{1}{y}\right) - \dfrac{5}{4}$.
11. $x^4 - 8x^3 + 19x^2 - 12x + 2$.
12. $(x+b)^4 - (a^2+c^2)(x+b)^2 + a^2c^2$.   13. $ab + cd$.

Exercise ix.

1. $2(1+3x^4)$, $2xy^3(3x^4 + x^2y^6)$.   2. $9b(a^2 + b^2 + ab^2)$, $b(27a^2 - 27ab + 7b^2)$.   3. $(x+y)^3$.   4. $8a^3$.   5. $8x^3$.
6. $8x^3$.   7. a^3.   8. $27x^3$.   9. $(2+x)^3$.   12. $8(x^2+y^2)^3$.
14. $(a^3+b^3)(x^3+y^3)$.   15. 0.   16. 0.

Exercise x.

1. $1 - 3x + 6x^2 - 7x^3 + 6x^4 - 3x^5 + x^6$, $a^3 - b^3 - c^3 - 3a^2(b+c) + 3b^2(a-c) + 3c^2(a-b) + 6abc$, $1 - 6x + 21x^2 - 56x^3 + 111x^4 - 174x^5 + 219x^6 - 204x^7 + 144x^8 - 64x^9$.

2. $-(x^9+18x^8+27x^7+29x^6-24x^5-36x^4+5x^3-3x^2-2)$.
5. 0.   6. $45x^6+168\Sigma x^4y^2-432x^2y^2z^2$.   7. $(ax+by+cz)^5$.

Exercise xi.

1. $x^6+6x^5y+15x^4y^2+20x^3y^3+15x^2y^4+6xy^5+y^6$,
$x^7+7x^6y+21x^5y^2+35x^4y^3+35x^3y^4+21x^2y^5+7xy^6+y^7$,
$x^8+8x^7y+28x^6y^2+56x^5y^3+70x^4y^4+56x^3y^5+28x^2y^6+8xy^7+y^8$, $x^{12}+12x^{11}y+66x^{10}y^2+220x^9y^3+495x^8y^4+792x^7y^5+924x^6y^6+792x^5y^7+$&c.

2. The signs will be alternately positive and negative.

3. $a^5-5a^4b+10a^3b^2-10a^2b^3+5ab^4-b^5$,
$a^4-8a^3b+24a^2b^2-32ab^3+16b^4$, same as last, terms in inverse order.   4. $1+6m+15m^2+20m^3+15m^4+6m^5+m^6$, $m^5+5m^4+10m^3+10m^2+5m+1$, $64m^6+192m^5+240m^4+160m^3+60m^2+12m+1$.   5. 120.

6. $x^8-4x^6y+6x^4y^2-4x^2y^3+y^4$, $a^5-10a^4b^2+40a^3b^4-80a^2b^6+80ab^8-32b^{10}$, $a^{18}-12a^{15}b^3+60a^{12}b^6-160a^9b^9+240a^6b^{12}-192a^3b^{15}+64b^{18}$.

7. $495a^8b^4-792a^7b^5$.

Exercise xii.

1. $1+x^3+x^4+x^6+x^{17}$.   2. $1+x+x^2+x^3+x^4+x^6+x^7+x^8+x^9+x^{15}$.   3. $x^4+2x^3-85x^2-86x+1680$, $2x^9-3x^6+4x^5+x^4+x^3-2x^2-x+2$.   4. $x^6-57x^4+266x^2-1$.   5. $18x^8+21x^7+8x^6+x^5+68x^3+96x^2+48x+6$.   6. $1-\tfrac{1}{2}x^2-\tfrac{1}{8}x^4$.   7. $6x^{12}-4x^9-5x_0^8-2x^7+9x^6-10x^5+x^4-5x^3+5x^2+x+4$.   8. $x^3+9x^2+10x+11$.
9. x^4+3x^3.   10. x^4-3x^3.   11. x^4+8x^3-8x.
12. (1), -1, (2), -1, (3) -4.   13. -1.

Exercise xiii.

1. $3x^3 - 2x^2 - 4x + 2$.
2. $5x^4 - 4x^3 + 3x^2 - 2x + 1$.
3. $a^4 + 2a^3 + 3a^2 + 4a + 5$.
4. $x^3 + 2x^2y + 3xy^2 + 4y^3$.
5. $a^3 + 3a^2x + 3ax^2 + x^3$.
6. $4x^2 + 8x + 7$, $-13x - 20$.
7. $10x^3 + 5x^2 + 1$, $10x + 10$.
8. $x^2 - xy + y^2$.
9. $x^2 - a^2$.
10. $x^4 + (1-a)x^3 + (1-a+b)x^2 + (1-a)x + 1$.
11. $3x^3 + 2x^2 + x + 1\frac{1}{2}$, $3\frac{1}{2}(x+1)$.
12. $5x^2 + 13xy + 12y^2$.
13. $6x^5 - x^4 - x^3 + x^2 - x + 6$, -1.
14. $2x^4 - 3x^3 + 4x^2 - 5x + 6$.
15. $a+b$.
16. $x+y+z$.
17. $10x^3$, $10(x^4 - 20)$.
18. $mx^3 + nx^2 + a$.
19. $1 + x - 5\frac{3}{4}x^2 - 3x^3 + 9x^4$.
20. 33.
21. -4.
22. -20.
23. $15y^4$.
24. $85x + 8$.
25. 755.

Exercise xiv.

1. $y^3 - 2y^2 - 4y - 9$, if $y = x - 1$.
2. $y^3 + 3y + 5$, if $y = x + 1$.
3. $y^4 + 81$, if $y = x - 2$.
4. $y^4 + 4y^3 - 48y^2 + 92y - 67$, if $y = x + 2$.
5. $3y^5 + 30y^4 + 119y^3 + 238y^2 + 249y + 106$, if $y = x - 2$.
6. $y^4 - \frac{59}{8}y^2 - \frac{91}{8}y + \frac{845}{256}$, if $y = x - 1\frac{3}{4}$.
7. $y^3 - \frac{19}{3}y + \frac{155}{27}$, if $y = x - \frac{1}{3}$.
8. $(x - 2y^3) - 3y(x - 2y)^2 - 18y^2(x - 2y) - 24y^3$.
9. $(x-y)^5 - 10y^2(x-y)^3 - 20y^3(x-y)^2 - 10y^4(x-y)$.
10. $(2x+y)^3 + 2y^2(2x+y) + 5y^3$.
11. $512y^3 - 3y - \frac{19}{144}$, if $y = \frac{1}{8}x - \frac{1}{16}$.
12. $y^4 - 24y^2 + 49y - 28$, if $y = x + 2$.

Exercise xv.

1. a^2b, $+ ab^2 + a^2c + b^2c + bc^2 + ac^2$,
   $(a-b)^2 + (b-c)^2 + (c-a)^2$, $a(b-c) + b(c-a) + c(a-b)$,
   $ab(x-c) + bc(x-a) + ac(x-b)$,
   $abc(a^2b + a^2c + b^2c + ab^2 + ac^2 + bc^2)$,

$(a+b)(c-a)(c-b)+(b+c)(a-b)(a-c)+(c+a)(b-c)(b-a)$.
$(a+c)^2-b^2+(b+a)^2-c^2+(c+b)^2-a^2$,
$a(b+c)^2+b(c+a)^2+c(a+b)^2$.

2. $abc+bcd+cda+dab$,
$a^2(b+c+d)+b^2(c+d+a)+c^2(d+a+b)+d^2(a+b+c)$,
$(a-b)+(a-c)+(a-d)+(b-c)+(b-d)+(c-d)$,
$a^2(a-b)+b^2(b-c)+c^2(c-d)+d^2(d-a)$.

13. x and y.   14. ax and by, x, y, z.   15. f and h.
16. x and y, also x and $-z$, and y and $-z$.
17. a, b and $-c$.   18. $x^2, -y^2$ and z^2.   19. b and c.
20. a and c.   21. a and b.   22. a^2 and $2ab$.
23. a^2b and abc.   24. a^2b, abc.   25. $x^5, x^4y,$ and x^3y^2;
same; x^4y, x^3y^2.   26. Not symmetrical.
28. $a^4, a^3b, a^2bc, abcd; a^4, a^2b^2$.   29. a^3, a^2b.

Exercise xvi.

1. $4(a^2+b^2+c^2)$.   2. $3(a^2+b^2+c^2)+2(ab+bc+ca)$.
3. $4(a^2+b^2+c^2+d^2)$.   4. $2(a^3+b^3+c^3)$.
5. $4(x^2+y^2+z^2+n^2)$.   6. $2(a^3+b^3+c^3)+6\Sigma a^2b - 12abc$.
7. $14(x^2+y^2+z^2)+2(xy+yz+zx)$.   8. $24abcmnr$.
9. $2abc(a+b+c)$.   10. $a^2b^2+b^2c^2+c^2a^2$.

Exercise xvii.

1. 115.   2. $pa^3-3qa^2+3ra-s$.   3. 2.   4. -17.3538.
5. $1, 2(3a^2+1)$.   6. 0 or $2y^n$, $2y^n, 0$.   7. 36.
8. $-(b^2+a^2)^3-(3b^2)^3$.   9. $-15a^4$.   10. $3888a^4b^4$.
11. $a^2b^2(a+b)$,   12. 0.   13. $2a^3-3ab(a-b)$,
$2b^3+6ab(a+b), 2(a^3+b^3)$.

Exercise xx.

1. 3.   2. 1.   3. $-1\pm 2\sqrt{-2}$.   4. 2.   5. 36.   6. 11.
7. $-1\frac{15}{27}$ 13. $p=-q, q=6$.   14. $p=-46, q=14$.

Exercise xxi.

1. $b = -8$, $c = 8\frac{3}{4}$, $d = -24$.   2. $c = -20\frac{1}{4}$, $d = -13\frac{1}{2}$, $e = 60\frac{3}{4}$.
3. $b = -8$, $c = -10$.   4. $a = 8$, $b = 0$, $c = -57$.   5. $a = -2$, $c = 24\frac{1}{4}$, $e = 0$.   6. $c = -106\frac{1}{2}$, $d = 202\frac{1}{2}$.   7. $a = 200$, $b = -810$, $c = 639$.   8. $a = 4$, $c = -27$, $d = 7$, $e = 30$.
9. 899.   10. $x^3 - (p+3)x^2 + (2p+q+3)x - (p+q+r+1)$.
11. $x^3 - (p-3)x^2 - (2p-q-3)x - (p-q+r-1)$.
12. $rx^3 - (3r-q)x^2 + (3r-2q+p)x - (r-q+p-1)$.
13. $x^3 - qx^2 + prx - r^2$.   14. $x^3 - (p^2-2q)x^2 + (q^2-2pr)x - r^2$.
15. $x^3 - 2qx^2 + (pr+q^2)x + r^2 - pqr$.   16. $rx^3 - (pq+3r)x^2 + (p^3 - 2pq + 3r)x - (pq-r)$, 33. -1.   34. 1.   35. -1.
36. 1.   37. -1.   38 and 39. $a+b+c+d$.   40. -1.

Exercise xxii.

1. $5b^4 + 15c^4$.   2. 6.   3. 3.   4. $-\{A(b+c+d) + \ldots + \ldots + \ldots\}$.
5. 0.   6. $5b^4 - 30ab^3 + 30a^2b^2 - 5a^3b$.   8. 0.   9. 0.   10. 0.   11. 1.
12. $(a+b+c+d)$.   13. -1.   14. $a+b+c+d$.
15. $(a+b+c)(a^2+b^2+c^2+ab+bc+ca)+abc$.
16. $(a+b+c)^2(a^2+b^2+c^2)+2abc(a+b+c)$.   17. $a+b+c+d$.
18. $(a+b+c+d)^2$.   19. $(a+b+c+d)\{(a+b+c+d)^2 - (ab+ad+ac+bc+bd+cd)\}+abcd$.   20. $a+b+c$.   21. 3.
22. -1.   23. 0.   24. 0.   25. 0.   26. $1+\frac{1}{2}x-\frac{1}{8}x^2+\frac{1}{16}x^3$.
27. $1-\frac{1}{2}x-\frac{1}{8}x^2-\frac{1}{16}x^3$.   28. $1+x+x^2+x^3$.
29. $1-2x+3x^2-4x^3$.   30. $1+\frac{1}{3}x-\frac{1}{9}x^2+\frac{5}{81}x^3$.

Exercise xxiii. (a)

1. $(p-p'+q)^2 = (p+1)(p^2-pp'-q)$.
8. $9(p^2-q)(r^2-qt) - (pr-t)^2 = 9\{3(p^2-q)(qr-pt) - (pq-r)(pr-t)\} \times \{3(pq-r)(r^2-qt) - (pr-t)(qr-pt)\}$
9. $x^n(4x^3+3px^2+3qx+r) \div (x^4+4px^3+6qx^2+4rx+t)$.

10. $-4p$, $(4p)^2-2(6q)$, $-(4p)^3+3(4p)(6q)-3(4r)$,
$(4p)^4-4(4p)^2(6q)+4(4p)(4r)+2(6q)^2-4t$,
$-(4p)^5+5(4p)^3(6q)-5(4p)^2(4r)-5(4p)(6q)^2+5(4p)t+$
$5(6q)(4r)$, $(4p)^6-6(4p)^4(6q)+6(4p)^3(4r)+9(4p)^2(6q)^2-$
$6(4p)^2t-12(4p)(6q)(4r)-2(6q)^3+6(6q)t+3(4r)^2$.

11. $s_0s_4-4s_1s_3+3s_2^2$, $s_0s_6-6s_1s_5+15s_2s_4-10s_3^2$, where s_0, s_1, &c., are the coefficients of the terms (taken in order) of the quotient in No. 10.

12. $x^n(4x^3-28x+1) \div (x^4-14x^2+x-38)$; $s_1=0$, $s_2=28$, $s_3=-8$, $s_4=544$, $s_5=-70$, $s_6=8683$; $\Sigma(a-b)^4=4526$, $\Sigma(a-b)^6=264122$.

Exercise xxiv.

1. $(3m+2)^2$, $(c^m-1)^2$.   2. $(y^3-z^3)^2$, $4y^2(2x+y)^2$.
3. $(3ab+2c)^2$, $4y^2(3x-y)^2$ 4. $(\frac{1}{2}x^2-4yz)^2$, $(\frac{1}{2}a^2-\frac{1}{3}b^2c^2)^2$.
5. $(a+b+c)^2$, $(3x^4-\frac{1}{4}y^2)^2$.   6. $(z-x+y)^2$, $\left\{\left(\frac{a}{b}\right)^m-\left(\frac{b}{a}\right)^m\right\}^2$.
7. $(x^2-z^2)^2$.   8. $(x-y)^4$.   9. $(a+b)^2$, $(\frac{2}{3}p^3-\frac{4}{9}q^3)^2$.
10. $(x-y)^2$.   11. $4(x^2+y^2)^2$.   12. $(x+y)^4$.
13. $\left\{\left(\frac{a}{b}\right)^m-\left(\frac{b}{a}\right)^n\right\}^2$ 14. $(a-b+c)^2$.
15. $(a^2-b^2-c^2)^2$.   16. $(2a-2c)^2$.   17. $(2a^2-8b+4c)^2$.

Exercise xxv.

1. $(7a+2b)(7a-2b)$.   2. $(3a+\frac{1}{2}b)(3a-\frac{1}{2}b)$.
3. $(3a-2b)(9a^2+4b^2)(3a+2b)$.   4. $(10x-6y)(10x+6y)$.
5. $5b(a+2xy)(a-2xy)$ 6. $(3x^3-4y^2)(3x^3+4y^2)$.
7. $(\frac{3}{4}c+1)(\frac{3}{4}c-1)$.   8. $(2y^2-\frac{2}{3}xz)(2y^2+\frac{2}{3}xz)$.
9. $(3a-1)(3a+1)(9a^2+1)$.   10. $(a-2b)(a+2b)(a^2+4b^2)$.
11. $(a-b)(a+b)(a^2+b^2)(a^4+b^4)(a^8+b^8)$.

12. $(a+b-c)(a-b+c)$. 13. $(a+2b-3x+4y)(a+2b-3x+4y)$.
14. $(x^2-y^2)^2$ 15. $(x+y+2z)(x+y-2z)$. 16. $16(x+1)(1-x)$.
17. $(x+y+z)(x+y-z)(z-x+y)(z+x-y)$.
18. $4xy(x+y)(x-y)$. 19. $(x-z+y)(x-z-y)(x+z+y)(x+z-y)$.
20. $4(a+c)(b+d)$. 21. $24x(1+2x^2)$. 22. $8ab(a+b)^2$.
23. $(a+b+c+d)(a+c-b-d)(a-b-c+d)(a+b-c-d)$.
24. $(x+y+z)(x-y-z)(x+y-z)(x-y+z)$.
25. $8a^3b^3(a^6-3a^3b^3+b^6)$. 26. $(a^3+b^3)(a^3-b^3)^3$.
27. $(x^2+y^2+z^2)(x^2+y^2+z^2-2xy-2yz-2zx)$.
28. $(x+2z)(x-2y)$. 29. $(a+b-c)(a-b+c)(b+c+a)(b+c-a)$.
30. $(x-y+z)(x+y-z)(x+y+z)(x-y-z)$.

Exercise xxvi.

1. $(x-7)(x+2)$, $(x-7)(x-2)$, $(x+4)(x+3)$.
2. $(x-3)(x-5)$, $(x-7)(x-12)$, $(x-12)(x+5)$.
3. $2(2x-5)(x+2)$, $3(3x-20)(x-10)$.
4. $\frac{1}{4}(x+12)(\frac{1}{2}x-3)$, $5(x+1)(5x+3)$, $(3x^3-4)(3x^3-5)$.
5. $(\frac{1}{3}x+4)(\frac{1}{4}x+3)$, $4(4x-5)(x+1)$.
6. $(x-a)(x+a)(x-b)(x+b)$, $\{2(x+y)-11\}\{2(x+y)+9\}$.
7. $(x^2+y^2-a^2)(x^2+y^2+b^2)$. 8. $(a+b-3c)(a+b+c)$.
9. $(x+y)(1+x+y)\{x+y+(x-y)^2\}$.
10. $(a+b)(1-a-b)\{a+b+(a-b)^2\}$.
11. $(x^2+xy+y^2+2x+y) \times \{x^2+xy+y^2-(x+2y)\}$.
12. $(a-5b+3c)(a+b-c)$. 13. $(x^3+y^3+a^3)^2-b^6 = \&c$.
14. $(x^2-10x-12)(x^2-10x+8)$.
15. $(x^2-14x+10)(x-9)(x-5)$. 16. $(x^2-y^2)^2$.
17. $(z+1)(z-1)(z^2-2)$, $(x^2-3)(x^2+1)$,

$(3x^4+5y^2)(3x^4-2y^2)$.   18. $(c^m+2)(c^m-1)$,
$(x^3-2)(x+1)(x^2-x+1)$, $(x^m-4y^n)(x^m+2y^n)$,
19. $(x^m-ay^n)(x^m+by^n)$.

Exercise xxvii.

1. $(x-by)(bx-y)$.
2. $3(x+2y)(2x-y)$.
3. $4(14x-5y)(x-y)$.
4. $4(14x+5y)(x-y)$.
5. $(14x-y)(x-20y)$.
6. $4(7x-5y)(2x-y)$.
7. $2(28x+y)(x-10y)$.
8. $4(14x-5y)(x+y)$.
9. $(8x-5y)(7x-4y)$.
10. $(8x+5y)(7x-4y)$.
11. $2(3x+y)(x-3y)$.
12. $(3x-2y)(2x+3y)$.
13. $2(28x+y)(x+10y)$.
14. $2(28x-5y)(x-2y)$.
15. $2(28x+5y)(x-2y)$.
16. $(56x-5y)(x-4y)$.
17. $2(4x-y)(7x-10y)$.
18. $4(14x+y)(x-5y)$.
19. $3(3x+y)(4x-5y)$.
20. $(8x+5y)(9x-8y)$.

Exercise xxviii.

1. $(5x-7)(2x+3)$.
2. $(5x+3)(2x-7)$.
3. $(5x-3)(2x+7)$.
4. $(2x-5)(3x-11)$.
5. $(4a+1)(3a-2)$.
6. $(3x-7)(4x-8)$.
7. $(3x+7)(4x+8)$.
8. $(5a^3-4b^2)(3a^2+5b^3)$.
9. $(4x+1)(3x-1)$.
10. $3y^2(x-y)(3x+2y)$.
11. $(2x+3y)(2x+y)$.
12. $x^2(3b+x)(2b-3x)$,
13. $(3x^2+7y^2)(2x^2-5y^2)$.
14. $(2x^2-9)(x^2+5)$,
15. $(2x+y)(2x-y)(x-3y)(x+3y)$.
16. $(2x+4+y)(2x+4-y)(x+2-3y)(x+2+3y)$.
17. $160xy$.
18. $(19y^2+60xy-6x^2)(35x^2-12xy+30y^2)$.
19. $2(4xy-3x^2-3y^2)(61x^2-49xy+61y^2)$.
20. $2(5x^2+4xy+10y^2)(x^2+10xy+2y^2)$.

Exercise xxix.

1. $(7x+6y+8)(x-y-z)$. 2. $(5x-5y-22)(4x+y+4)$.
3. $(3x^2+4y^2+13)(x^2-y^2-1)$. 4. $(4x+5y)(5x-4y+7)$.
5. $(9x+8y-20)(8x-y-1)$. 6. $(x+8y)(x-4y-5)$.
7. $(4x+3y-z)(2x+3y+z)$. 8. $(3x-2y-2z)(2x-3y+4z)$.
9. $(3x^2-2y^2+5z^2)(2x^2+5y^2-5)$.
10. $(15x^2+8y^2+5z^2)(x^2-2y^2+3z^2)$.
11. $(2a-5b-7c)(2a+3b+3c)$.
12. $(a-b+c)(a+b-c)(a+b+c)(a-b-c)$.

Exercise xxx.

1. $x^2+\tfrac{7}{3}\pm\tfrac{3}{2}\sqrt{5}$, $2x^2+\tfrac{7}{3}\pm\tfrac{3}{2}\sqrt{5}$. 2. $x^2+\tfrac{7}{2}y^2\pm\tfrac{3}{2}y^2\sqrt{5}$,
$\tfrac{1}{12}(6x^2+5y^2\pm y^2\sqrt{13})$. 3. $\tfrac{1}{4}(4x^2+5\pm\sqrt{13})$,
$\tfrac{1}{12}\{6(x+y)^2+5z^2\pm z^2\sqrt{13}\}$.
4. $(x^2+\tfrac{1}{2}y^2)(x^2+6\tfrac{1}{2}y^2)$, $(x^2+\tfrac{11}{2}y^2)(x^2+\tfrac{3}{2}y^2)$.
5. $(2x^2+4\tfrac{1}{2}y^2)(2x^2+\tfrac{1}{2}y^2)$, $\tfrac{1}{4}\{4(a+b)^2+5c\pm\sqrt{13}\}$.
6. $\tfrac{1}{12}(6x^2+5y^2)(6x^2+11y^2)$, $(6x^2+5)(6x^2+11)$.
7. $\tfrac{1}{5}(5x^2+10\pm 8\sqrt{10})$, $(2a^2+3\pm 2\sqrt{2})$.
8. $\{2(x+y)^2+(3\pm 2\sqrt{2})z^2\}$;
$\tfrac{1}{5}\{10x^2+(10\pm 8\sqrt{10})y^2\}\{10x^2+(20-6\sqrt{10})y^2\}$.
9. $\tfrac{1}{9}(9x^2+7\pm\sqrt{13})$, $\tfrac{1}{2}\{2x^2+(6\pm\sqrt{16})(y+z)^2\}$.
10. $\tfrac{1}{2}(2x^2+6\pm\sqrt{6})$, $\tfrac{1}{7}(7x^2+20\pm\sqrt{85})$.
11. $\tfrac{1}{2}\{4x^2+(9\pm\sqrt{23})y^2\}$.
12. $\tfrac{1}{7}\{7(a-b)^2+8c^2\pm c\sqrt{29}\}$, $\tfrac{1}{8}\{3a^2\pm b^2\sqrt{3}\}$.
13. $\tfrac{1}{3}\{3x^2+(3\pm\sqrt{3})y^2\}$, $\tfrac{1}{3}\{8(a+b)^2+(3\pm\sqrt{3})(a-b)^2\}$.
14. $\{7a^2+(6\pm\sqrt{14})b^2\}$, $(5m^2+9n^2)(5m^2+3n^2)$.
15. $\{7(m+n)^2+(6\pm\sqrt{14})(m-n)^2\}$.

Exercise xxxi.

1. $(x^2 \pm 2xy + 3y^2)$, $(x^2 \pm xy - y^2)$, $(x^2 \pm xy + y^2)$.
2. $(x^2 \pm 2xy + 2y^2)$, $(4x^2 \pm 3xy + y^2)$, $(\frac{1}{2}x^2 \pm xy + y^2)$.
3. $(x^2 \pm \sqrt{2}x + 1)$, $(x^2 \pm \sqrt{6}xy + 3y^2)$, $(1 \pm 2y - 4y^2)$.
4. $(x^2 \pm 3x + 1)$, $x^2 \pm \sqrt{6}x + 3$, $\frac{1}{2}x^2 \pm 2xy + y^2$.
5. $(y^2 \pm \frac{11}{2}x^2 \pm \frac{5}{2}x^2 \sqrt{5})$, $(x^4 + 2y^4 \pm 2x^2y^2)$, $x^2 + 4 \pm 2x$.
6. $(2x^2 + y^2 \pm \frac{7}{2}xy)$, $(x^2 + y^2 \pm \frac{1}{4}xy\sqrt{39})$, $(2x^2 + 1 \pm 2x)$.
7. $(x^{2m} + 8y^{2m} \pm 4x^m y^m)$, $x^{2m} + 2y^{2m} \pm 2x^m y^m$, $(\frac{1}{2}x^2 - \frac{3}{4}y^2 \pm xy\sqrt{5})$.
8. $(2x^2 - 1 \pm 2x)$, $-(\frac{1}{2}x^2 - 6y^2 + xy)(\frac{1}{2}x^2 - 6y^2 - xy)$, $(x^2 + a^2y^2 \pm axy\sqrt{2})$.
9. $mx^2 - ny^2 \pm xy\sqrt{p}$, $x^{2m} + 2^{m-1}y^{2m} \pm 2^m y^m$.
10. $4x^2 - 3 \pm x$, $2x^2 - 2 \pm 2x\sqrt{2}$, $-(3x^2 - 2y^2 + xy)(3x^2 - 2y^2 - xy)$.
11. $2x^2 \pm \frac{4}{5}xy - 3y^2$, $x^2 \pm 2x + 5$.
12. $2(a^2 + ab + b^2)^2$, $(2a^2 + a + 1)^2$.
13. $\{(x+y)^2 + 3(x+y)z + z^2\}\{(x+y)^2 - 3(x+y)z + z^2\}$
14. $(a+b)^2 + \frac{7}{2}c^2 \pm \frac{3}{2}c^2\sqrt{5}$.
15. $\{4a^2 + 5a(b-c) + 2(b-c)^2\}\{4a^2 - 5a(b-c) + 2(b-c)^2\}$.
16. $4(a^2 + 5ab - 2b^2)(b^2 + 5ab - 2a^2)$.
17. $\{(x^2 + y^2 - xy)^2 \pm 3(x^2 + y^2 - xy)(x+y) + (x+y)^2\}$.
18. $\{(a^2 + ab + b^2) + \frac{7}{2}(a-b)^2 \pm \frac{3}{2}(a-b)^2\sqrt{5}\}$.
19. $(4a^2 \pm 2a + 1)$, $x^2 \pm 7x + 4$.
20. $(x^2 \pm 9xy + 9y^2)$, $(1 \pm 3z + 5z^2)$.
21. $4(3x^2 - 2x + 1)(x^2 - 2x + 3)$.

Exercise xxxii.

1. $(x^2 + 3)(x+3)(x-1)$.   2. $2(x^2 + 3)(x^2 + x - 3)$.
3. $(x^2 + 4)(x+4)(x-1)$.   4. $(x+2)(x-2)(3x^2 + x + 12)$.

ANSWERS.   xiii

5. $(x^2-3)(5x^2+4x+15)$.   6. $(x^2+6)(10x^2+5x-60)$.
7. $(\frac{1}{2}x^2+\frac{1}{10})(\frac{1}{2}x^2+40x-\frac{1}{10})$.   8. $(5x^2-1)(5x^2-8x+1)$.
9. $(5x^2-8)(7\frac{1}{2}x^2-6x-12)$.
10. $(3x^2-4)(21x^2-13x-28)$.   11. $(18x^2+1)(45x^2+\frac{9}{5}x+\frac{5}{2})$.
12. $(11x^2+1)(22x^2-3x-2)$.   13. $(\frac{1}{2}x^2-\frac{2}{5})(\frac{1}{2}x^2+\frac{1}{5}x+\frac{2}{5})$.
14. $8(x^2-2y^2)(10x^2-4xy+20y^2)$.
15. $(2x^2-5y^2)(12x^2-6xy+30y^2)$.
16. $(x^2-16y^2)(2x^2+\frac{1}{2}xy+32y^2)$.
17. $(x^2-\frac{2}{5})(11x^2+10x+\frac{6.6}{5})$.   18. $10(x^2+2)(4x^2+3x-8)$.
19. $(x^2-6y)^2(13x^2-12xy+78y^2)$.
20. $(x^2+4y^2)(3x^2+3xy-12y^2)$.
21. $(x^2-3y^2)(5x^2+4xy+15y^2)$.
22. $2(x^2-2y^2)(2x^2-7xy+2y^2)$.
23. $(x^2+\frac{1}{5}y^2)(x^2+80xy-\frac{1}{5}y^2)$.   24. $(x^2-6y^2)(2x^2-xy+12y^2)$.

Exercise xxxiii.

1. $(x^2+3x+27)(x^2-9x+27)$.   2. $x^2+x(1\pm\sqrt{3})+4$.
3. $\{x^2+1+\frac{1}{2}(1\pm\sqrt{5})x\}$.   4. $x^2+1\ \ x(2\pm\sqrt{5})$.
5. $2x^2+2-3x\pm x\sqrt{23}$.   6. $(x^2+15x-5)(x^2-x-5)$.
7. $(4x^2-2)(4x^2-6x-2)$.   8. $(x^2+8x+4)(x^2-3x+4)$.
9. $(x^2+7x-2)(x^2-x-2)$.
10. $(x^2+5xy+3y^2)(x^2-xy+3y^2)$.
11. $(x^2+10x-1)(x^2+2x-1)$.
12. $(x^2+7xy+y^2)(x^2-3xy+y^2)$.
13. $2x^2+xy-5y^2\pm xy\sqrt{46}$.   14. $(x^2+7xy-y^2)(x^2-xy-y^2)$.
15. $x^2+2y^2+3xy\pm xy\sqrt{3}$.
16. $(3x^2+10xy-2y^2)(3x^2-4xy-2y^2)$.
17. $\frac{1}{11}\{11x^2+22y^2+5xy\pm\frac{2 5}{11}xy\sqrt{11}\}$.

ANSWERS.

Exercise xxxiv.

1. $(y-z)(x^2-y)$.   2. $(by+c)(ax+by-c)$.
3. $(z^2+a)(x+a)(x-a)$.   4. $(2x-a)(x-2b)$.
5. $(x+3a)(x+2b)$.   6. $(x-b^2)(x-a)(x+a)$.
7. $(x-b)(x+b)(x-a)(x^2+ax+a^2)$.   8. $(2x+3a)(4x+5b)$
9. $(a+bx)(a-bx+cx^2)$ 10. $(a-bz)(a+bx+cx^2)$.
11. $(ax-d)(bx^2+cx-f)$.   12. $(px-q)(x^2-x-1)$.
13. $(a-b-c)(a+2b+3c)$.   14. $(x+a)(x^2+x+1)$.
15. $(mx-n)(px^2+qx-r)$.   16. $(x-a)(x-b)(x-c)$.
17. $(x+a)(x-b)(x-c)$.   18. $(x+a)(x+b)(x-c)$.
19. $(a^2+z)(x-ay)(x^2-y)$.   20. $(abx+cdy-cfz)(ax+by)$.
21. $(ax+c)(ax^2-bx+c)$.   22. $(x-y)(x+y)(mx-ny+rz)$
23. $(mx-ny)(ax+by+cz)$.   24. $(mx+n)(ax-bcx+a)$.
25. $(c^2-xz)(b^2-yz)(a^2-xy)$.   26. $(x^3-m^2x^2-a)(x^2-n+n^2)$.
27. $(1+x-x^2)(1-ax+bx^2-cx^3)$.
28. $(ax-dy)(ax-by)(ax+cy)$.   29. $(mx+q)(px+n)(m^2x-n)$.
30. $(mx+ny)(mx-ny)(p^2x^2+q^2y^2)(x+1)$.

Exercise xxxv.

1. $(a+x)(a-b)$.   2. $(ax+by)(bx-ay)$.
3. $(x-a)(x+a)(x^2+ax+a^2)$.   4. $x(a+x)(a^2+ax+x^2)$.
5. $(ax-b)(cx+d)$.   6. $(5x^2-1)(5x^2-x+1)$.
7. $(a-b)(a+b+x-c)$.   8. $(a^2+b)(a+b)$.
9. $(x-y)(x+y)^3$.   10. $(x-y+1)(x^2+xy+y^2)$
11. $(b-2x)(2+bx)$.   12. $(x-1)(x+2)^2$.
13. $(p-q)(p^2-2q^2)$.   14. $(a-1)(a^2+2a+2)$.
15. $(ab^2-1)(3ab^2+1)$.   16. $(y-1)^2(y+2)$.

ANSWERS.

17. $(a+b)(2a^2-3ab+2b^2)$.   18. $(b^m-1)(b^{2m}+2b^m+2)$.
19. $(y^n+z^n)(y^{2n}-3y^nz^n+z^{2n})$.   20. $(a-b)(a^3+ab-3b^2)$.
21. $(a^m-c^n)(a^m-2c^n)$.   22. $(ax-b)(x^2-ax-b)$.
23. $(5x^n-3a^2)(7x^n+3a^2)$.   24. $(ab+bc-ca)(ab-bc+ca)$.
25. $(m-b)(m+b)(a-m)$.   26. $(\frac{1}{2}-3a^2)(1-3a)(1+3a)$.
27. $(x-y-z)(x^2-2xy+y^2+z)$.   28. $(6m-7n)(4m^2+n^2)$.
29. $(x^n+y^n)(x^n+y^m)$.   30. $(x^2+xy+ax+y^2)(x^2+xy-ax-y)^2$.

Exercise xxxvi.

1. $(x-y)(x+y)(x^2+xy+y^2)(x^2-xy+y^2)$, $(x-1)(x^2+x+1)$,
   $(x+2)(x^2-2x+4)$, $(2a-3x)(4a^2+6ax+9x^2)$,
   $(2+ax)(4-2ax+a^2x^2)$.
2. $(x-a^2)(x^4+x^3a^2+x^2a^4+xa^6+a^8)$,
   $(3a-4)(9a^2+12a+16)$, $(a^3-b^2)(a^3+b^2)(a^6+b^4)$,
   $(x^2-2y)(x^8+2x^6y+4x^4y^2+8x^2y^3+16y^4)$.
3. $(a-b)$.   4. $x+4y$.   5. $(x+y)(x^2+y^2)(x^4+y^4)$
6. $5(y^2-x^2)(7x^4-11x^2y^2+7y^4)$, $(a^2-2b)(a^2+2b)(a^4+4b^2)$.
7. $y(x-y)(y+1)$.   8. $(x-a)(x^2+ax+a^2)(a+b)$.
9. $(a+b)(m+a)(m^2-am+a^2)$.
10. $(x^2+xy+y^2)(x^2-xy+y^2)(x^2+2xy-y^2)$.
11. $(a^2+bc)(a^4-4a^2bc+7b^2c^2)$.
12. $(x-a+b)\{(x-a)^2-(x-a)b+b^2\}$.
13. $(x^2-2xy+4y^2)(x+2y+4xy)$.
14. $(2x+3y)(2x-3y)^2$.   15. $(1-2x)(1+4x^2)$.
16. $(a^2+abc+b^2c^2)(a+bc)(a^2-abc+b^2c^2)$.

Exercise xxxvii.

1. $3(x+y)(y+z)(z+x)$.   2. $(a-b)(b-c)(a-c)$.
3. $3(a^2-b^2)(b^2-c^2)(c^2-a^2)$.   4. $(x+y)(y+z)(z+x)$.

xvi   ANSWERS.

5. $3(a+b)(b+c)(c+a)$.   6. $(a+b+c)(a-b)(b-c)(c-a)$.
7. $(a+b)(b+c)(c+a)$.   8. $(a^2-b)b^2-c)(c^2-a)$.
9. $(a+b)(b+c)(c+a)$.   10. $(a-b)(b-c)(c-a)$.
11. $(x^2+y^2)(y^2+z^2)(z^2+x^2)$.
12. $(a^2+b^2+c^2-ab-bc-ca)(a-b)(b-c)(c-a)$.
13. $(a^2+b^2+c^2)(a+b+c)$.
14. $-(c-b^3)(a-c^3)(b-a^3)$.   15. $(x^2-y^2)(y^2-z^2)(x^2-z^2)$.
16. $(x+y+z)(x-y+z)(y-z+x)(z+y-x)$.
17. $(a-b)(b-c)(a-c)$.   18. $8(a+b+c)^3$.
24. $(a-b)(b-c)(a-c)(a^2+b^2+c^2+ab+bc+ca)$.

EXERCISE xxxviii.

1. $(a-2)(a^2-7a+2)$.   2. $(x-2)(x-3)(x-4)$.
3. $(x-3)(x-2)^2$.   4. $(x-2)^2(x+4)$.
5. $(x+1)(x^2+2x+3)$.   6. $(x^2+2x+3)(x^2+2x+3)$.
7. $(x+2)(x-1)^2$.   8. $(x^2+2x+3)(x^2-2x+3)$.
9. $(m-n)(m^2-2mn-2n^2)$.   10. None.
11. $(m-n)(m-2n)^2$.   12. $(b+3c)(b^2-2bc+18c^2)$.
13. $-(m-n)^2(m^2-mn+n)$.   14. $(a+2b)(a-2b)(a^2-7ab+4b^2)$.
15. $(x-5)(x-3)^2$.   16. $(x+2)(x^2+3x+1)$.
17. $(a-1)(a^2-2a-195)$.   18. $(p+2)(p-1)(p_s+4)$.
19. $(a-1)^2(a+2)(a+3)$.   20. $(a^{2n}-1)(a^{2n}-2)(a^{2n}-3)$.
21. $a^2+4b^2\pm7ab$.   22. $(a-b)^2(a^2+2ab+2b^2)$.
23. $(p-2)(p^2-2p+2)$.   24. $(x^n-1)(x^{2n}+5x^n+5)$.
25. $(y-2)(y^3-3y^2+2y+4)$.   26. None.
27. $(a-b)(a^2+2ab+3b^2)$.   28. $(a^n+1)(2a^{2n}-3a^n+2)$.
29. $(x-2)(x-3)(x-6)(x-7)$.   30. $(x-y)(x-2y)(x-3y)^2$.

Exercise xxxix.

1. $2(x-1)(x^2-9x+10)$, $(x-2y)^2(x-3y)$.
2. $(4x+3y)(3x^2-xy+y^2)$, $(x-1)(4x-2)(2x+3)$.
3. $(x-5a)(3x^2+a^2)$, $(2x+3y)(x^2+3xy-y^2)$.
4. $(b+c)(b-4c)(2b^2-bc+c^2)$, $(5a+4b)(3a^2+7ab-8b^2)$.
5. $(2p+q)(2p+3q)(p^2+q^2)$.
6. $(10x-9y)(15x+16y)(x^3-5xy+8y^2)$.
7. $(2x-3y)(2x+3y)(3x+4y)(3x-5y)$.
8. $(5x-2z)(2x^3-3x^2y+8xy^2+12y^3)$.

Exercise xl.

1. $1+x^2$. 2. $(x^2-1)^2$. 3. $(x^4+a^2x^2+a^4)(x^8-a^4x^4+x^8)$
4. $(x+2y)(x^2+8y^3)$. 5. $1-2x+3x^2$.
6. $(a-x)(a+x)^2$. 7. $x^2+y^2+z^2+xy+yz-zx$.
8. $(a+b)(3a+b)$. 9. $(x-y)(2x+3y)$.
10. $a^2-b^2+c^2$. 11. $7a^2-8ab+2b^2$.
12. $a-7$. 13. $(a-b)(b-c)(a-c)$.
14. $(x-a)^2-b(x-a)+b^2$. 15. $x^2+y^2+z^2+1$.
16. $x(x^2-ax+b)$. 17. x^2+y^2.
18. $(x-y)(x^2+y^2)$. 19. $a^2-b^2+c^2+1$.
20. $a^3-b^3-c^3$. 21. $a+x$. 22. $(c-b)(a+b+c)$.
23. $ab-ca-bc$. 24. $x^2+y^2+1-xy+x+y$.
25. $(x^3-2)(x+1)$. 26. a^2+5a+3.
27. $(2x-y)a^2-(x+y)ax+x^3$. 28. $a(x^2+x+1)-(x+1)$.

Exercise xli.

1. x^2-3. 2. $x+a$. 3. x^2-x+1. 4. ax^2+bx+c.
5. None. 6. c^a+c^b. 7. $(a-b)(x+a)$. 8. $b(x+y)$.

xviii ANSWERS.

9. $(a-b)(b-c)(c-a)$. 10. $a^{2m}+1$.
12. $5(a-b)(b-c)(c-a)$. 13. $(y-1)(x-1)$.
15. $(x+1)(x^2+1)(x-1)^3$. 16. $(x+1)(x+2)(x+3)(x+4)$. 17. 5.
18. Same as given quantity. 25. $(a-b)(b-c)(c-a)$.
29. x^4+x^2+2x+1.

EXERCISE xlii.

1. $(x-1) \div (x^2+4x+16)$, $x(8y-7) \div y(7y-8)$.
2. $(x^2-ax+a^2) \div (x^2-a^2)$, $(x+4) \div (x-1)^2$.
3. $(x-1)(x+2) \div (x^2+5x+5)$, $(x^2+2x+3) \div (x^2-2x-3)$.
4. $1 \div (b-2x)$, $1 \div (x^2-2x+2)$.
5. $5a^3(a+x) \div x(a^2+ax+x^2)$, $(4x^2+1) \div (5x^2+x+1)$.
6. $(x-y) \div (x+y)$. 7. $(3ax^2+1) \div (4a^3x^4+2ax^2-1)$,
   $(ax+by) \div (ax-by)$. 8. $-1 \div abc$.
9. $-(a+b+c) \div (a-b)(b-c)(c-a)$. 11. $5 \div 7(x^2+xy+y^2)$

EXERCISE xliii.

1. $(4-x) \div (5-x)$, $(a^2+b^2) \div 2ab$.
2. x, $2a \div (a^2+1)$. 3. $a(1+a) \div (1+2a+3a^2)$, x.
4. $b^2 \div a^2$, $(b+1) \div ab^2$. 5. $(ac-bd) \div (ac+bd)$, $b \div a$.
6. $1+6x^2yz(y+z) \div \{y^2z^2-x^2(y+z)^2\}$. 7. $(a^2+b^2+c^2) \div abc$.
8. 1. 9. $-(a^4+a^2b^2+b^4) \div ab(a-b)^2$.
10. $(a+b+c)^2 \div 2bc$. 11. $\left(\dfrac{1-x}{1+x}\right)^2$, $4a^2x^2 \div (a^2+x^2)$.
12. $(x+y) \div (x-y)$. 13. $(a-b)^3 \div (a+b)^3$.
14. $(x+y) \div (x-y)$ 15. $1 \div x^3$.
16. $1 \div n$. 17. $\pm (1-b) \div (1+b)$. 18. $1 \div c$.

ANSWERS.

Exercise xliv.

1. $(x-a) \div 5$. 2. $a+b$. 3. $16a^5x \div (a^4 - x^4)^2$.
4. 0. 5. $1 \div (x+2)$. 6. $1 \div (a^4-x^4)$.
7. $12xy \div (9x^2 - 4y^2)$. 8. $(4x^2+2) \div x(16x^4-1)$.
9. $1 \div (x+1)(x+2)(x+3)$. 10. $4(x^4+4x^2y^2+y^4) \div (x^4-y^4)$
11. $(a-b)^3 \div (x+a)^2(x+b)^2$. 12. $2a \div x$.
13. $(236-77x) \div 18(11x-8)$. 14. $1 \div (a-b)$.
15. $15a(3a-x) \div (9a+2x)(a+3x)$.
16. $(10x-7) \div (x-1)(2x-5)-1 \div (2x-7)(x-4)$. 17. 2.
18. $y^n(y^n-x^n)$. 19. $(a-b)^{2n}+2$.
20. 0. 21. $4x^2 \div (x^{12}-1)$.
22. $-(a^2+b^2)(a^2-ab+b^2) \div (a^2-b^2)(a^2+ab+b^2)$.

Exercise xlv.

1. $x-y$. 2. $a+b$. 3. 0. 4. 0. 5. 0.
6. $\{(a+b)(c+a)x^2 + 2(ab+bc+ca)ax - 2a^2bc\} \div$
   $(a+b)(a+c)(x+a) \times (x+b)(x+c.)$ 7. 1.
8. $a+b+c$. 9. 1. 10. x^3-y^3. 11. 0.
12. $(a-b)(b-c)(a-c) \div (a+b)(b+c)(c+a)$.
13. $x^3 \div (x-a)(x-b)(x-c)$. 14. 1. 15. 0.
16. $\{b(x+a-b)+ax\} \div \{ab+(b-a)(x-b)\}$.

Exercise xlvii.

1. $(a-b)^2+4c^2=0$. 2. 8. 3. 10. 4. a^2+b^2.
5. $m=2$, $n=1$. 6. $2x^2$, or 5. 7. $m=-5$, $n=6$.
8. ± 12. 9. $(a^2+b^2)(c^2+d^2)$. 11. $-3bc-4c^2+b^2c^2-4b^2$.
12. $(x^2-4x+3)(x^2-4)$, also $(x^2-3x+2)(x^2-x-6)$.
13. $\frac{1}{3}(-1 \pm \sqrt{5})$. 15. $a \div c = d^2 \div e^2$, $a \div b = f^2 \div e^2$,

$b \div c = d^2 \div f^2$.   17. $ac^3 = b^3 d$ and $9ad = bc$.
19. $4p^3 + 27q = 0$.   24. $p = 2m^3 q \pm 2mq \sqrt{(m^2+1)}$.
25. $4(p-3) = q$.

Exercise xlix.

1. $5, 3\frac{1}{4}, a, -3$.   2. $-4\frac{1}{4}, -a, 2, 10$.
3. $a+b, c-a, b-c, 3$.   4. $-2, 6, -5, 12$.
5. $-14, a-3b, 2a-3b, 5b-3a$.   6. $7, 4, a, b$.
7. $\frac{1}{3}c, 5 \div a, 0, 1$.   8. $-1, \{(a+b)^2 - a\} \div b, a+b$.
9. $(b-a), a+b$.   10. $1 \div a - b, 1 \div (a-b), 1 \div (a^2+b^2)$.
11. $2b, a$.   12. $a+b, c \div (a+b), b \div (a-c)$.
13. $(b-c) \div (a-b), b+c$.   14. $a+b, a^2+ab+b^2$.
15. $a^2 - ab + b^2, 1$.   16. $-1, (a+b) \div (a-b)$.
17. $(e+b)(e-b), 2 \div 15, 3 \div 14$.   18. $-1 \div 12, b \div ac, a \div b$.
19. $(a^2+b^2) \div a^2 b^2) \div a^2 b^2, a(b^3+c^3) \div bc$.
20. $10, 12, 4, \frac{1}{2}$.   21. $1000, \frac{3}{5}, \frac{3}{5}$.   22. $9\frac{9}{10}, ab, bc \div a$.
23. $b^2 \div ac, c(a+b), b(a+b) \div a$.
24. $a \div b, (a-b) \div (a+b), -(a+b)^2 \div (a-b)^2$.
25. $-1, -1$.   26. $(a^2-c^2) \div (a+b)^2, 2, 3\frac{1}{4}$.
27. $ab, b \div a, ac \div b, 12$.   28. $12, -ac \div b$.
29. $9, 2$.   30. $12, 1$.   31. $3, 1$.   32. $(2a-1)(2a+2), 0$.
33. $1 \div m$.   34. 1.   35. $(ab+bc+ca) \div (a^2+bc+c^2)$.
36. $(a^2+b^2+c^2) \div (ab+bc+ca)$.   37. $a+b+c$.   38. 1.
39. 1.   40. 1.   41. 1.   42. 15.   43. $16\frac{1}{4}$.   44. 6.
45. 5.   46. $(npqa + pqb + qc + d) \div mnpq$.   47. $-\frac{1}{6}$.
48. 0.   49. $-25 \div 136$.   50. 1.

ANSWERS.   xxi

EXERCISE 1.
1. 2, 3.   2. $\frac{1}{2}$, $\frac{1}{3}$.   3. ± 2, $1\frac{1}{4}$.   4. 1, $1\frac{1}{2}$.   5. $\pm\frac{2}{3}$, $\pm(a+b)$, a.   6. 4, 5, 2, $2\frac{1}{2}$.   7. -3 or 2; 4, -3; $2\frac{1}{3}$, $-1\frac{1}{3}$.   8. 1; $\frac{2}{3}$ or $\frac{3}{2}$; $\frac{1}{3}$ or 3.   9. $-\frac{2}{3}$ or $\frac{3}{2}$, $\frac{1}{6}$ or 6; $\frac{4}{3}$ or $-\frac{3}{2}$.   10. -1, 2, $-\frac{1}{2}$, 1.   11. 0, $-b$, $3b$.
12. a, $\pm a\sqrt{-1}$.   13. 1; $\frac{1}{2}(-1+\sqrt{5})$.   14. $\pm a$.
15. $\pm bc$, $-(b+c)$.   16. $a+2b$.   17. b or $\pm a$.
18. $-2ab$, $\frac{1}{3}ab(1\pm\sqrt{7})$.   19. a, b, $-(a+b)$.   20. a, b.
21. a or $1-a$.   22. $-a$, $-b$, $a-2b$.
23. a, b, $b(1-b)\div(1+a-b)$.   24. $x^3-6x^2-37x+210$.
25. $x^4-4ax^3-13a^2x^2+64a^3x-48a^4$.
26. $x(x-1)(x+2)(x-4)=0$.   27. $x^4-4x^3+x^2+6x+2=0$.

EXERCISE li.
1. 4.   2. $-7\frac{6}{7}$.   3. -107.   4. 8.   5. $3a$.   6. $\frac{31}{135}$.
7. $50\frac{30}{59}$, 17.   8. 22, $46\frac{1}{3}$.   9. 7, 3.   10, 10, 10, 11.
11. 0 or 11; 33.   12. $3956\div3971$.   13. $\frac{1}{7}(15\pm\sqrt{190})$.
14. 3.   15. 3.   16. 4.   17. $1\frac{3}{8}$.   18. $1\frac{1}{2}$.   19. $3\frac{1}{4}$.
20. 4.   21. ± 3.   22. 11.   23. 2 and $-1\pm\sqrt{-3}$.
24. $2\frac{1}{2}$.   25. 0.   26. $3a$.   27. $\frac{3}{4}$.   28. $\frac{16}{15}$.   29. 3.
30. 10.   31. 0, 1, or $(-5\pm\sqrt{-23})\div 8$.   32. $102\frac{3}{5}$.
33. $(-11\pm\sqrt{4681})\div 20$.   34. 2, $\frac{1}{3}$, $\frac{7}{4}$.   35. -4.
36. 0 or $\pm\sqrt{(a^2+b^2)}$.

EXERCISE lii.
1. $(1-a)\div(1+a)$, $a(m+1)\div(m-1)$, $b(m+1)\div a(m-1)$.
2. $a-b$, 0, 0.   3. b, $ma\div b$, $b\div ca$.   4. 1, -1, 0.
5. $-\frac{3}{2}$ or -1.   6. $(c-b)(b^2+c^2)\div 2abc$.   8. 14, $4\frac{1}{2}$.
9. 2, $6\div 295$.   10. $73\div 210$, $(a+b+c+d)\div(m+n)$.

ANSWERS.

11. $b \div a$. 12. $b \div a$. 13. a or 0.
14. $\pm\sqrt{a^2+1} \div 2$. 15. $\frac{1}{2}$. 16. $\frac{14}{13}$. 17. 0 or 4.
18. $c \div ab$. 19. $88\sqrt{(2x-1)} = 100\sqrt{(3x-3)}$.
20. $75 \div 52$. 21. 8. 22. $84\frac{49}{144}$.
23. $1 \div n(n-1)$. 24. $ac \div (b-a)$. 25. 4, $8\frac{3}{4}$ or $13\frac{1}{2}$.
26. $a^2b^2 \div (a-b)^2$, 3. 27. $4a^2 \div (1+a)^2$, $b(a+b)^2 \div (a-b)^2$.
28. $(1+b^2) \div 2ab$. 29. $\sqrt{(1-x)} = 2 \div (a+1)^2$.
30. $-a \pm a\sqrt{\{(1+b+b^2) \div 2b\}}$.
31. $\left(\frac{x+1}{x-1}\right)^5 = \left(\frac{a+1}{a-1}\right)^5$.

Exercise liii.

1. 8. 2. 0. 3. 3. 4. $(\sqrt{m}+\sqrt{n})^2$.
5. $ab \div (1-2\sqrt{b})$. 6. $4 \div 7$. 7. $1 \div (a-2)$.
8. $18962 \div 12393$. 9. $\sqrt{a} \div (\sqrt{a}+2)$.
10. $(c^4 - 2bc^2) \div (2c^3 - 2b)$. 11. $\frac{1}{2}$. 12. $18a$.
13. $x^2 = 80 \div 81$. 14. $\pm \frac{10}{3}\sqrt{\frac{17}{21}}$. 15. $\pm\frac{4}{11}\sqrt{-11}$.
16. $\pm \sqrt{\left\{a^2 - \frac{(b-2a)^3}{27b}\right\}}$. 17. 0. 18. $\frac{9}{16}a$.
19. $(c-a-b)^3 = 27abc$. 20. $x^2 = a^2(n-1)^2 \div (2n-1)$.
21. $16xy = (n-4x-y)^2$. 22. 0, $-\frac{24}{25}$.
23. $\left(\frac{a^2}{2a-2} - 1\right)^2$, 0. 24. $2\sqrt{(1-m^2)} \div m\sqrt{(4-m^2)}$.
25. $(a^2-1)\{a^2+2 \pm \sqrt{(a^2+1)}\} \div a^2$.
26. $(cn-an+c)^2 \div b(n-1)^2$. 27. ± 5.
28. $2\sqrt{(3x^2+10)} = (17\sqrt{17} - 3\sqrt{3}) \div 7$. 29. ± 5.
30. $\pm \sqrt{(3b-2a)}$. 31. $\sqrt{\frac{3}{2}(a^2-b^2)}$.

ANSWERS. xxiii

32. $(2y+2z-2x)^3+216xyz=0.$ 33. $\frac{2}{5}a\sqrt{6}.$
    $a(n^2-4n+8) \div (2n-4).$ 35. $a^2+2a.$
36. $\pm\sqrt{(3a^2+b^2)} \div \sqrt{3}.$

Exercise liv.

1. $-(a^2+b^2)\div a.$ 2. $(2a^2+b^2)\div 2a.$
3. $\{(a-b)a^2 - 2c(a^2+ab+b^2)\}\div\{a^2 - 2c(a^3-b^3)\}.$
4. $-b.$ 5. $a+b+c.$ 6. $ab\div(b-a).$
7. $x^2-3ax-a^2=0$, &c. 8. $a.$ 9. $\frac{1}{3}(a+b+c)$ 10. $1\div abc.$
11. $1\pm(a+b+c).$ 12. $(a-b)(ac-2b)\div(a+b)ac.$ 13. $-c.$
14. $(\sqrt[3]{a}+\sqrt[3]{b})^3.$ 15. $\pm 2.$ 16. $c\div(a-b).$
17. $(a-b)\div(a+b).$ 18. $\frac{5}{8}a.$ 19. $\pm 2.$ 20. ± 2, &c.
21. $\frac{1}{2}(a+c)\div(a-c).$ 22. $a, (3ab-3b^2-a)\div(1+3a-3b).$
23. $a.$ 24. $a, b, 2b.$ 25. $a, (c^2+6ab)\div 6b.$ 26. $\frac{1}{6}(c+6a).$
27. $\frac{1}{2}a.$ 28. $a+b.$ 29. $(ab+bc+ca)\div(a+b+c).$
30. $\pm b, \pm a.$ 31. $\sqrt{\{1\div(a\ 1)\}}.$
32. $\{6(a-b)-4c(c-b)\}\div\{4c-3b-a\}.$
33. $(c^2-ab)\div(a+b-2c)$ 34. $\frac{1}{2}(-29\pm\sqrt{37}).$
35. $(x+a)^2=2b^2-a^2.$ 36. $\sqrt{(b^2-\frac{1}{3}ab)}.$ 37. $\frac{1}{2}(b-a).$
38. $3\frac{1}{2}, \frac{4}{5}.$ 39. $x^2-6x=a.$ 40. $1\pm\sqrt{19}.$ 41. $b, b-a.$
42. $(a^2+b^2)\div(a+b).$ 43. $x=-5\div 2.$ 44. $\frac{1}{3}(5\pm\sqrt{3}).$
45. $-2a, \frac{6}{7}a, \frac{3}{8}a.$ 46. $-3a.$

Exercise lv.

1. $bc\div(a+c).$ 2. $(a^2+b-2ab)\div(a+b^2).$
3. $(ad-bc)\div(a-b).$ 5. $\dfrac{a^2+bc}{b+c}.$ 6. $c.$ 7. $\frac{1}{2}(a+b)$
8. $a+b.$ 9. 0. 10. 0. 11. $abc.$
12. $(a^2+b^2+c^2)\div(a+b+c).$ 13. $(a+b+c)\div(a^2+b^2+c^2).$

14. $(a^2+b^2+c^2) \div (ab+bc+ca)$ 15. $\dfrac{a+b}{a-b}$. 16. $\dfrac{ab}{a+b}$.

17. $4\frac{1}{2}$. 19. 4. 20. -140. 21. 17. 22. 10. 23. a.

24. $\dfrac{e^2(a^2-b)}{a^2d+c^2-bd}$. 25. $8\frac{1}{4}$, 0. 26. $8\frac{11}{21}$. 27. $(ab-c^2) \div (a+b)$.

28. $-b, a$, 29. 0, 0. 30. $\frac{1}{2}(a+b-c)$. 31. $\dfrac{ab}{a+b}$.

32. d. 33. $ab \div (a^2-b^2)$. 34. $-3\frac{2}{3}$. 35. $\frac{8}{9}$. 36. $-3\frac{2}{3}$

37. Infinity. 38. 10. 39. $abc \div (ab+bc+ca)$.

40. $(ab+bc+ca-ad-bd-cd) \div (a+b+c-3d)$.

41. $a(b+c)^2 \div (b^2+c^2-ab+bc-ca)$.

42. $bc(d-a)+(a-b)(b-c)(c-d) \div (ab+bc+cd-ad-b^2-c^2)$.

43. $bc^2-b^2c-ac^2+b^2d-abd+acd \div (ab+bc-ac-b^2)$.

44. $-(a+b+c)$. 45. $a+b+c$. 46. $(ab+bc+ca) \div abc$.

47. $-\frac{1}{2}(b+c)$. 48. $(ab+c) \div 2a$. 49. 9. 50. 2. 51. 7.

52. 4. 53. $\frac{1}{27}(5 \pm \sqrt{785})$. 54. 4, $(am-nb) \div (n-m+a-b)$.

55. $\frac{1}{12}$, $b(a+c) \div (a^2+ab+b^2)$. 56. 0, $-\frac{1}{3}$, $\frac{1}{4}$. 57. 10.

58. $(apnq-cmpq) \div (apm^2+cqm^2)$; $\dfrac{nq(ap+mb)-mp(cq+nd)}{apm^2+mn^2b-m^2cq-m^2nd}$.

59. $ab \div (b-c)$, $c\{a^2+(b-c)a-bc\}+a(b^2-c^2) \div (a^2+b^2-c^2+ab-bc-ac)$. 60. $b \div (a+b)$.

61. $mpcq+apnq \div (apn^2-cqm^2)$.

62. $\{bm(a-c)+cn(b-a)+ap(c-b)\} \div \{m((a-c)+n(b-a)+p(c-b)\}$. 63. $(a^2+b^2) \div ab$, 0, $\frac{31}{13}$.

64. $(ap-cm) \div (an-bm)$, $\dfrac{d(n-q)-q(b-d)}{a(n-q)-m(b-d)}$. 55. $\frac{1}{2}$, $\frac{1}{4}$.

66. 100. 67. 13, 111. 68. 11, 7.

69. $(a+b-m-n)$. 70. $\dfrac{a^2+2ac+ad+2bc+2ab}{a-d}$.

71. $\frac{1}{2}(a+b) \pm \sqrt{\{\frac{1}{4}(a-b)^2 - \frac{1}{6}c^2\}}$, a or b.   72. 0.   73. $a+b+c$.

74. $\dfrac{ab+bc+ca}{abc}$.   75. $a+b+c$.   76. $a+b+c$.

77. $(ab+bc+ca) \div (a+b+c)$.   78. $b^2+a^2-c^2$.

79. $c-a-b$.   80. 0.   81. 0 or 11.

Exercise lvi.

1. $A=0$, or $B=0$.   2. $A=0$, or $B=0$, or $C=0$.
3. $x=0$, or $a-b=0$.   4. $x=0$, or $y=0$.
5. In the first case either $x-5y=0$, or $x-4y+3=0$, in the second case both conditions hold.   6. $x=0$, or $x=a$.
7. $x=0$, or $x=-b$.   8. $x=a$, or $x=c \div b$.
9. $x=0$, or $x=3$.   10. $x=0$, or $x=a+b$.
11. $x=0$, or $x=\pm a$.   12. $x=0$, or $x=b \div a$.
13. $x=0$, or $x=a$.   14. $x=0$, or a.
15. $x=0$, or $x=a+b$.   16. $x=0$, or $a+b$.
17. $-(2ab) \div (a+b)$.   18. $x=a$, or b.
19. $x=a$, or b, or c.   20. 5.   21. 1.   22. 21.
23. $x=\frac{1}{3}$, $x=3$.   24. $x=9$, $x=4$.
25. $x=1$, or 3.   26. $(ab) \div (a+b)$.
27. $x=a$, or b.   28. $x=(a^2+b^2) \div (a+b)$, $x=b+a$.
29. $(2ab) \div (a+b)$.   30. $x=a$, or b.
31. $x=1$, or $(1+a) \div (1-a)$.   32. $x=a$.
34. $x=a-b$, or $\frac{1}{2}(b+c)$.   35. $x=a+b$, or $\frac{1}{2}(a+c)$.
36. $x = \dfrac{a}{a+b+c}$, or 1.   37. $a+b-c$.
38. $x=a$, or $\frac{1}{3}(4b-a)$.   39. $x=-c$, or $a+b+c$.

40. $x=1$, or $\dfrac{m-n}{n-p}$.

41. $x = \dfrac{nc-pb}{mc-ap}$.

42. $\dfrac{p(a-b)-c(m-n)}{m(c-b)-a(n-p)}$.

43. $x=\tfrac{1}{2}(a+b)$, or $\tfrac{1}{2}(b-a)$.

44. $x=2a-b$, or $3b-2a$.

45. $x=a+c-b$, or $x=\dfrac{4a+4c-2b}{3}$.

46. $x=a+b$, or $\dfrac{17a-7b}{5}$.

47. $x=4a+b$, or $a+b$.

48. $x=\dfrac{a-b}{b-c}$, or $\dfrac{a}{c}$.

49. $(a-b)(b-c)x^2 - (a^2+b^2+c^2-ab-bc-ca)x + (a-c)(a-b) = 0$.

50. $x=\pm 3$, or ± 2.

51. $x=\pm 6$, or ± 2.

52. $x=3$, or $\tfrac{1}{3}$.

53. $x=\dfrac{a+b}{a-b}$, or $\dfrac{a-b}{a+b}$.

54. $x=\dfrac{a}{b}$, or $-\dfrac{b}{a}$.

55. $x=b-2a$, or $a-2b$.

56. $x=\dfrac{2a+3b}{5}$, or $\dfrac{3a+2b}{5}$.

57. $x=(mb+na)\div(m+n)$, or $(ma-nb)\div(m+n)$.

58. $x = \sqrt{\{(m+2n)-\sqrt{(m-2n)}\}} \div \sqrt{\{(m+2n)+\sqrt{(m-2n)}\}}$.

59. $a\left\{\dfrac{\sqrt{(c+1)}+\sqrt{(c-1)}}{\sqrt{(b+1)}-\sqrt{(c-1)}}\right\}$.

60. $a\{\sqrt{(3c-2)}+\sqrt{(2-c)}\} \div \{\sqrt{(3c-2)}-\sqrt{(2-c)}\}$.

61. $a\{\sqrt{(2c-1)}+1\}\div\{1-\sqrt{(2c-1)}\}$.

62. $\tfrac{1}{3}(a+2b)$.

63. $\tfrac{1}{2}(a+b)-(a-b)\sqrt{(m-2n)}\div\sqrt{(m+2n)}$.

64. $\dfrac{-a^2+b^2}{2b}$, or $\dfrac{a^2+b^2}{2a}$.

65. $2ab\div(a+b)$.

66. $x=\dfrac{3a+5b}{8}$, or $\dfrac{3b-5a}{8}$.

67. $x=2a\{\sqrt{(c+4)}-\sqrt{(c-4)}\}\div\{\sqrt{(c+4)}+\sqrt{(c-4)}\}$.

ANSWERS.   xxvii

68. $x = 4$, or 3.
69. $\frac{1}{2}\{a \pm \sqrt{a^2 - 4m}\}$ where $m = a^2 \pm \frac{1}{2}\sqrt{(c + a^4)}$; 3 or 1.
70. $a, b.$ 71. $x = \frac{1}{2}[a + b \pm \sqrt{\{(a+b)^2 - 4(ab+t)\}}]$,
    where $t = \frac{1}{2}(a-b)^2 \pm \frac{1}{2}\sqrt{\{(a-b)^4 + 4r\}}$.
72. $x = 0$, or a, or $\frac{1}{2}a(1 \pm \sqrt{-3})$; $x = 4$, or 2.
73. $x = 0$, or $a + b$, or $\frac{1}{2}\{(a+b) \pm \frac{1}{2}\sqrt{(a-b)^2 - 4ab}\}$.
74. $x^2 - (a-b)x + ab = \&c.$ 75. $x = \frac{1}{2}(3a - b).$ or $\frac{1}{2}(3b - a).$
76. $x = 3a - 2b$, or $3b - 2a$.
77. $y^2 - m^2 = 0$, where $y - m = x$ and $2m = a + b$.   See Key.
78. $y^2 - m^2 = 0.$ 79. $y^2 - m^2 = 0.$ 80. $y^2 - m^2 = 0.$
81. $y^2 - m^2 = 0.$ 82. $y^2 - m^2 = 0.$ 83. $y^2 - m^2 = 0.$
84. $(y^2 - k^2)(5y^2 + 7k^2) = 0$, (where also $k = \frac{1}{2}(a-b)$.
85. $k^4 - y^4 = c.$ 86. $k^5 + 10k^3y^2 + 5ky^4 = c(k^4 - y^4)$ &c.
87. $sy \pm k\sqrt{(k - 3c \pm r)} = 0$, where $s^2 = 3k + c$, and $r^2 = (k - 3c)^2$
    $+ (k - c)(3k + c).$ 88. $-3 \pm \sqrt{(9 \pm 12\sqrt{24})}.$
89—102. Work with a variable w such that $wx = x^2 + 1$.
89. $w = (a \pm s) \div b$, where $s = a^2 + 2b^2$.
90. $w = (3a + 2b \pm s) \div 2(a - b)$ where $s = \pm 5\sqrt{(a^2 + 2ab + 4b^2)}$.
91. $w = (3 \pm s) \div (1 \pm s)$ where $s = (b - 4a) \div b$.
92. $(w + 1)^2 = a \div (a - b).$ 93. $w^2 = 2a \div (b - a).$
94. $(x + 1) \div (x - 1) = a \div (a - 8b).$
95. $(w + 2) \div (w - 2) = \frac{1}{2}(1 \pm s)$ where $s = (16a + b) \div b$.
96. $w^2(4a - b) \div (a - b).$ 97. $w^2 = (4a - 3b) \div (a - b).$
98. $w = (b \pm s) \div 2a$ where $s^2 = b^2 + 16a^2$.
99. $w = (a + b \pm s) \div 2(a - b)$ where $s = (a + b)^2 + 8(a - b).$

100. $w = (a+b\pm s) \div 2(a-b)$ where $s^2 \div (a-b)^2 = \{(a+b)^2 + 4(a-b)^2\} \div (a-b)^2$.

101. $(w+2)\pm(w-2) = \pm s \div (4\pm 3s)$ where $s^2 = 2a \div (a+b)$.

102. $(w+2)\div w = \pm\sqrt{\{5a \div (a+4b)\}}$.

103. $\frac{1}{3}(2a+b)$, $\frac{1}{3}(a+2b)$.   104. $2a-b$, $\frac{1}{2}(a+b)$, &c.

105. 1, 2, 4, 5.   106. ± 1, 2, 4.   107. 1, 2, 3, 4.

108. $-\frac{1}{2}$, $-\frac{1}{4}$, 1, $\frac{5}{4}$.   109. -1, 3, 4.   110. $-a$, $5a$, $5a$.

111. 15, 20.   112. $2\frac{1}{4}a$.   113. 4, -1.

114. 7, -1.   115. $\frac{1}{2}(bc\div a + ca \div b + ab \div c)$.   116. $\pm a \div m$, &c.

117. $2s(s-a)(s-b)(s-c) \div \sqrt{\{s'^2-a^2)(s'^2-b^2)(s'^2-c^2)\}}$ where $2s = a+b+c$, $2s^1 = a^2+b^2+c$.

118. $(2ab+2ac^2+2bc^2-a^2-b^2-c^4)\div 4c^2$.   119. a, b, $\frac{1}{3}(a+b)$.

120. $\pm a$ or $\pm ja\sqrt{3}$.   121. a, b, $\frac{1}{2}(a+b)$.

Exercise lvii.

1. $x, 7; y, 9$.   2. $x, 2; y, 1$.   3. $x, 8; y, 1$.
4. $x, 9; y, 5$.   5. $x, -10\frac{1}{2}; y, 5\frac{1}{2}$.   6. $x, -2; y, \frac{1}{2}$.
7. $x, -1; y, 1$.   8. $x, -2; y, -3$.   9. $x, -\frac{3}{4}; y, \frac{1}{2}$.
10. $x, -\frac{1}{4}; y, \frac{3}{4}$.   11. $x, 12; y, 8$.   12. $x, 8; y, -9$.
13. $x, 10; y, 12$.   14. $x, 12; y, 15$.   15. $x, 18; y, 13$.
16. $x, \cdot 3; y, \cdot 2$.   17. $x, 7; y, 9$.   18. $x, 7; y, \cdot 3$.
19. $x, 7; y, 3$.   20. $x, 2; y, 3$.   21. $x, 3; y, 4$.
22. $x, \frac{4}{5}; y, \frac{9}{10}$.   23. $x, \cdot 3; y, \frac{1}{7}$.   24. $x, 12; y, 15$.
25. $x, \frac{7}{10}; y, \frac{3}{10}$.   26. $x, 8; y, 9$.   27. $x, 3; y, 1$.
28. $x, 7; y, 8$.   29. $x, 11; y, 7$.   30. $x, 17; y, 13$.
31. $x, 5; y, -4$.   32. $x, -\frac{31}{10}; y, -\frac{19}{10}$.   33. $x, 18; y, 10$.
34. $x, 4\frac{2}{5}; y, 3\frac{3}{10}$.   35. $x, 11; y, 6$.   36. $x, 7; y, 5$.

ANSWERS. xxix

37. x, 2; y, 3.
38. x, 5; y, 3.
39. Equations not independent.
40. x, 3; y, 1.
41. x, 7; y, 5.
42. $x=0=y=0$.
43. 0, 0.
44. $x=0$ or 13; $y=0$ or $\frac{39}{11}$.
45. x, 17; y, 20; z, 5.
46. x, $1\frac{231}{130}$, y, $1\frac{234}{130}$, z, $\frac{247}{130}$.
47. 11, 7, 9.
48. 21, 22, 23.
49. $-15, -6, -8$.
50. 3, 4, 5.
51. 12, 15, 10.
52. 5, 3, 1.
53. $\frac{5}{6}$, $1\frac{1}{2}$, $\frac{2}{3}$.
54. 3, 5, 7.
55. 11, 13, 17.
56. 5, 3, 1.
57. 9, 7, 3.
58. $7\frac{1}{4}$, $8\frac{1}{4}$, $9\frac{1}{4}$.
59. $3\frac{1}{7}$, $2\frac{1}{7}$, $1\frac{1}{7}$.
60. 2·3, 3·4, 4·5.
61. 30, 20, 70.
62. $88 \div 59$, $1098 \div 59$, $1004 \div 59$.
63. 30, 12, 70.
64. 6, 12, 20.
65. 5, 2, 0.
66. 1, 1, 1.
67. 11, 9, 7.
68. 5, 3, 1.
69. 2, 3, 1.
70. 3, 4, 5.
71. $\frac{1}{3}$, $\frac{1}{2}$, $\frac{1}{4}$.
72. 5, 4, 3.
73. 7, 3, 1.
74. 2, 3, 1.
75. 1, 3, 5.
76. 0, 1, 2.
77. $1755 \div 698$, $360 \div 349$, $-15705 \div 698$.
78. $\frac{1}{5}$, $\frac{1}{3}$, 1.
79. 5, 4, 1, 3.
80. $4\frac{2}{5}$, $3\frac{3}{10}$, $2\frac{2}{10}$, $1\frac{1}{10}$.
81. 31, 41, 51, 21.
82. 7, $4\frac{1}{2}$, 4, $8\frac{1}{2}$.
83. 20, 10, 0, 30.
84. $11 \div 24$, $\frac{1}{4}$, $1 \div 24$, $\frac{1}{4}$.
85. $270 \div 117$, $-52 \div 117$, $15 \div 117$, $-126 \div 117$.
86. Each 210.

EXERCISE lviii.

1. $(a'c-ac') \div (a'b-ab')$. 2. $b(cn-dm) \div (ad-bc)$.
3. $b(d-c)(d-a) \div d(b-c)(b-a)$, $c(d-a)(d-b) \div d(c-a)(c-b)$.
4. $y = cz + du + ew + ax$, $z = du + ew + ax + by$,
   $u = ew + ax + by + cz$, $w = ax + by + cz + du$.
5. $x = \frac{1}{2}m(a-b+c)$, &c. 6. $x = \{p(a^2-b) - m(ab-1) +$

$n(b^2-a) \div \{a^3+b^3-3ab+1\}$, &c.

7. $x=(l-am+abn-abcp+abcdr) \div (1+abcde)$, &c.
8. $1=a \div (1+a)+b \div (1+b)+c \div (1+c)$.
9. $1=ab+bc+ca+2abc$.

Exercise lix.

1. $(nc-bd) \div (na-bm)$, $(mc-ad) \div (mb-na)$.
2. $(nc+bd) \div (an+bm)$, $(mc-ad) \div (bm+an)$.
3. $c(n-b) \div (an-mb)$, $c(m-a) \div (bm-am)$.
4. $(b-c)a \div (b-a)$, $b(a-c) \div (a-b)$. 5. $ab \div (a+b)$, y, same
6. $ab^2 \div (a^2+b^2)$, $a^2b \div (a^2+b^2)$. 7. $ac \div (a+b)$, $bc \div (a+$
8. $(a^2-b^2) \div (am-bn)$, $(b^2-a^2) \div (bm-an)$.
9. $a+b-c$, $c+a-b$. 10. $a+c$, $b+c$.
11. $a(cn-dm) \div (bd-ac)$, $b(cn-dm) \div (ad-bc)$.
12. $y=\{93(a^2-c^2)-b(b+2a)\} \div \{(a-b)^2-c^2+4bc\}$.
13. $a+b-c$, $a-b+c$. 14. $a+b-c$, $c+a-b$.
15. $(m-a)(n-a) \div (b-a)$, &c. 16. $1 \div (a-b)(a-c)$, &c.
17. $(m-bc)(l-a) \div (c-a)(a-b)$, &c.
18. $x=p \div (pl+mq+nr)+a$, so y and z.
19. $p\{1-(la+mb+nc)\} \div (pl+mq+nr)+a$, &c.
20. $(m^2+2a^2-b^2-c^2) \div 3a$, &c. 21. $y=a-b+c$, &c.
22. $x=(ab+bc+ca)(b+c-2a)(2b-a-c) \div \{(a-c)(b+c-2a)+(b-c)(2b-a-c)\}$. Corrected equation, $x=\frac{1}{2}(b+c)$, &c.
23. $ma \div (a+b+c)$, &c. 24. $npr \div (anp+bmp+cmq)$.
25. $1 \div (b-c)$, &c. 26. $\frac{1}{2}(b+c-a)$, &c.
27. $au \div d$, &c. 28. $z=1 \div (a+b-c)$.
29. $a+b$, &c. 30. $1 \div 2a$, &c.

ANSWERS. xxxi

31. $(m^2+n^2-l^2) \div 2mn$, &c. 32. $\frac{1}{2}(a+c-b)$, &c.
33. $l(m^2+n^2) \div 2mn$, &c. 34. $1 \div (b+c-a)$, &c.,
35. $bc \div (b+c)$, &c. 36. $b+c-a$, &c.
37. a, b, c. 38. b^2-c^2, &c.
39. $\frac{1}{8}(a+2b-c+3d)$, &c. 40. $\frac{1}{4}(4a+b+3c-2d+5e)$.

Exercise lx.

1. $a+b, a-b$. 2. $\frac{1}{2}(a^2+b), \frac{1}{2}(a^2-b)$.
3. $(3n^2+m^2) \div 5m, (2n^2-m^2) \div 5m$.
4. $a \div (a-b), b \div (a+b)$. 5. $1 \div (a-b), 1 \div (a+b)$.
6. $a+b-c, a-b+c$. 7. $a+b-c, a-b+c$.
8. $(a^2+ab+b^2) \div (a+b), (a^2-ab+b^2) \div (a-b)$.
9. $(ab-1) \div (a-1)(b-1), (a-b) \div (a-1)(b-1)$.
10. $(1+a) \div (ab-1), (1+b) \div (ab-1)$.
11. $(a+1)(b+1) \div (ab-1), (a-b) \div (ab-1)$.
12. $a(a+b), b(a-b)$. 13. $a\{b(a+b)-c(a-c)\} \div (a^2-bc)$, $a\{b(a-b)+c(a+c)\} \div (a^2-bc)$. 14. $-(a+b)$. ab.
15. $\frac{1}{4}(b+c)$, &c. 16. $(a-2b+3c) \div 38$, &c.
17. $2 \div (b+c)$, &c. 18. $a+b$, &c. (by symmetry).
19. b^2-c^2, &c. 20. b^2-c^2, &c.
21. $\frac{1}{4}abc, (1-a)(1-b)(1-c), (2-a)(2-b)(2-c)$.
22. $2abc \div (ab+bc-ca)$. 23. $1, 1, 1$.
24. $ar \div (ma+nb+pc+qd)$, &c.
25. $u=0$, or $\left(\dfrac{d}{a}-1\right) \div \left(\dfrac{d^2}{b^2}-\dfrac{d^2}{c^2}\right)$.
26. $(b+c-a) \div (a+b+c), y=(b-c-a) \div (a-b-c)$.
27. $\frac{1}{2}(a-b+m-n)$, &c.

28. $(4a+2c-d-3b)$, $y+z$ by symmetry.
29. $-(a-b+c+d)$, $(ab+bc, \&c.)$, $-(abd+\&c.)$, $abcd$.
30. $\frac{1}{2}(a-b+c-d+e)$, others by symmetry.
31. $x = (a-lb+lmc-lmnd+lmnpc) \div (1+lmnpq)$, the others by symmetry. 32. $x = b+c-e$, &c.
34. $y = (a+5b+3c-7d+9e) \div 22$, &c.
35. $z = \frac{1}{2}(a+c)$, then symmetry. 36. $z = c+d+e$, &c.
37. $x = a-2b+3c-2d+e$, then by symmetry.

Exercise lxi.

1. $x = (2ab+a+b+r) \div 2(a-b)$ where $r = 4a(b^2+b+1) \div (3a-b)(3b-a)$.
2. $x = (ar+1) \div (ar-1)$ where $r^2 = (b^2-1) \div 3(a^2-b^2)$.
3. $x = \{\sqrt{(1+a)(1+b)} - \sqrt{(1-a)(1-b)}\} \div \{\sqrt{(1+a)(1+b)} + \sqrt{(1-a)(1-b)}\}$.
4. $(ab-\alpha\beta) \div (ab+\alpha\beta)$, $(a\beta+b\alpha) \div (a\beta-b\alpha)$.
5. $x = \{\sqrt{(a+b+c)(a+b-c)} + \sqrt{(b+c-a)(a+c-b)}\} \div \{\sqrt{(a+b+c)(a+b-c)} - \sqrt{\ldots}\}$.
6. $(a+b) \div (1-ab)$. 7. $x = (\alpha\beta-ab) \div (a\beta+b\alpha)$.
8. $\left(\dfrac{x+1}{x-1}\right)^4 = \dfrac{(a+m)(b+n)}{(a-m)(b-n)}$.
9. $x = \{a\sqrt{(1-b^2)} - b\sqrt{(1-a^2)}\} \div \sqrt{(a^2-b^2)}$.
10. $x = (b+c-a) \div \{(\sqrt{b+c-a})(c+a-b)(a+b-c)\}$.
11. $x = (b+c) \div \sqrt[3]{(a+b)(b+c)(c+a)}$.
12. $x = \sqrt{(a+b+c)} \div a$, &c.
13. $x = \{b^2+c^2-a(b+c)\} \div \sqrt[3]{2(a^3+b^3+c^3-3abc)}$.
14. $(b+c-a)$, &c. 15. a or $(a^2-b) \div (1-ab)$.

ANSWERS.

16. $x+y = \sqrt{(a+b)(a+2b)} \div \sqrt{(a-b)}$,
    $x-y = \sqrt{(a+b)(a-2b)} \div \sqrt{(a-b)}$, &c.

17. $(x+y)^2 = \frac{1}{2}\left(3b - \frac{a}{b}\right)$, $(x-y)^2 = \frac{1}{2}\left(\frac{3a}{b} - b\right)$.

18. $(x+y) \div (x-y) = \sqrt{(a+3b)} \div \sqrt{(a-b)} = m$ suppose.

19. $y^2 = m \div (am^2 - m + 1)$ where $m = 1 \div a + b \pm \sqrt{(a^2 - b^2 + 1)} \div (a+b)$.

20. $a, b.$ 21. $x = \{\sqrt{(a-c)} + \sqrt{c}\}y \div \{\sqrt{(a-c)} - \sqrt{c}\}$.

22. $x = (a+c)y \div (a-c)$, &c.

23. $x+y = (ab-1) \div (a-b)$, &c.

24. $xy = \{\sqrt{(b+2)} - \sqrt{(b-2)}\} \div \sqrt{\{(b+2) - \sqrt{(b-2)}\}} = p$ suppose, $x \div y = \{\sqrt{(a+2)} + \sqrt{(a-2)}\} \div \{\sqrt{(a+2)} - \sqrt{(a-2)}\}$.

25. $x^2y^2z^2 = \frac{1}{2}(a+b-c)(b+c-a)(c+a-b)$, &c.

26. $x = (ab-bc-ca) \div 2\sqrt{abc}$.   27. $a - x^2 = \pm m$, where m is the value of v in the equation $4ca - 4(c+a)r + 4v^2 = (ca - ab - bc)^2 + 4b(ca - ab - bc)v + 4b^2v^2$.

28. $x = \frac{1}{2}\sqrt{(abc)}\left(\frac{1}{b} + \frac{1}{c}\right)$, y and z by symmetry.

29. $a(b^2 + c^2) \div (b^2 + c^2) \div (b^2 + c^2)$, &c.

30. $c\{\sqrt{(a+b)} + \sqrt{(a-b)}\} \div \sqrt{(a+b)}$.

31. 0 or $a(b+c) \div 2bc$, &c.

32. $z = -1$ or $a\sqrt{(a^2-1)} \div \sqrt{(a^2-b^2)}$, &c.

ANSWERS.

EXAMINATION PAPERS.

I.

1. $(a+b+c)(x+y+z)$.
2. $(x^2-4y^2)^3$, $2(a^2b^2+b^2c^2+c^2a^2)-a^4-b^4-c^4$.
3. $x^4-6x^3+13x^2-12x+4$, $x^2+9x^{-2}+3$,
   $x^{n(n-1)}+x^{n(n-2)}+x^{n(n-3)}\ldots$
4. $2x^{2m}-\tfrac{1}{3}x^{3m}$, $\dfrac{a}{b}-\dfrac{b}{c}-\dfrac{c}{a}$. 5. $\pm 14\sqrt{-19} \div 27$, 0 or $\tfrac{4}{3}$.
6. 16, 7. 199, 8, AB 37, CA 52, BC 45.

II.

1. $(a+b)^3$. 2. $(a+b+c)(a^3+b^3+c^3-3abc)$. 3. $1 \div x^3$.
4. $(a^{2m}+2a^m+2) \div (a^m+2)$, $(a+b+c) \div (a-b-c)$.
7. $-\tfrac{7}{8}$, 4 or 6. 8. $p^2q \div (r-pq)$.
9. 584. 10. $(am+bn) \div (a+b)$, (by common rule).

III.

1. -117, $a^2(z-x)+(x-y)ab+(y-z)b^2$.
2. $a+bx+cx^2$, $3-4x+7x^2-10x^3$
3. $2\tfrac{1}{2}$; 6, 9, 12. 4. 160 eggs.
5. 40, 35. 6. $\sqrt{\tfrac{1}{2}(1+a)}+\sqrt{\tfrac{1}{2}(1-a)}$.
7. 5 or $\tfrac{1}{4}$; 4. 9. $(a+b+c)(a-b)(b-c)(a-c)$.

IV.

1. $(a+b+c)^2$. 2. $a \div b$. 3. $(4x^2-9y^2)(4x^2-4y^2)$. $1+\sqrt{x}$.
4. $\tfrac{1}{3}\sqrt{3}-\tfrac{1}{6}\sqrt{6}$, 17. 5. 4. 6. 1. 7. 0 or 4; 8, 1.
8. $7\tfrac{1}{2}$, 12. 9. 4 or 6. 10. $\tfrac{1}{2}(-3\pm\sqrt{-39})$.

V.

1. $8x^3+1 \div 125$. 3. $16a^2b^2$. 4. $b \div 8a$. 5. 15, 12.
6. $\tfrac{1}{2}a$, $\tfrac{1}{4}a$; 4 or -9, 7. 6, 7, 8, or $-6, -7, -8$.

VI.

1. (a^2+b^2).   2. $a^2+b^2+c^2+ab+ac-bc$.
3. $x^2+y^2\pm\frac{3}{2}xy$, $(7x+6y-9)(x-y+4)$.   4. $-20, 0$.
5. b or $1\div b$.   7. $x=4$ or 9, $y=9$ or 4; $1, 2, 3$;
  $x=\frac{1}{2}(-7\pm\sqrt{33})$.   8. $-1, 0, 1$, or $5, 6, 7$, or $-\frac{11}{7}, -\frac{4}{7}, \frac{3}{7}$.
9. $x(x^2+3)(x^2-2x-1)=0$.   10. $x\div y=3\div 4$.

VII.

2. $-\cdot 01$.   4. x^2+x-1.
5. $(36x^2+18x+9)\div(16x^4-81)$, $(x^2-ab)\div(x^2+ab)$.
6. $9; 3; x=\frac{1}{2}(a-b)$.

VIII.

2. $(x^3-y^3)^2(x-y)$; $a=p+q$, $b=pq$.
3. $(x^2-xy+y^2)^2$; $a^4+b^4-a^2b^2$.
5. $ax^3+bx+cx^{-1}$; $\frac{1}{2}x^{\frac{3}{4}}+\frac{1}{3}x^{\frac{5}{6}}-\frac{1}{4}x^{\frac{7}{12}}$.
6. x^2+px+p^2; $50(x+5)(x-4)(c-6)$.
7. $a=0=b$ or $a=1$, $b=2$.
8. 3 or $-43\div 7$ satisfies the equation $2-\sqrt{\ldots}$ &c.

IX.

4. $axy+b$.   7. $\pm\sqrt{ab}$; $\sqrt{(a+b)}\div\sqrt{(a-b)}$, and $y=$ reciprocal of this.   8. $P(22a-21b)\div 20a(a-b)$.   9. $7, 15, 48$.

X.

1. ax^2+by^2+2cxy.   3. $x-2$.   4. 24.
5. $5x^2-3ax+4a^2$; $\dfrac{x}{y}+\dfrac{y}{x}-\sqrt{\frac{1}{2}}$.   7. $3; \frac{1}{2}$.   8. $4, 6$.

XI.

1. 2.   2. $6, 8$.   3. $(4x^{\frac{1}{2}}-9y^{\frac{1}{2}})^3$, $\frac{1}{15}x^4+\frac{4}{81}y^4$.
4. x^2+3+9x^{-2}, $x^2-(a+b)x-c$.   6. 0.
7. 0 or $-2\frac{1}{2}$; $y=\pm 3$ or $\pm\sqrt{-9}$, &c.

XII.

2. $a^{2^n} - b^{2^n}$. 3. $1+x+2x^2+4x^3+8x^4$, $2^r x^{r+1}$.
5. $(1-3x-4x^2)(1-2x-13x^2+38x^3-24x^4)$.
6. $m^2 - m = p$, $q = 0$.
7. $x^2 = (a-2) \div (a+4)$, $\sqrt{x} = \{b(n-1) - cn\} \div \sqrt{a(n-1)}$,
8. $ab \div \{(a-1))b-1)-1\}$. 9. 87.

XIII.

1. 2. 3. $1 + c_1 x + c_2 x^2 + \&c.$, where c_1, c_2, &c., represent the combinations of a_1, a_2, ... taken *one, two*, &c. at a time.
4. $4000. 5. $\{m(b'c - bc') - n(ca' - c'a)\} \div (ab' - a'b)$,

$$\frac{a}{a'} = \frac{b}{b'} = \frac{c}{c'}.$$ 7 $(\alpha+\beta)^{\frac{2}{3}} + (\alpha-\beta)^{\frac{2}{3}} - 2a^{\frac{2}{3}} = 0$. 9. 65.

XIV.

1. $12abc$. 4. 8 miles. 5. $am \div n$; $x+y = \pm 5$ or ± 1, $x-y = \pm 1$ or ± 5. 6. 20.

XV.

2. $\frac{1}{8}$. 3. $c^2(c^2 - bd) + d^2(b^2 - ac) + ac(ad - bc)$. 4. $2 \div (m-n)$. If $m-n$ is negative x is neg. which shows that they were together *before* noon. If $m-n = 0$, x is infinite, *i.e.*, they are *never* together.
5. $x^2(x^2 - a^2)(x-2a)$; $(x^2 - a^2)(x^2 - y^2)$.
6. $(a+b+c+3d) \div (a+b+c+d)$: $(x+y+z) \div (x-y+z)$; $3a \div (a+b)$.

XVI.

4. $2x^4 - 3x^3 + 4x + 3$. 5. See paper XIX., prob. 4.
6. $\frac{1}{2}\{1 \pm \sqrt{(a^{\frac{1}{n}} + 4b^{\frac{1}{n}})} \div \sqrt{4b^{\frac{1}{n}}}$. 8. $(b_2 c_1 - b_1 c_2) \div (a_1 b_2 - a_2 b_1)$ $= (b_3 c_2 - b_2 c_3) \div (a_2 b_3 - a_3 b_2)$. 9. $ac \div (a-b)$.

XVII.

1. $\{2ab-(a+b)\} \div ab.$ 2. $x^4 - 2x^2(a+b^2)+(a-b^2)^2$; $8a^2 - 2ab - 10ac - 3b^2 + 2c^2 + 5bc.$ 6. $m^{\frac{2}{3}} - n^{\frac{2}{3}} = 4.$
8. $(m+n)(bq-pc) \div (mq-pn),\ (p+q)(mc-bn) \div (mq-pn).$

XVIII.

1. $1-x^{16}.$ 2. $1.$ 4. $x-a-b.$ 6. $(x-y)(x-z).$
7. $1.$ 8. $9a$; $\pm 1 \div \sqrt{2}$ or $\pm \tfrac{1}{2}\sqrt{6}.$ 9. $10, 11.$

XIX.

1. $\tfrac{1}{4}(4x-a+1)^2.$ 3. $5x^2-1$; $(x^2+xy+y^2)^2.$
4. $(a-c)(a-d)(b-c)(b-d) \div (a+b-c-d)^2.$
6. $(a+2)^2 \div 4(a^2+a).$ 7. $(7b^2-a^2) \div 12a.$ 8. $18, 22, 50.$

XX.

1. $(1+m)x+(1-n)y.$ 3. $x^2-1.$
4. $(x+y-z)(y+z-x)(z+x-y) \div (x+y+z)^2$; $(a^2b-2ab^2-a^3+ab^3+b^3) \div (a^4-b^4).$
5. $1.$ 6. $a''(cb'-bc')+b''(ac'-a'c)+c''(a'b-ab')=0.$
7. $\sqrt{\left\{n^3 - \left(\dfrac{m^3-2n}{3m}\right)^3\right\}}$; $2, 3, 4.$
8. $a(b-c) \div (b-a),\ b(c-a) \div (b-a).$ 9. $8.$ 10. $2000.$

XXI.

1. $8.$ 2. $24abc$; $2a^8+43a^6x^2+62a^4x^4+44a^2x^6+18x^8.$
4. 0; $a^7+16.$ 5. 16; $x+2a^{\frac{1}{2}}x^{\frac{1}{2}}-2a.$ 7. 3377 oz. of gold, 783 oz. of silver.   9. $-(8\pm 4\sqrt{3}) \div (3 \pm 2\sqrt{3})$; $x=\pm 2$ or $\sqrt{-1}$, $y=\mp 1$ or $\mp 2\sqrt{-1}$; $8, 4.$
10. $y =$ cost of 2nd bale $= 60 \pm 20\sqrt{7}.$

XXII.

1. $\cdot 02997$, a^3+aq+p^3.   2. $ab^2-b^2+c=0$,
   (a). $(a-b)(8a-3b)$. (b). $x(x-1)(a-b)(b-c)(a-c)$.
3. $b^2=4ac$. (a). $(a+b)^2$.   5. (a) $(ax-by)\div(ax+by)$
7. (a) $\frac{1}{3}(a+b+c)$, (b), $\frac{2}{5}$, $\frac{2}{3}$, 2.
   (c). $b-c$, $c-a$, $a-b$.   (d). $-1\pm\sqrt{2}$.

XXIII.

1. $\frac{1}{3}x^5+\frac{4}{3}x^4-x^3+\frac{1}{3}x^2+\frac{16}{13}+(95x^3+21x^2-40x+42)\div$
   $(3x^4-21x^3+9x-6)$; -382.   4. $x=a+2c$, $y=b+3c$.
5. (2). $\frac{1}{32}$.   7. (1). $a+b+c$. (2), 1, 2, 3, 4. (3), 0 or $\frac{7}{12}$.
8. $\frac{1}{24}(u^2-1(3n+2))$.   11. $\begin{Bmatrix} a_1b_1c_1d_1 \\ a_2b_2c_2d_2 \\ a_3b_3c_3d_3 \\ a_4b_4c_4d_4 \end{Bmatrix} = 0$.

12. Coll. to Newmarket 63 miles.

XXIV.

1. $1\div\{(x-y)(y-z)(z-x)$.   2. $8-m-n$; 0.   3. 5, 3, $4\frac{1}{2}$.
4. (1). $A=bc\div 2a$, $B=ac\div 2b$, $C=ab\div 2c$; (2). $a^2+b^2=c^2$.
11. $y=\pm 2$, $x=\pm 5$, &c.; $x^2-10x=-19$ or -16,
   $\therefore x=8$ or 3, &c.; $(b^{-\frac{1}{2}}\pm a^{-\frac{1}{2}})^{\frac{4pq}{p-q}}$

XXV.

1. $(a^2-3x)\div(a^2-x^2)$; 1.   2. $\dfrac{1}{a}-\dfrac{b}{ax}-b^2(x+a)\div a^2x^2$.
5. $x^2-2x=2$, $x=3$, $y=2$; $y=53\div 24$, &c.   6. 4 miles, 3 do.
7. $(x^n-1)\div(x^2-1)x^{n-1}$.   8. $R^n\left(P-\dfrac{p}{r}\right)+\dfrac{p}{r}$.   9. 7.

10. $1+\dfrac{1}{1}+\dfrac{1}{1\cdot 2}+\dfrac{1}{1\cdot 2\cdot 3}+$ &c. $= 2\cdot 71828$ approximately;

$\{(5x)^r \div 3^{r+\frac{1}{2}}\}\{1\cdot 6\cdot 11 \ldots (5r-4)\} \div \underline{|r}$. 11. $\tfrac{3}{2}$ and $-\tfrac{7}{2}$.

XXVI.

2. $a - \sqrt{(ab - a^2)}$; 4. 8. 2, $\tfrac{1}{2}$, or $\tfrac{1}{2}(-8 \pm \sqrt{5})$; $x+y+z = \sqrt{(a^2 + 2b^2)}$, $\therefore z = c + \sqrt{(a^2 + 2b^2)}$, &c.; $\tfrac{1}{3}(-4 \pm \sqrt{76})$.

4. 3.   7. $p^2 \div 4 + r^3 \div 27 = 0$.   10. $(1+x) \div (1-x)^2 - x^{n-1}\{3x - 1 + 2n(1-x)\} \div (1-x)^2$; $n \div (15n+9)$, $\tfrac{1}{15}$.

XXVII.

1. $(lm - nk)^2$.   2. $(a+b+c)(a-b)(b-c)(a-c)$.
3. $(a-b)(b-c) + (b-c)(c-a) + (c-a)(a-b)$; $(a-b)^2 + (b-c)^2 + (c-a)^2$.   4. $x = a \div (a^2 + b^2 + c^2)$, &c.
5. $\sqrt{ab}$; 18 or -2; $\tfrac{3}{4}+3j$.   7. $\tfrac{1}{2}(a+b)$ &c.; $x = \{1 + a^2 - b^2 \pm \sqrt{(1-a-b)(1-a+b)(1+a-b)(1+a+b)}\} \div 2a$; $x \div y = (1+a)(1-b) \div (1-ac)$, &c.
8. -8; $(p^4 - 4p^2q + 8pr) \div (p^2 - 4q)$.
9. $(m+n) \div 2mn$, $(n-m) \div 2mn$.

Note.

Owing to the pressing demand for the issue of the present edition of this work, the Authors have been unable to furnish solutions to the problems in appendix, in time for insertion. Those wishing to get such solutions, will be supplied gratis, either direct from Publishers, or through local Booksellers. The Publishers will feel grateful to teachers should they kindly point out errors in this, or any other of their publications. W. J. GAGE & Co., Educational Publishers.

XXII.

1. $\cdot 02997$, a^3+aq+p^3.   2. $ab^2-b^3+c=0$,
   $(a). (a-b)(8a-3b)$. $(b). x(x-1)(a-b)(b-c)(a-c)$.
3. $b^2=4ac$. (a). $(a+b)^2$.   5. (a) $(ax-by) \div (ax+by)$
7. (a) $\tfrac{1}{3}(a+b+c)$, (b), $\tfrac{2}{3}$, $\tfrac{1}{3}$, 2.
   (c). $b-c, c-a, a-b$. (d). $-1 \pm \sqrt{2}$.

XXIII.

1. $\tfrac{1}{3}x^5 + \tfrac{4}{3}x^4 - x^3 + \tfrac{1}{3}x^2 + \tfrac{16}{13} + (95x^3+21x^2-40x+42) \div$
   $(3x^4-21x^3+9x-6)$; -382.   4. $x=a+2c$, $y=b+3c$.
5. (2). $\tfrac{1}{32}$.   7. (1). $a+b+c$. (2), 1, 2, 3, 4. (3), 0 or $\tfrac{7}{12}$.
8. $\tfrac{1}{24}(u^2-1(3n+2))$.   11. $\begin{cases} a_1 b_1 c_1 d_1 \\ a_2 b_2 c_2 d_2 \\ a_3 b_3 c_3 d_3 \\ a_4 b_4 c_4 d_4 \end{cases} = 0.$

12. Coll. to Newmarket 63 miles.

XXIV.

1. $1 \div \{(x-y)(y-z)(z-x)\}$.   2. $3-m-n$; 0.   3. 5, 3, $4\tfrac{1}{4}$.

10. $1 + \frac{1}{1} + \frac{1}{1\cdot 2} + \frac{1}{1\cdot 2\cdot 3} +$ &c. $= 2\cdot 71828$ approximately;

$\{(5x)^r \div 3^{r+1}\}\{1\cdot 6\cdot 11 \ldots (5r-4)\} \div \lfloor r$. 11. $\frac{3}{2}$ and $-\frac{7}{9}$.

XXVI.

2. $a - \sqrt{(ab - a^2)}$; 4. 3. 2, $\frac{1}{2}$, or $\frac{1}{2}(-3 \pm \sqrt{5})$; $x+y+z = \sqrt{(a^2 + 2b^2)}$, $\therefore z = c + \sqrt{(a^2 + 2b^2)}$, &c.; $\frac{1}{3}(-4 \pm \sqrt{76})$.

4. 3. 7. $p^2 \div 4 + r^3 \div 27 = 0$. 10. $(1+x) \div (1-x)^2 - x^{n-1}\{3x - 1 + 2n(1-x)\} \div (1-x)^2$; $n \div (15n+9)$, $\frac{1}{15}$.

XXVII.

1. $(lm - nk)^2$. 2. $(a+b+c)(a-b)(b-c)(a-c)$.
3. $(a-b)(b-c) + (b-c)(c-a) + (c-a)(a-b)$; $(a-b)^2 + (b-c)^2 + (c-a)^2$. 4. $x = a \div (a^2 + b^2 + c^2)$, &c.
5. $\sqrt{ab}$; 18 or -2; $\frac{3}{4} + 3j$. 7. $\frac{1}{2}(a+b)$ &c.;
$x = \{1 + a^2 - b^2 \pm \sqrt{(1-a-b)(1-a+b)(1+a-b)(1+a+b)}\} \div 2a$; $x \div y = (1+a)(1-b) \div (1-ac)$, &c.
8. -8; $(p^4 - 4p^2q + 8pr) \div (p^2 - 4q)$.
9. $(m+n) \div 2mn$, $(n-m) \div 2mn$.

www.ingramcontent.com/pod-product-compliance
Lightning Source LLC
Chambersburg PA
CBHW030745250426
43672CB00028B/787